# ABOUT ISLAND PRESS

Island Press, a nonprofit organization, publishes, markets, and distributes the most advanced thinking on the conservation of our natural resources—books about soil, land, water, forests, wildlife, and hazardous and toxic wastes. These books are practical tools used by public officials, business and industry leaders, natural resource managers, and concerned citizens working to solve both local and global resource problems.

Founded in 1978, Island Press reorganized in 1984 to meet the increasing demand for substantive books on all resource-related issues. Island Press publishes and distributes under its own imprint and offers these services to other nonprofit organizations.

Support for Island Press is provided by Geraldine R. Dodge Foundation, The Energy Foundation, The Charles Engelhard Foundation, The Ford Foundation, Glen Eagles Foundation, The George Gund Foundation, William and Flora Hewlett Foundation, The John D. and Catherine T. MacArthur Foundation, The Andrew W. Mellon Foundation, The Joyce Mertz-Gilmore Foundation, The New-Land Foundation, The J. N. Pew, Jr., Charitable Trust, Alida Rockefeller, The Rockefeller Brothers Fund, The Rockefeller Foundation, The Tides Foundation, and individual donors.

# THE ENERGY-
# ENVIRONMENT
# CONNECTION

# THE ENERGY-ENVIRONMENT CONNECTION

EDITED BY *Jack M. Hollander*

FOREWORD BY *Reverend Theodore M. Hesburgh*

ISLAND PRESS

WASHINGTON, D.C. ☐ COVELO, CALIFORNIA

Grateful acknowledgment is made for permission to include the
following previously published material:

"Acid Rain," by John Harte, from *The Cassandra Conference:
Resources and the Human Predicament*, edited by Paul R. Ehrlich
and John P. Holdren, published by Texas A&M University Press,
1988. "Toward a Sustainable World," by William D. Ruckelshaus,
*Scientific American*, September 1989. "Energy in Transition," by
John P. Holdren, *Scientific American*,
September 1990.

Library of Congress Cataloging-in-Publication Data

The energy-environment connection / edited by Jack M. Hollander ;
    foreword by M. Hesburgh.
        p.    cm.
    Includes index.
    ISBN 1-55963-119-8 (acid-free paper).—ISBN 1-55963-118-X
(pbk. : acid-free paper)
    1. Energy development—Environmental aspects.   2. Energy
conservation.   3. Energy policy.   I. Hollander, Jack M.
TD195.E49E543   1992
333.79'14—dc20                                        92-8634
                                                         CIP

Printed on recycled, acid-free paper

Manufactured in the United States of America

10   9   8   7   6   5   4   3   2   1

Funding to support the publication of this book
was provided by The Energy Foundation.

# CONTENTS

# FOREWORD

POSSIBLY THE MOST COMPELLING PHOTOGRAPH of all time is the portrait of planet earth, a jewel shining in the darkness of space, blue and brown and flecked with white clouds, the human habitat of incomparable beauty.

As we contemplate earth from the moon, we see no sign of the many divisions that have separated humans here, no indication of the factions that have spawned centuries of wars and injustices, no trace of the religious, political, or ethnic differences that still abound. We see but one earth, beautiful and tranquil as it orbits around the sun, not too close to the sun and yet not too far away either.

Is not this vision laden with philosophical and theological implications: that humankind, like planet earth, is one and potentially more beautiful than our history has yet indicated; that we must survive together or perish together in this unitary habitat, sharing the same hospitable climate, the air, the water, the land, and, more significantly, the yearnings for knowledge, freedom, and a civilization marked by justice and peace rather than injustice and war? This vision of our only human habitat underscores the interdependence that characterizes our globe and our lives upon it.

Earth is a great gift of God to each and all of us. It provides an incredible abundance of natural resources that we have harnessed to power our technological civilization and make life more meaningful and interesting. Earth, with its bounty, is ours to use and maintain

with responsible and reverent stewardship, to pass on undiminished to succeeding generations of humans of every sort, whatever their race, religion, nationality, or language. The greatest imperative of our time is to create a spiritual unity of human understanding and justice that matches the physical unity and beauty of the earth, so that we will use and not abuse this precious gift, so that earth will continue to nourish and sustain us and those who follow us for all time.

Reverend Theodore M. Hesburgh
*President Emeritus*
*University of Notre Dame*

# INTRODUCTION

ENVIRONMENTAL DAMAGE HAS BEEN ASSOCIATED WITH ENERGY USE at least since Plato. Scanning the barren hills of Greece, he mourned the lost forests sung of by Hesiod and Homer, fed to the forges and fireplaces of a prospering civilization. For centuries human societies accepted such impacts as a small price to pay for the development of the human habitat. The resulting affluence has freed hundreds of millions of people from disease, poverty, and ignorance. In recent decades, though, we have recognized that the immense scale of our energy use may be responsible for great and irreversible damage to the habitat that sustains us. Will the world's peoples master these impacts and achieve a healthful and sustainable environment? Or will the human capacity prove inadequate to meet this crucial test?

Sharpening the challenge is the fact that some of these problems are regional or global in scale and, moreover, are expected to manifest themselves mostly in the future. Carbon dioxide, an unavoidable product of fossil fuel combustion, may contribute to significant and undesired future changes in the world's climate. Chlorinated fluorocarbons (CFCs), commonly used chemicals, seem to be eating away the protective stratospheric ozone layer, which may increase the incidence of skin cancer. Sulfur and nitrogen oxides from combustion are causing widespread acid precipitation, which may harm human health and damage natural vegetation and crops. The list goes on and on. While there is consensus in the scientific community about the reality of these energy-environment connections, great uncertainty enshrouds their magnitudes and timing.

The tightening connections between energy use and environmental damage seem to present the entire world with a dilemma. How much time do we have before some of these problems become irreversible? How do we evaluate the rights of future generations in making decisions that affect them? What kinds of political arrangements will be necessary to mitigate the impacts? What will the costs be, and who will pay them? Will poor countries be forced to choose between economic development and environmental preservation? Will rich countries summon up the financial resources and diplomatic skill to curb global impacts, while helping the developing world preserve its opportunity for better living standards? There is no consensus. It is clear, though, that new standards of goodwill and global cooperation will be needed, along with new science and new technology.

In one area the human community has made progress—dramatic progress—that offers hope. The superpower nations have halted the nuclear arms race, which might have exploded in an exchange of nuclear missiles that would have instantly devastated the human habitat. This achievement gives hope that humankind will have the vision to confront realistically and humbly the more subtle but equally compelling challenges of the energy-environment dilemma, make the required effort to narrow the scientific uncertainties, and find approaches to sustainable solutions that are technically feasible, politically acceptable, and just.

Connections between human use of energy resources and the environment extend back to the earliest uses of wood for shelter and warmth. Primitive societies were, however, so small and fragile in comparison with the forces of nature with which they contended that their impacts were minimal. When people began, several thousand years ago, to cultivate the land for agriculture and later replaced the digging stick with the plow, their use of land became more intensive and they had to learn respect for the soil, including crop rotation, to maintain its productivity. Significant connections between resource use and the environment surfaced again in the eighteenth century, when agrarian, subsistence-level groups in northern Europe began to make the transition to industrial society with its higher productivity and living standards. Many factors, including the development of machines, helped make the industrial transition possible. Among the most crucial was the availability of inexpensive and abundant energy resources.

One such resource was coal. In early industrial Britain, coal was readily at hand, apparently limitless in abundance, and convenient to mine, transport, and use. With significant advantages over wood, coal powered the industrial revolution and transformed that island nation into the world's first manufacturing society. But coal had a serious disadvantage. It is not a clean fuel—mining coal is inherently messy, and burning coal is inherently dirty. Terrible scars began to appear around the coal mines in what had been Britain's lovely rural landscape, and entire regions became covered with dense smoke from the coal-burning factories. Miners and factory workers and their families were deeply affected by these environmental insults, but probably rationalized them as part and parcel of the new employment opportunities provided by industrialization.

Charles Dickens, in his 1854 novel *Hard Times*, describes the industrial city of Coketown:

> It was a town of red brick, or of brick that would have been red if the smoke and ashes had allowed it; but, as matters stood it was a town of unnatural red and black like the painted face of a savage. It was a town of machinery and tall chimneys, out of which interminable serpents of smoke trailed themselves for ever and ever, and never got uncoiled. It had a black canal in it, and a river that ran purple with ill-smelling dye, and vast piles of building full of windows where there was a rattling and trembling all day long, and where the piston of the steam-engine worked monotonously up and down, like the head of an elephant in a state of melancholy madness.

Coketown industrialist Josiah Bounderby observes, "First of all, you see our smoke. That's meat and drink to us. It's the healthiest thing in the world in all respects, and particularly for the lungs."

Not only were the costs of the workers' slums, the sweatshops, and the miners' lung diseases high, in damage to people and the environment, but these costs were not paid by the industries that generated the damages. Nonetheless, the costs were not avoided—they were either paid in nonmonetary terms, such as ill health by the people directly affected, or postponed to future generations as a legacy of environmental damage. In today's economic jargon such costs are called externalities—real but unevaluated costs not included in standard economic calculations. Only in the last few decades have societies developed a rudimentary understanding of the external costs of energy-environment connections, and even today most energy-related environmental costs are not internalized into the bills paid by those who reap the associated benefits.

Although the negative aspects of the industrial revolution were serious, and gave rise to a host of political repercussions (including Marxism), a positive side was developing as well. A broad middle class emerged, and the general standard of living in the industrializing countries steadily rose, so that by the late nineteenth century the quality of life in those countries far outpaced that in the nonindustrial societies. Eventually the availability of mass-produced consumer goods at affordable prices, especially the automobile, brought to ordinary people leisure and mobility theretofore enjoyed only by the affluent. More important, the new middle class was able to provide sanitation, health care, and education, which ensured that the growth in well-being would not only continue but accelerate and that more of the young would survive.

In the United States, environmental quality was not an issue of general interest during the period of heavy industrialization in the late nineteenth and early twentieth centuries. The affluent were able, as always, to provide their own clean, comfortable environments of beautiful homes, neighborhoods, and country estates. The middle class was gaining a better living environment, as well, and America's cities were becoming more comfortable and enjoyable. The availability of the commons grew, too, as urban, state, and national parks were created, so that all had access to environmental beauty and recreational possibilities. Even society's poorest could enjoy these shared amenities to some extent.

To nourish the economic development that made these amenities possible, the industrial societies intensively exploited their natural resources. In the United States, energy and mineral resources were extracted from the earth's surface and depths; forests were used for wood and wood products; land was used for farms, factories, homes, roads, and railways; rivers and lakes were used for movement of people and materials and as dumping grounds for wastes. The environment was already under moderate stress by the mid-nineteenth century, but signs of heavy stress were relatively few and mostly localized. In any event, when local resources were diminished, spoiled, or depleted, frontier-oriented Americans were motivated to move on to new lands. "Go west, young man," said Horace Greeley.

This practice of exploiting and moving on suggests that a basic presumption about the environment was widely shared—that the earth, with its vast untapped resources and virgin lands, is extremely large compared with human numbers and activities and hence can be expected to support human development essentially without limit. For a brief period in the industrial era, this idea appeared to be correct. Contrary predictions that resource exhaustion would limit population growth, such as those of the nineteenth-century British economist Thomas Malthus, were never borne out, as resource substitution and new technology came to the rescue. And the winds and waters seemed robust enough to cleanse the earth of industry's effluents. Pollution could even be seen as a blessing. This writer recalls that in the depression years of the 1930s, the ubiquitous gray cloud over the steel mills of northeastern Ohio was actually welcomed. Even though one choked on the smoke and one's clothes were instantly soiled by the soot, they meant that the mills were operating and fathers and husbands had jobs and one had a home to live in and food on the table. The rivers were so terribly polluted with iron residues and refuse that the water was almost hidden, but no matter—only an hour away beautiful forests, bubbling brooks, and chirping birds could be found. To me, as presumably to my counterparts in Britain nearly two centuries earlier, the price of prosperity seemed small.

A few farsighted individuals had concerns for the environment earlier than most. In the first decades of this century, visionaries such as John Muir and Gifford Pinchot convinced the nation it should set aside in perpetuity great tracts of virgin lands as national parks, forests, and wilderness areas. These pioneers of the environment must have sensed that the relentless onslaught of human development might leave no acre untouched, and they believed fervently that some of the earth's natural environment should be preserved for all time. What these environmental pioneers did not know, and probably could not have known in their own time, is that merely withdrawing land from development is not enough to save it from environmental insults such as acid rain that affect extensive areas and cross political boundaries without hindrance.

Although the remarkable achievements of these early preserva-
tionists have stood the test of time, the next major chapter in protecting the environment had to await the experiences of recent decades,
including several energy crises, environmental disasters, and wars.
These experiences include a significant increase in chronic energy-
related environmental costs: metropolitan air pollution, once only
mildly irritating but now virtually intolerable in some cities; acid
precipitation, once barely detectable but now known to damage
vegetation across vast regions; and marine oil spills, which disrupt
sensitive aquatic ecosystems. Other costs have been catastrophic: the
chemical spill at Bhopal, the nuclear reactor accident at Chernobyl,
and Iraq's deliberate torching of the Kuwaiti oil fields. Now civilization confronts the possibility of enormous costs associated with
changes in the earth's climate over the next century induced by
increasing concentrations of atmospheric greenhouse gases (mostly
carbon dioxide), the legacy of three centuries of fossil fuel burning.

These experiences have given rise to a new awareness that the
earth does not have unlimited capacity to absorb, dilute, and flush
away the by-products, waste products, and accidental products of
energy use. The scale of human activities has reached global dimension, and the impacts and costs extend far into the future. One of the
most pressing issues of our time is the need for civilization to decide
how to deal with critical energy-environment problems that affect
not only the entire earth but unborn generations as well.

The energy-environment connection can thus be seen as a complex set of interactions surrounding the benefits and costs of supplying and using energy, in the present and in the future. Energy's
*benefits* have been appreciated since the time when energy resources
fueled the remarkable economic transition to an affluent,
consumption-oriented society. Energy remains a prime ingredient of
economic development and its benefits are no less important
today—to sustain the quality of life of the developed nations and to
make possible the vitally necessary future transition of the other 80
percent of the world's people from poverty to a good life.

The *costs* of energy use are more complex and less well understood.
Most of today's energy issues, especially the energy-environment
connections, revolve around energy's costs. A new and profound
energy transition began in the 1970s—from an era of low-priced but
socially costly energy resources and technologies to a future era of
resources and technologies that are higher-priced but lower in social

cost. (An example of the former is coal; an example of the latter might be higher-priced electricity produced from less polluting photovoltaic cells.) For several decades, society has been grappling with how to define, understand, and measure environmental costs, especially those involving the future, and how to develop technologies, behavior patterns, and political arrangements to reduce environmental costs in ways that are socially equitable, economically prudent, and ethically just. The energy-environment connection challenges civilization to find a balance in this complex web of energy costs and benefits and, moreover, to discover relationships among energy, environment, and development that are not only consistent with the reasonable aspirations of the world's peoples for a better life but globally sustainable as well for future generations who are not able to share in the decisions we make today.

This book is a compendium of essays that address the most significant connections between energy and the environment. It is not meant as a guide to specific day-to-day actions that people can take to safeguard the environment or save energy or money; a number of such guides exist. Rather, its purpose is to inform the reader about the factual basis of energy and environment problems and solutions, to convey a sense of context about these facts, and to outline the directions this country and the world must follow in order to create a livable and sustainable environment.

The book is organized around three themes: the environmental impacts of major energy-supply sources, including fossil fuels, nuclear power, and biomass; the environmental and economic benefits of efficient energy use; and policy statements and comprehensive issues, including questions of environmental ethics and Third World economic development. Part I considers the environmental impacts of energy use. In Chapter 1, Umberto Colombo provides a broad overview of the book's major themes. Two major challenges to the global community in the coming decades, Colombo writes, are the protection of the environment and the development of the Third World. Until recently, the development issue and the environment issue were treated quite separately, but there is a growing realization that they are strongly linked and that energy—the motor of economic growth as well as the main cause of global environmental degradation—is the crucial link between them. Energy requirements must be reconciled with the needs of a livable and sustainable environment. The main hindrance to such reconciliation, Colombo points out, is the domination of the world's energy system—especially in developing countries—by short-term economic considerations and a deep conservatism, whereas concern for the global environment requires a long-term view both in philosophy and in concrete actions. We need to think in terms of a "green economy, one that attributes monetary value to the quality of such common goods as air, water, and the whole environment and adjusts market forces according to environmental requirements."

For 200 years, fossil fuels—mostly coal, petroleum, and natural gas—have been the main sources of energy supply for development of the world's industrial societies. The benefits and costs of fossil fuels have been mentioned in the preceding pages. Because the environmental costs, in particular, are so pervasive, this volume offers separate discussions of the major environmental issues associated with fossil fuels: air pollution, acid rain, global climate change, and oil spills.

Air pollution was once thought of as a local, and mostly urban, problem. Today it is increasingly understood that the atmosphere is a complex and subtly balanced chemical system and that noxious emissions from power plants, industrial factories, and automobiles can degrade air quality and affect health, structures, and crops far from the sources of emissions. In Chapter 2, Duncan Brown and I show how the links between energy use and air pollution are tightening, as global energy use continues to increase and U.S. energy use resumes its climb after a decade of decline in the 1970s. Most growth in energy consumption in the next few decades is expected to take place in developing countries, few of which have adequate environmental controls. Of special concern in these countries are rapid growth in the numbers of automobiles and plans for heavy use of coal—an especially prolific source of air pollution—to meet development needs in China and India. But even in the United States, which pioneered many aspects of air pollution control, health-related air quality goals have proved elusive and stiffer regulations are being imposed in most large urban areas. Chapter 2 focuses on descriptions of the major air pollutants and their sources, national and state regulatory efforts to control these sources and emissions, and the ambitious plan that one region—Southern California—has developed to restore the quality of its air in the coming decades. As environmental costs continue to escalate, "pollution abatement will give way more and more to pollution avoidance, as humanity learns new and more energy-efficient ways of managing its industry, meeting its transportation needs, and planning its cities."

Production of strong acids in the atmosphere is a major chemical consequence of sulfur-oxide and nitrogen-oxide emissions from fossil fuel combustion. These acids can be transported long distances through the atmosphere and deposited on earth with rainfall, often called "acid rain." In Chapter 3, John Harte describes the complete cycle of acid rain—from the chemistry of the production process to the consequences of the acidity after deposition in the form of rain, snow, fog droplets, or dry particles. These consequences are pervasive: lakes have been rendered biologically unproductive ("dead"), important fish species and other organisms have suffered reproductive losses, trees and forests have been severely damaged, and the quality of human life has been degraded by visual damage to ecosystems and by loss of essential ecosystem services such as maintenance of clean air and water. Although quantitative evidence detailing these damages is in many cases lacking, Harte urges stronger control measures without delay and argues that "a prudent society ought not make regulatory action contingent on understanding every last nuance of the problem."

Even more dramatic is the issue of global and regional climate changes that may occur as a result of increasing concentrations of so-called greenhouse gases (such as carbon dioxide, methane, and chlorofluorocarbons) in the atmosphere. This is an example of a scientific issue with critical policy implications where the current state of knowledge is surrounded by great uncertainties. These uncertainties become Stephen Schneider's starting point in Chapter 4: a summary of the ranges of scientific estimates of climate changes in the twenty-first century as the atmosphere's carbon dioxide doubles. State-of-the-art "general circulation models" vary widely in their projections of average global temperature, sea level, and average rainfall. Even greater uncertainties arise from the lack of scientific consensus over the effects of cloud cover, the potential for intensification of storm (and heat-wave) frequency and magnitude, and the time delays (perhaps many decades) in realizing the predicted changes. One fact, however, is certain: previous climate changes of the projected magnitudes occurred over thousands of years (at the end of the last ice age), while this man-made change may occur in less than a century. Since animal and bird species can migrate fairly rapidly, but ecosystems cannot, rapid climate changes "could tear apart the fabric of current ecosystems with hard-to-predict consequences." The policy dilemma is whether society should invest resources now to reduce the buildup of greenhouse gases as a hedge against potential change, or whether ecosystems should be adapted to such changes as they occur (by ecological engineering). Because measures to reduce the greenhouse gas buildup—such as increasing energy efficiency and banning CFCs—have other beneficial effects, Schneider opts decisively for the first approach.

The world's annual consumption of oil has been steadily increasing, and during the decade of the 1980s averaged about 2.9 billion tons. About one part in a thousand, or 3.2 million tons, enters the sea each year as a pollutant from marine transportation, municipal and industrial wastes and runoff, natural sources, atmospheric fallout, and offshore oil production. In Chapter 5, Mary Hope Katsouros reviews the state of knowledge of the chemical and physical processes involved in the disposition of oil in the oceans, the biological and environmental impacts of that oil, and the effectiveness of human and natural response mechanisms. Although Katsouros concludes that accidental oil spills and chronic oil pollution do not rank among the first tier of long-term environmental problems, she points out that several large oil spills have wrought catastrophic damage to coastal ecosystems—and that the best response is prevention. Among the many interesting facts that emerge from this chapter is that the 1989 *Exxon Valdez* spill was far from the world's largest in terms of oil flow (36,000 tons); this dubious honor goes to the *Atlantic Empress*, which spilled 257,000 tons near the West Indies in 1979. The greatest spill of all time was not from a tanker, but from Iraq's opening the valves of Kuwait's major oil-loading terminals in January 1991, which released three times as much as any tanker has spilled. The Persian Gulf spill will also prove to be the most damaging because of the gulf's extremely fragile ecosystems.

Nuclear power, despite its controversial history, remains the second-largest contributor to the U.S. electricity system (after coal). But today the United States finds itself with a de facto moratorium on nuclear power more extreme than that of any other major nuclear country. Critics and proponents alike are asking what went wrong with nuclear power and why. In Chapter 6, Christoph Hohenemser, Robert Goble, and Paul Slovic address these questions. Nuclear science, the ancestor of nuclear power, is about one century old, and its first major commercial application, medical X-rays, is generally considered benign, even lifesaving. But the public's real initiation into the nuclear age came with the explosions of two atomic bombs in World War II; since then, most people have distinguished only poorly, if at all, between military and commercial uses of nuclear energy. In the United States, badly conceived government policies and ambiguous safety studies encouraged opposition to nuclear power, but the point of no return came with the nuclear accidents at Three Mile Island (1979) in the United States and Chernobyl (1986) in the former Soviet Union. These accidents could not have been more unlike: no one was killed at Three Mile Island, and its main impact was an economic disaster for the U.S. utility industry; Chernobyl, however, was truly a "worst-case scenario" with serious environmental and health consequences.

A dilemma today for environmentalists is that, even in an energy-efficient economy, nuclear power can substitute for fossil-fueled electricity generation and reduce its most undesirable environmental impacts—acid rain and greenhouse gas emissions. Hohenemser and his colleagues discuss the conditions necessary for a revival of nuclear energy. Technological improvements to reactor safety, which are likely to be remarkable, will provide some of these conditions. But the bottom line, as with all major technologies, is public confidence. Given the public's juxtaposition of military and commercial nuclear images, even the most well-intentioned efforts to correct the built-in errors of the present system may not suffice to rebuild public confidence and fuel a rebirth of civilian nuclear power in the United States.

On a global scale, the most widely used energy-supply alternatives to fossil and nuclear systems make use of biomass. The traditional biomass resources, including fuelwood, charcoal, and various forms of agricultural and animal wastes, provide about half the energy used in developing countries and are especially relied upon by rural households and the urban poor. Even though most current biomass uses are relatively primitive, biomass technologies are widely perceived to be environmentally superior to other energy sources and hence form the core of many proposed "alternative energy" strategies.

In Chapter 7, Janos Pasztor and Lars Kristoferson examine a number of critical environmental issues surrounding increased use of biomass energy. Biomass systems can have serious effects on land use and the quality of air and water. Poorly conceived programs can alter the distribution of wealth in undesirable ways. Such impacts can be eliminated or reduced by proper agricultural and forestry practices. Without such management, bioenergy systems can cause more environmental harm than equivalent fossil-based systems.

On balance, the potential impacts of biomass energy systems are clearly positive. The key to achieving these environmental benefits is integration of many technical and socioeconomic components. Often the positive impacts of a bioenergy system—for example, a newly planted forest that benefits wildlife, local climate, soil erosion, and aesthetics—are difficult to quantify in economic terms, but these factors are important and must be part of the equation. Pasztor and Kristoferson conclude that biomass systems, while posing many significant environmental challenges, have the potential to "form part of a matrix of energy supply that is environmentally sound and therefore a significant contributor to sustainable development."

Amidst a sea of conflicting viewpoints about energy and environment connections, one axiom commands widespread and consistent support: the fastest, cheapest, and surest way to ease energy's environmental damages is to increase the efficiency with which energy is produced, transported, and used. Energy efficiency is a "resource" that can be mined indefinitely to produce energy services of the quality and quantity that people want with less resource use and less environmental impact. In Part II of this volume, energy efficiency in the major end-use sectors—buildings, transportation, and industry—is the focus of three chapters.

Buildings account for more than 35 percent of all primary energy consumed in the United States and 66 percent of all electricity, at a total annual cost of $200 billion. In Chapter 8, Arthur Rosenfeld and Ellen Ward point out that half of this energy—and $100 billion—could be saved annually by energy-efficiency improvements in existing buildings. About 250 million tons less carbon (as $CO_2$) would be injected into the atmosphere—nearly as much as is emitted each year by all the country's automobiles. To convert this "technical potential" into reality is a major challenge to society that will require a social consensus and development of incentives for millions of owners and occupants of buildings to invest in energy-efficiency improvements. The authors advocate a national strategy, with strong government leadership at all levels, to compensate for the short-sightedness of consumer behavior.

Transportation accounts for about 25 percent of U.S. primary energy consumption and 70 percent of U.S. oil consumption. In Chapter 9, Marc Ross and Marc Ledbetter concentrate on automobiles and light trucks, which use about two-thirds of the transportation oil and are responsible for injecting 280 million tons of carbon into the atmosphere annually, as well as carbon monoxide emissions and other important urban air pollutants. Great progress has been made in increasing the energy efficiency of autos and light trucks since 1973. Without these improvements, these vehicles (which now consume 6.5 million barrels of oil per day) would be consuming an additional 4 million barrels per day—more than twice Alaskan oil production and more than half of current U.S. imports. The authors also point out that the 15-year trend of improvement has ended. The average fuel economy of new cars and light trucks has dropped, the number of highway vehicle-miles traveled continues to grow, and the fraction of light trucks—less efficient than cars—in the fleet is increasing. Yet it is feasible, say Ross and Ledbetter, to improve new car fuel economy to over 40 mpg within about 10 years. The chapter outlines the technological, economic, and policy challenges standing in the way of such improvements and quantifies the environmental benefits of overcoming these challenges.

Industry accounts for almost 40 percent of the primary energy resources consumed in the United States, so this sector is at least as important in the energy-environment equation as transportation and buildings. It is simpler to consider industry's use of energy, however, because the market system works better there than in the other two sectors; in industry, decisions about producing, buying, and selling are related more closely to costs. Energy use in U.S. industry has become steadily more efficient for many decades. In Chapter 10, Dan Steinmeyer points to several factors underlying this continuing improvement. First, efficiency in industry, with respect to energy as well as other production factors, evolves naturally from the competitive drive to lower production costs. Second, there is a continuing trend toward reduced use of materials in manufactured goods; a $10,000 computer uses far less material than a $10,000 automobile. Third, consumers are recycling materials more than they used to, a practice that reduces energy use in remanufacturing items such as aluminum cans. Looking to the future, Steinmeyer notes that the major driving force behind improvements in manufacturing is the desire for competitive advantage, which can be achieved both through routine production-cost reductions and through technological breakthroughs that produce new or better products (for example, in the biotechnology industries). Energy-efficiency improvements will usually happen as by-products of other advances, especially in periods of rising energy prices. Similarly, the environmental consciousness of industry is also tied mainly to costs. The key to environmentally beneficial changes in industry is the quantification by society of the costs of pollution, so that industry can internalize these costs into the production process.

The presumption that industrialization and development inevitably lead to a better life for future generations has been supported by so much evidence that it has become virtually a moral imperative of Western culture. Even major environmental issues such as the threat of global climate change have not seriously challenged the basic premises of development but have instead focused attention mainly on the economics of mitigation. Economists debate whether mitigation is a good investment—whether its benefits over the long run, discounted to the present, justify incurring large mitigation costs in the next few decades amidst serious scientific and technical uncertainties. In Part III we consider energy in transition.

In Chapter 11, Richard Norgaard and Richard Howarth hold that the basic issue is ethical rather than economic. The real question, they say, is not whether climate mitigation will provide a suitable return as an investment for us but "What kind of a world do we want to leave to our children and what is the best way to provide it?" Framing the question as an investment decision, they say, implies that current generations have the *right* to exploit the atmosphere, because future generations can be expected inevitably to be better off. Norgaard and Howarth argue that the environmental rights of future generations involve issues of intergenerational equity, not economic efficiency, and must be approached as moral issues first. In political debates about civil rights, the authors note, economic calculations are inadmissible because equity or moral criteria must take precedence. They conclude that "it is better economics for society to make moral decisions about the climate rights of future generations and let the economy follow than to construct arguments for or against such moral decisions using reasoning and data rooted in an implicit assumption that current generations hold all rights."

Among the original imperatives for human survival was sustainability. Preindustrial societies had to sustain the ecosystems upon which they depended for food, shelter, and warmth, for their only alternatives were to move or perish. As a consequence they developed a consciousness about nature that was essentially spiritual; humans considered themselves to be part of the ecosystem, shared with the animals, plants, and the earth. Only later, with industrial development, did humans assert their dominance over nature and custom tailor their environment by intensive exploitation of natural resources and environmental services. Today the idea of sustainability is regaining its importance as people ponder the future of energy and the environment—for, as pointed out earlier in this introduction, the scale of human activities is beginning to overwhelm the environment.

According to William Ruckelshaus in Chapter 12, a new creed of sustainability must be developed on the basis of three fundamental beliefs. First, the human species depends for sustenance on a finite natural world and must preserve the natural systems that regenerate this world. Second, economic activity must account for the environmental costs of such activity rather than "stealing" wealth from future generations. Third, the maintenance of a livable global environment depends on the sustainable development of the entire human family. If 80 percent of the world's people remain poor, there will be no peace; but if the poor try to attain affluence by emulating the past practices of the affluent, sustainability will be lost for poor and rich alike. Ruckelshaus suggests that a sustainable world economy will require people to redefine concepts of political and economic feasibility, and to listen more closely to the words of the earth, upon which we utterly depend.

Most students of energy agree that the so-called energy crisis of the early 1970s marked society's first general awareness of a permanent change in the terms on which we use energy. In Chapter 13, John Holdren notes that while civilization will not run out of energy resources in an absolute sense, the most accessible and inexpensive stocks of oil and natural gas are already gone except in the Middle East, whose stocks are likely to be depleted in a matter of decades. Since the more abundant energy resources that could in principle replace oil and gas—coal, solar energy, fission and fusion fuels—are all expensive to develop and use in acceptable ways, the transition to high-cost energy will be permanent. High costs for energy will extract a toll on civilization, and particularly on oil-poor Third World countries.

The energy transition is characterized by depletion of another finite resource: earth's capacity to absorb energy-related disruptions of the environment. Civilization will eventually have to pay for the loss of services provided by damaged ecological and geophysical processes such as building and fertilizing soil, regulating water supply, controlling pests and pathogens, and maintaining a tolerable climate, whether or not it develops technologies to replace these services. The transition, then, is to an energy system in which these costs are more fully reflected in the monetary costs of energy. Holdren makes the case for a two-pronged strategy: "no regrets" and "insurance policies." No-regrets actions provide leverage against the potential problems of the energy transition but also provide benefits even if the problems do not materialize; examples include reducing the environmental costs of existing energy sources, increasing the efficiency of energy use, and accelerating research on long-term energy alternatives. Insurance-policy actions offer leverage at modest cost, so that even without sure benefits they are a good investment. A plan to reduce global carbon emissions by 20 percent per decade would be an important insurance policy against global climate change.

A crucial factor bearing on future energy-environment connections is population. Today's world population of 5.3 billion people will surely grow to at least 9 billion, and some believe the number could reach almost twice that. No matter how difficult controlling population may be, the difficulty of providing energy, food, water, and shelter to a world of 15 billion people would be far greater. Holdren concludes with a reminder that "none of the preceding measures, nor all of them together, will be enough to save us from the folly of failing to stabilize world population."

\*    \*    \*

Environmental decisions increasingly determine our energy options. In this way, energy use reflects many of our deepest values and social goals. Future generations, examining the options we have given them, will judge those values by how we defined and pursued the goal of a sustainable world.

As we recognize the global scope of environmental problems, and the potential for irreversible harm to our environment, we will move toward that goal. Science and technology have given us many of the tools we will need. But those tools must be put to work by societies acting strategically and cooperatively. As the challenge of geopolitical struggle fades, nations face an unprecedented opportunity to rise to the challenges of a new sustainable world. These challenges are illuminated in informative and often provocative ways by the contributors to this book, who have inquired into many of the issues that will shape the lives of generations to come.

Jack M. Hollander

# PART I

# THE ENVIRONMENTAL IMPACTS

CHAPTER 1

# DEVELOPMENT AND THE GLOBAL ENVIRONMENT

*Umberto Colombo*

THE DEVELOPMENT OF THE THIRD WORLD and the protection of the environment are two major global problems humankind must deal with in the next decades. These two problems are strictly interconnected. Energy—the motor of economic growth and the main cause for degradation of the global environment—is the connection. Only an appropriate response to the energy issue can assure compatibility between the need for development and the preservation of the environment, including global climate.

Until recently these issues were treated quite separately—more as regional and local problems than with a planetary vision in full awareness of their interconnectedness. If one compares the discussions taking place in meetings of high-level experts in the fields of energy and environment, one cannot fail to be puzzled by the sharply divergent views of these two circles.

When energy experts meet—for example, on the occasion of the World Energy Conferences—they usually produce forecasts on global energy supply and demand indicating substantial growth in consumption as an extrapolation of the trends that have so far prevailed. Fossil fuels, as we are all aware, now supply almost 90 percent of the total commercial energy demand worldwide, and their consumption has risen from 500 million tons a year at the beginning of the century to the present 7 billion tons. In 1989 oil contributed 43 percent to the total supply of fossil fuels. A reduction in this high share has been in the agenda of policymakers ever since the energy crises of the 1970s. The questionable security of energy supplies has made replacement of oil by alternative fuels, mainly gas and coal, a recurring objective. Coal is the most abundant fossil fuel on earth, and its repenetration into the world energy system—which it dominated prior to the advent of oil—is seen by many energy experts as the most appropriate solution to our problems.

Such a solution is seen under a very different light by environmentalists, whose awareness of the potential risk of global warming— caused largely by increasing emission of carbon dioxide resulting from fossil fuel burning—is now a major feature of the environmental debate. A serious threat to the survival of humankind has been identified in the climate change induced by increased emissions of greenhouse gases, principally $CO_2$. The message that comes ever more loudly from global climate experts is that action is urgently needed: first to check and then substantially to reduce these emissions in order to halt the global warming trend.

Let us now examine in greater detail this three-knotted nexus, starting with the environment. In its global dimension, the environment involves first and foremost the world's climate. Climate has always been in a state of change. Smaller and greater variations have occurred repeatedly in the earth's history with different cyclical frequency. At the height of the last glaciation, about 18,000 years ago, Europe and North America were covered by a layer of ice about 2000 m thick. Sea level was about 100 m lower than it is now, and the landscape was quite different. Then, about 10,000 years ago, the climate began to heat up, the glaciers melted, and the sea level rose to about today's level. In short, the climate became highly favorable to humans and fostered the beginnings of agriculture worldwide.

Periods of minor glaciation recur in millennial cycles, the last of which took place in the second half of the seventeenth century. There is also some evidence of a two-century cycle of alternating warmth and cold provided by such indicators as the size and carbon-14 content of tree growth rings. These climatic changes have characterized the history of our planet, and they are likely to continue to occur in the future.

But today there is a difference. Since the nineteenth century, with the industrial revolution, the human population and its economic activities have grown to such an extent as to interfere seriously with natural phenomena by complex feedback mechanisms we do not wholly understand—mechanisms that could accelerate and intensify climate changes. Human society, in other words, has become a force capable of influencing our global environment. Urbanization, industrialization, the development of overintensive agriculture, deforestation, and, perhaps most important, consumption of ever growing amounts of energy in the form of fossil fuels (first coal, then, increasingly, oil and gas) with the release of huge quantities of carbon dioxide in the atmosphere: all of these anthropogenic processes are making us finally reflect on the fact that we have begun playing with the climate on a planetary scale—and may soon have to pay a high price for it unless we make full use of our ever growing knowledge with longsightedness and, may I add, humility.

Let me now briefly review some of the global environmental issues that have come to the fore in recent times. First, ozone. We know that the ozone present in the stratosphere, roughly between altitudes of 12 and 25 km, protects us from incoming ultraviolet radiation. After years of debate, scientific evidence has shown that two classes of chemical compounds, namely chlorofluorocarbons (CFCs) and halons (chlorinated and brominated organic products), are responsible for the observed depletion of the ozone layer. These organic compounds are produced by a few chemical firms and used in different industrial and consumer sectors for a variety of more or less useful purposes. Although scientific debate on the depletion of the ozone layer has gone on for over a decade, only in 1987 was an international protocol signed in Montreal to reduce production of CFCs and halons.

The Montreal Protocol was a historical landmark, but it proved not to be enough. In fact, conclusive scientific evidence of the destruction of stratospheric ozone by CFCs and halons has recently been gathered, and commitments for a more drastic reduction of their production were undertaken at the 1990 London Conference. Replacement products and technologies that will allow us to live comfortably without CFCs are gradually coming to the fore and should make a total ban of these obnoxious products less painful. One issue still on the floor is the need to distribute fairly the economic burdens deriving from a ban of CFCs, especially with respect to developing countries, some of which have invested heavily in CFC-related technologies.

Another environmental problem of international, if not global, dimensions is acid rain, which is largely due to emissions of sulfur dioxide and nitrogen oxides in the combustion of fossil fuels—for example, in power plants, domestic heating, and automobile engines. Acid rain exerts its deleterious effects on the ecology of water systems, on forests, and on historical and cultural artifacts. The acid rain produced by some countries' emissions often happens to fall on other countries. The problem was underrated until the evidence of its importance became overwhelming. Suitable measures are now being taken in a number of countries and regions, including North America and the European Community, but we are still lacking worldwide coordination and adequate action to prevent growth of acid rains, which are increasingly taking place in many developing countries.

But the most important of all global environmental problems is greenhouse warming, which could induce climatic changes of geological proportions even before the end of the next century. The temperature of the earth's surface increased about 0.6°C over the last 100 years, and as a consequence sea level is estimated to have risen by perhaps 20 cm. Humankind is contributing with a great many economic activities to the increase in the atmospheric concentration of various greenhouse gases—gases that absorb heat which would otherwise escape into space and then return some of it to the earth's surface, thus raising the temperature of the planet. Carbon dioxide from fossil fuel combustion, methane from increased human activity, CFCs (which, as we have seen, are also depleting the ozone layer), deforestation—all contribute to the greenhouse effect. Most scientists agree that there is a cause–effect relationship between the observed emission of greenhouse gases and global warming. Furthermore, they predict that if the atmospheric concentration of greenhouse gases continues to increase, as present trends of fossil fuel consumption indicate they will, the earth's temperature may increase in the next century by another 2 and perhaps 4°C. If this prediction comes true, the sea level could rise between 30 and 60 cm before the end of the twenty-first century. The impact on coastal settlements could be dramatic; there could be a displacement of fertile zones for agriculture and food production toward higher latitudes; and the decreasing availability of fresh water for irrigation and other essential uses could seriously jeopardize the survival of entire populations.

We are dealing here with nonlinear phenomena, and it is possible that relatively small perturbations caused by human activity may trigger dramatic and unforeseeable excursions in global and regional climates. The mechanism of the greenhouse effect is itself complicated by feedbacks that can be negative or positive—thus compensating or enhancing, respectively, the primary warming due to increased atmospheric concentrations of $CO_2$ and other greenhouse gases. Among the negative feedback processes might be higher rates of photosynthesis and greater $CO_2$ absorption by seawater, respectively resulting from higher temperatures and from greater and deeper ocean circulation.

But positive feedbacks—linked, for example, to increased plant respiration (accelerated metabolism)—might actually deplete the stock of biological carbon and discharge even more $CO_2$ into the atmosphere, thus accelerating global warming. Analysis of the air bubbles trapped in ice cores in the Antarctic indicates that over the past 160,000 years temperature increases have been regularly associated with higher atmospheric concentrations of $CO_2$.

Thus we still have much to learn about the complex problems that link the geological carbon cycle, with all its physical, chemical, and biological components, to the climate. But the issue is not only scientific. It involves the political world, as well, for sooner or later policy decisions will have to be made. The heart of the matter is whether we can afford to give priority to enhanced scientific research while postponing measures aimed at counteracting possible greenhouse warming—or whether we should start to act now with determination despite the scientific uncertainty. This second option is now clearly prevailing in Europe, as the preliminary conclusions of the Intergovernmental Panel on Climate Change (IPCC) become available.

Let us now take a look at the question of energy. Since 1900, annual world consumption of primary energy has risen from 600 million tons of oil equivalent (TOE) to 8 billion in 1989. In particular, as noted earlier, annual world consumption of fossil fuels rose from 500 million to 7 billion TOE. This led to a substantial increase in the atmospheric concentration of $CO_2$. Measurements taken at the Mauna Loa observatory in the Hawaiian Islands show an increase from 315 ppm in 1958 to 350 ppm in 1989. In the preindustrial age, the concentration was in equilibrium at 280 ppm.

The world's population has grown from about 1.5 billion people in 1900 to 5.3 billion today. More than three-quarters of them live in the Third World, which accounts for only one-quarter of total energy consumption. Per capita energy consumption in the Third World is over ten times lower than in the industrial countries. The United Nations reports that per capita income in as many as forty countries is below the poverty level. Adding in China and the lower classes of India's population, where per capita income is also very low, it is evident that an enormous number of people (about half of humankind) are living in poverty. Life expectancy in the poorest countries lags far behind the world average, notwithstanding the "Green Revolution" that has made countries like China, India, the whole of South and Southeast Asia, and a good part of Latin America self-sufficient in food production. It must be kept in mind that world population will probably about double once again in the next century before leveling out—and that it will take a huge amount of resources of every kind, especially energy, if the Third World is to emerge from poverty.

Taking into account demographic trends, most current projections of world energy consumption in the decades to come indicate quite conspicuous increases both in the South and in the North of the planet. A group of thirty experts who carried out a study for the 1989 World Energy Conference in Montreal predicted for the year 2020 a rise in global energy consumption of 50 to 75 percent in the period starting from 1988. Thus even in the lowest economic scenario (one assuming a yearly rate of economic growth of 1.5 percent in the North and 2.8 percent in the South), the global energy demand is estimated to reach 12 billion TOE by the year 2020. The present low prices of fossil fuels, in spite of what many commentators expected from the Gulf War, does not encourage energy conservation or favor the penetration of renewable energy sources now competitive only in very special circumstances. Now that the Gulf War has come to an end, let us hope that policymakers will take advantage of expected low prices of energy for investing in R&D on renewable sources and conservation.

As far as nuclear power is concerned, this source is in a quasi-stalemate situation. Unless major advances in technology are achieved in order to assure substantial progress in safety and an acceptable answer to the issue of radioactive waste disposal, for the time being it is unlikely that its share in world energy supply will increase.

This is the energy background against which one should consider the need to reduce emissions of $CO_2$ and other greenhouse gases. The 1988 Toronto Conference on global environmental security recommended that we should cut global emissions of carbon dioxide by 20 percent by the year 2005. For this reduction to take place, one must act on several fronts. First and foremost, the efficiency of energy production and use must be increased, both in the developed and developing countries. Even in all our economies, we are still very far not only from the maximum theoretical efficiency but also from what would be achievable through the use of currently available technologies. Second, there must be more research and development of renewable energy sources, in particular wind, photovoltaic solar, and biomass. Of these, wind power is closer to competitiveness (indeed, it is already competitive in favorable geographic areas), but it has a lower potential for long-term market penetration than either photovoltaics or biomass.

For conservation and renewables to take off, it would be essential to set in place economic instruments—for example, a special tax—to force a rise in the price of fossil fuels. This solution, sometimes referred to as a "carbon tax," would also encourage a structural shift in the mix of fossil fuels toward use of more gas and less coal, desirable not only from the global warming viewpoint, given that the combustion of gas results in a lower $CO_2$ emission than that of oil and coal, but also for the control of acid rain and particulate emissions.

Achievement of the Toronto objective would also be favored by fostering the "dematerialization" of society, spurred by the ongoing technological revolution in the North of the planet. This process is taking place as new materials and technologies requiring less energy to perform the same functions replace conventional ones. Promoting a "recycle economy," aimed at reducing the amount of waste produced everywhere, is also part of this strategy.

This whole problem may seem intractable. However, our industrial economies can undertake—without undue sacrifice—an effort to substantially reduce greenhouse gas emissions. Energy conservation is at hand with the use of advanced technologies and, in fact, in our industrial economies there is already a trend to decouple the growth of GDP from the need for more energy. The more we become a service-centered economy, the less energy and raw material we need to produce an additional unit of GDP.

Things are quite different, however, in developing countries, where socioeconomic progress is, and for some time will be, severely conditioned by the amount of energy available to fuel the economy. It is therefore indispensable that we make available advanced technologies to the Third World, where at present a large fraction of the primary energy used is wasted.

Thus improved energy efficiency worldwide and incentives to use renewables are the key elements of a global energy strategy aimed at reducing $CO_2$ emissions. To these components we must add the imperative of halting deforestation and putting in place a worldwide reforestation plan. All of this of course requires major financial resources that could be raised, for example, by implementing a "Global Fund for the Atmosphere" using the revenues of a carbon tax to be levied on fossil fuels. Even a small tax, a few cents per gallon of gasoline, would mean the availability of many billions of dollars every year. The World Bank has created recently a "Global Environmental Facility," a step in the right direction.

We can thus conclude that the objective of reconciling energy needs with environmental requirements, difficult as it may be, is not unattainable. What worries one is that the two expert circles—energy and environment—do not speak the same language. The energy system is dominated by short-term economic considerations and marked by a deep conservatism. Change is slow, given that significant penetration of a new energy source takes many decades. Concern for the environment and particularly for long-term climate change, on the other hand, requires policy actions that reach far beyond these time frames. Conventional economics is often dismissed by hard-core environmentalists as being apparently incapable of treating environmental parameters appropriately. Patterns of resource exploitation and industrialization often overshoot ecological thresholds and, up to now, environmental costs have been only marginally regarded as parameters in decision making. We have therefore to think of ways to bridge these incompatibilities between economic and environmental exigencies—not only in individual nations, both developed and developing, but globally.

Leaving energy for a moment, we should remark that our perception of environmental problems and their relationship with development changed drastically in the 1980s. The concept of "sustainable development" gained general consensus after the Brundtland Report entitled "Our Common Future" was published in 1987. It calls for a pattern of economic growth that, while satisfying the needs of the present generations, does not jeopardize those of future generations. This principle requires us to think in strategic rather than tactical terms.

Today we know that it is not a presumably limited amount of nonrenewable resources that poses constraints on development. New resources can be "invented" using science and technology. Renewable energy sources can be developed to the extent that, a few decades from now, they may constitute a significant proportion of energy supply. The economy can be dematerialized by using more functional and efficient materials and processes, by replacing goods with information and services, and by reorganizing production to reduce and recycle waste. Quality can be improved and production diversified to better satisfy the needs of customers and society as a whole. High productivity can be linked to quality.

What can indeed limit growth are changes in the environment and climate. We can hope to expand these limits only if we pursue new technologies and solutions in the perspective of sustainable development. Conditions enabling a change of course do exist in the North, and it is here that we ought to begin setting a concrete example. We must accept, however, that more quantitative growth is still required over the greater part of the planet. Developing countries lack the most basic infrastructure and have yet to satisfy the fundamental needs of rapidly growing populations. Socioeconomic growth and an equitable income distribution are themselves a precondition for halting population explosion, as the experience from industrial countries proves.

The growth patterns hitherto adopted in the developing countries have relied on highly energy-intensive and materials-intensive processes and technologies imported from the industrialized world.

These have carried with them the same problems of pollution and wasteful resource use already experienced in the past in industrial countries. Moreover, deforestation has often been seen as a way to increase the amount of land available for cash crops and as a vital source of hard currency from the sale of timber.

It is feasible for North and South to work hand in hand so that Third World development can benefit from the whole array of advanced energy-efficient and materials-efficient technologies, exploiting them straight off in farming, in manufacturing for local consumption and export, and in building efficient infrastructure. In this connection one must admit that, by and large, aid-to-development policies carried out by most industrialized countries have not promoted environmentally benign practices in developing countries. Rather, industrial and other projects have been based on conventional, sometimes even obsolete, approaches characterized by high environmental impact and wasteful use of resources. Thus a drastic revision is needed to orient aid-to-development toward the buildup of an endogenous system of scientific and technological research capabilities in the Third World, emphasizing the use of renewable energy and encouraging a decentralized and balanced pattern of development.

In the same vein, it is absolutely essential to reverse the present pathological flow of financial resources from a heavily indebted South to a much richer North. Over the last decade, throughout the South debt service has risen as a percentage of exports. According to World Bank figures, by 1988 debt service had reached 28 percent for sub-Saharan Africa, 38 percent for North Africa and West Asia, and over 40 percent for Latin America. Yet resources generated in the developing world must be husbanded and reinvested if sustainable development is to be achieved. This is an issue that has been discussed at the international level since the mid-1980s, yet very little has been achieved. Per capita income growth actually became negative for both Africa and Latin America during the decade, and expectations for development are not much brighter. Meanwhile, despite periodic bursts of self-questioning when the oil price surges as a result of events in the Middle East, the North continues to squander energy and resources.

To cope with the inherent risks of the future—those generated by human activities and those caused by natural phenomena and disasters—we must develop rational resource use and flexible strategies incorporating all our scientific and technological learning, as well as the strength we derive from the wealth of our culture. Being flexible and adaptive means learning what nature teaches us with its extraordinary wealth and variety of plant and animal species and the mechanisms that enable them to respond to emergencies. Nature does not consider one species or one solution inherently better than another, but fosters precisely the varieties whose properties enable them to adapt to hazardous situations. For this reason, too, it is important to safeguard the diversity of the world's species, many of which are threatened by stresses on the world's major ecosystems.

Following nature, we must readjust our systems, giving greater emphasis to environmental requirements. A step in the right direction would be a process aimed at internalizing environmental costs and introducing into government thinking (and industrial decision making) parameters that are not strictly economic in the traditional sense, for a healthy environment can no longer be entered in any economic equation as a "free good." We are beginning to realize that conventional patterns of resource exploitation and industrialization have overshot ecological thresholds, yet conventional economics seems incapable of treating environmental parameters appropriately. We must devise alternative economic approaches to analyzing the global problems now before us and evaluating the costs and benefits of proposed solutions. We should indeed start thinking in terms of a "green economy," one that attributes monetary value to the quality of such common goods as air, water, and the whole environment and adjusts market forces according to environmental requirements.

Decisions are being forced upon us. A conscious change in direction toward an ethical stance with regard to the environment and development will mean a process of selection sifting the positive aspects of our traditions from those that place too great a burden on nature. This will be a chastening experience, for we are used to seeing the world through the optic of our own centrality, history as the succession of human generations. There are bigger forces at work, forces that are altering the scale of our perspective and humbling our overreaching arrogance.

Environmental concern is not going to be a transient phenomenon. It will continue to dominate the political and cultural arena into the next century. Development has rightly been seen as the greatest challenge facing the world. The precept that development should hinge on sound management of natural resources and the environment no longer has merely local or regional significance. Our planet is smaller than it used to be, but our responsibilities are not diminished thereby. To the contrary, each and every one of us ought to feel a sense of ethical commitment as a citizen of the world.

CHAPTER 2

# AIR POLLUTION

*Jack M. Hollander and Duncan Brown*

THE PROGRESS OF INDUSTRIAL CIVILIZATION in the past 200 years, especially the worldwide combustion of fossil fuels, has exacted a heavy
toll on the atmosphere. Mexico City is plagued almost year-round by
lung-biting, eye-stinging, crop-damaging photochemical smog, produced by high concentrations of fossil-fuel-burning vehicles and
industry in a sun-baked basin rimmed by high hills. At Shenandoah
National Park, sulfur and nitrogen oxides from coal-fired power
plants and ozone from nearby cities and highways have damaged
vegetation and cut visibility to an average of 15 miles. Germany's
Black Forest, bathed in pollution from the region's coal-fired power
plants, is turning into a thicket of stumps and snags. A century of
industrial pollution has damaged cultural treasures such as Greece's
Parthenon, Italy's Venice, and the great medieval cathedrals more
than all the previous centuries' wars and weather. The World Health

The authors wish to thank Barbara Finleyson-Pitts and Tihomir Novakov for their critical
reading of this chapter.

Organization estimates that 70 percent of the world's population breathes air that contains unhealthy concentrations of suspended particulates at least part of the time (World Health Organization and U.N. Environment Program, 1987).

Until quite recently air pollution was thought of as a local problem afflicting cities and industrial areas. Noxious emissions were often simply vented upward through tall smokestacks in the belief that the vast ocean of air would absorb and dilute them. Today we realize more and more that the atmosphere is a complex and subtly balanced chemical brew. Surprisingly small changes in its minor constituents can have major economic and ecological effects, not only locally, but regionally and globally. Sulfur dioxide, for example, rarely exceeds concentrations of 50 ppb, but it forms acids in the air that corrode stone and metal and reduce visibility over large areas downwind of its emissions. At similar concentrations, nitrogen oxides ($NO_x$), mainly nitric oxide (NO) and nitrogen dioxide ($NO_2$), form acids with similar effects and also help produce photochemical smog. The entire world may be warming, owing to the buildup of carbon dioxide ($CO_2$) from fossil fuel burning (at 350 ppm, a 50 percent increase from preindustrial levels) as well as several other gases. Lead in the air is considered a serious health hazard at concentrations of a few millionths of a gram per cubic meter. Ozone is a major urban pollutant, largely arising from automobile use, that damages vegetation and can harm lungs at concentrations of about 100 ppb. Chlorofluorocarbons (CFCs) from air conditioners, refrigerators, foam-blowing agents, and spray cans are eroding the stratosphere's ozone layer (a vital shield against ultraviolet radiation) and, with methane, nitrous oxide, and carbon dioxide, may be amplifying the global greenhouse effect at concentrations measured in parts per *trillion* (Graedel and Crutzen, 1989).

Pollutants interact with one another in complex and unexpected ways, reflecting the chemical complexity of the atmosphere. One ironic example is that thinning of the ultraviolet radiation shield provided by stratospheric ozone could boost ozone formation at lower levels, where it is a noxious and health-threatening pollutant (because ultraviolet light helps form ozone and photochemical smog). Global warming from the atmospheric buildup of carbon dioxide and other gases (including CFCs) could amplify this effect. A simulation of this interaction in Nashville, Philadelphia, and Los

Angeles suggests that stratospheric ozone depletion, combined with greenhouse warming, could increase smog formation as much as 50 percent (Whitten and Gery, 1986).

The more we learn about the atmosphere, the more important the subtle balance of its components appears. Tens of billions of dollars are spent each year, worldwide, on abating air pollution from human activities. Hundreds of billions more are spent on health care, maintenance of structures, and lost crops due to air pollution—not to mention the immeasurable costs of shortened lives and irreparable damage to civilization's treasures. As the world's fossil fuel consumption continues rising, these costs will grow heavier. In the future, pollution abatement will give way more and more to pollution avoidance, as humanity learns new and more energy-efficient ways of managing its industry, meeting its transportation needs, and planning its cities.

## PROGRESS (AND UNMET GOALS) IN THE UNITED STATES

The United States has made much progress over the two decades since the Clean Air Act of 1970 put air pollution on the national agenda. Between 1972 and 1988 the nation doubled its spending on air pollution prevention—in real dollars—to $31.9 billion (CEQ, 1991:272). These investments have had impressive payoffs. From 1970 to 1989—a period in which GNP nearly doubled and energy use grew 20 percent—lead emissions decreased by 96 percent (mainly from the phasing out of leaded gasoline); carbon monoxide emissions decreased by 40 percent and volatile organic compounds (such as gasoline vapors) by 31 percent, owing mainly to requirements for catalytic converters and other automotive engine improvements. Emissions of total suspended particulates (smoke, dust, chemical particles, and the like) from vehicles, power plants, and industry have fallen 61 percent. Sulfur dioxide emissions have been reduced by 26 percent through the use of flue-gas scrubbers and lower-sulfur coal by electric power plants and factories. Of the six "criteria pollutants" that the U.S. Environmental Protection Agency (EPA) regulates for the sake of air quality, only nitrogen oxides show rising emissions—an 8 percent increase since 1970 (EPA, 1991:1–14).

Still, in 1989, some 84 million Americans lived in areas that failed to meet at least one air quality standard (EPA, 1991). Since these standards are set on the basis of risk to human health, this means that, in most major population centers, susceptible people—the aged and those with asthma or diseases of the lung or heart, for example—may be at risk. The standard for ozone—a main ingredient of smog and a serious threat to the lungs—is by far the most widely violated; of those 84 million people, 67 million live in ozone "nonattainment areas." Some 33 million and 27 million people live in areas that violate the standards for carbon monoxide and particulate matter, respectively (EPA, 1991).

Moreover, a recent EPA study found that a wide variety of toxic chemicals—carcinogens, mutagens, and other poisons—are being emitted into the air in billions of pounds by large manufacturing facilities each year (CEQ, 1991). (The study did not look at smaller manufacturers, firms other than manufacturers, government, or homes.) The 1990 Clean Air amendments imposed controls on most of these emissions for the first time.

The link between energy use and air pollution is growing tighter. U.S. energy use began to increase in the 1980s, after a decade of decline, as Americans drove more and residential and commercial energy use began to rebound. Per capita energy use is back at the level reached in the mid-1970s, and the nation's total energy use is at an all-time high.

The U.S. Congress, recognizing these trends, passed a sweeping series of air pollution measures in 1990 intended to regain lost ground. The 1990 Clean Air Act amendments established new, more distant, deadlines for meeting the ozone, carbon monoxide, and particulate matter standards. At the same time, the amendments tightened enforcement of all three standards in many ways. They require states to make steady, measurable progress toward attainment in the form of specified percentage reductions of emissions. They impose a variety of new restrictions on emissions from vehicles and stationary emission sources. New penalties for failure to meet deadlines include fees and withdrawal of highway funds. The new restrictions include, in areas furthest from attaining the standards, restrictions on driving and requirements for fleet owners to use vehicles powered by natural gas, electricity, or other so-called clean fuels.

Besides these measures to secure attainment of the ozone, carbon monoxide, and particulate standards, the amendments have a variety of other provisions. They impose tighter standards on vehicle-tailpipe emissions of nitrogen oxides (a key ingredient in smog and acid rain), hydrocarbons, and carbon monoxide. To help clean Southern California's polluted air, they require a special clean-fuel program in California. They require the use of "maximum achievable control technology" for toxic emissions from industry.

The new legislation also contains a special acid-rain reduction program. Although the health-based air quality standards for sulfur dioxide and nitrogen oxides are violated in few areas—thanks to measures required by the original Clean Air Act of 1970—these substances have been implicated in regional acid precipitation downwind of emission sources. Both are to be reduced substantially. Sulfur dioxide emissions, currently about 20 million tons per year, are to be cut by about 8 million tons. This requirement breaks new regulatory ground by establishing a system of emission allowances—in essence, licenses to emit specified amounts of sulfur dioxide—that are tradable both within and among utilities. The intention is to give utilities incentives to adopt the most cost-effective measures to reduce their sulfur emissions; utilities that manage to cut emissions below the level set by their allowances may sell the excess allowances to those that find it more costly to reach those levels. This policy is expected to bring about the desired emission reductions at the lowest cost.

The acid rain program's nitrogen oxide reductions, in contrast, do not follow a market-based approach. Annual emissions—today about 20 million tons—must be reduced by about 1 million tons by the year 2000 through specific emission limitations on certain types of utility and industrial boilers. Market trading of emissions is not likely to be extended beyond sulfur dioxide because the chemistry of most other pollutants is so complex that unintended side effects could occur.

Measures required by the Clean Air Act amendments of 1990 may cost as much as $25 billion a year by the year 2000, according to initial White House projections (Rosewicz, 1990). But until detailed regulations implementing the new law are issued, it will be impossible to make a firm estimate.

## RISING ENERGY USE WORLDWIDE

World energy use, after years of slow growth, began rising rapidly in the mid-1980s. Between 1980 and 1986, energy use rose at an annual rate of only 1.5 percent (mainly because of near-zero growth in the major industrialized countries). But 1987 saw a 3.1 percent increase, and in 1988 the figure was 3.7 percent (U.S. Bureau of the Census, 1990). Most of the growth in energy demand for years has been in developing countries. Between 1973 and 1987, their energy use grew at an annual average of 5.1 percent (Levine et al., 1990). Their share of the world's commercial energy use rose from 14 percent in 1970 to 23 percent in 1985 (Sathaye, Ghirardi, and Schipper, 1987). Developing countries are expected to continue to account for most of the growth in energy use. The newly industrializing countries of Southeast Asia show the fastest growth of all: an 11.4 percent increase in 1988. Japan's energy use rose by 6.2 percent in that year, and that of the United States rose by 4 percent, reaching its highest level ever (World Resources Institute, 1990:141ff).

In 1988, oil accounted for 38 percent of world energy use, coal for 30 percent, natural gas for 20 percent, hydropower for 7 percent, and nuclear fission for 5 percent (World Resources Institute, 1990:141ff). This does not count the so-called traditional fuels—wood, charcoal, and crop residues—that make up a large part of Third World energy use. (The health effects of indoor and outdoor air pollution from the traditional fuels have not been thoroughly studied but are known to be significant in many places—notably among women and children who spend long hours in smoky kitchens.)

Natural gas is widely expected to be the world's fastest-growing energy source in the 1990s; many nations intend to use gas to reduce oil dependence and the environmental problems associated with other fuels (World Resources Institute, 1990:141ff). (Gas produces nearly no sulfur dioxide when burned and only half the carbon dioxide of an equivalent amount of coal.) But coal, while it may be dirty, is extremely abundant and widespread and hence will see substantial growth too, especially in developing countries. China and India, for example, both have substantial coal reserves and populations demanding higher standards of living. High growth in coal use can also be expected in the former Soviet Union and many Eastern European countries.

Growth in automobile use presents special energy and environment issues. There are about 500 million vehicles on the world's roads, and their population is increasing faster than the human population (Davis, 1990). Again, most of the growth is in developing countries. U.S. cities show how difficult it can be to control health-threatening pollutants due to automobile traffic, pollutants such as carbon monoxide and photochemical smog. In Third World cities, especially those in the tropics and subtropics, these problems will be even greater because of the more intense sunshine that induces photochemical smog formation and because pollution-control resources may be limited. Mexico City should be a warning.

Keeping the environmental impacts of the expected growth of energy use within reasonable bounds will challenge governments everywhere. The fact that most energy growth is taking place in developing countries without effective air pollution regulations is worrisome. Worrisome too are reports of other dangerous situations—high concentrations of the carcinogen benzene from petrochemical plants in São Paolo's industrial suburbs, for example, and uncontrolled lead pollution from Taiwan's battery factories.

The developing countries do have the benefit of the industrialized world's experience, however. As their energy use grows, they can choose the most effective regulatory measures and the most efficient control technologies. But these approaches are often costly and require investments of scarce capital that would otherwise go into high-priority economic development efforts. The industrial countries have a major stake in an environmentally safe future worldwide, and it is in their self-interest to provide needed assistance for Third World development that protects the environment.

THE INDUSTRIALIZED WORLD

Many advanced industrial countries have followed the U.S. lead in air pollution control and in some cases have surpassed the U.S. example. Since 1983 Germany—spurred by widespread forest damage attributed to acid rain—has put into force what may be the world's strictest air pollution regulations: every power plant is expected to install state-of-the-art pollution control systems by 1993, at an estimated cost of 21 billion deutschmarks, and emission standards for sulfur dioxide and nitrogen oxides will be progressively tightened as control technology improves. Japan, beginning in the early 1970s, reduced its average power plant's emissions of sulfur dioxide from nearly 7 grams per kilowatt-hour (g/kWh) to less than half a gram and cut nitrogen oxide emissions from about 1.2 to 0.5 g/kWh. Japan has also strictly limited concentrations of carbon monoxide, ozone, and particulates; pollution has been cut further by moves to reduce dependence on oil in favor of cleaner power sources such as natural gas and nuclear fission. Canada has strict limits on air concentrations of sulfur and nitrogen oxides and has taken recent steps to achieve additional curbs on both these pollutants as part of a program to stem acid rain. In the past few years, every major industrialized nation except the former Soviet Union has adopted tougher tailpipe pollution controls for automobiles (World Resources Institute, 1990).

At the same time, in Japan, Western Europe, and America people are driving more miles in less fuel-efficient cars. In the United States, energy use is rising again after a decade of declines. And coal—of all fuels, the most prolific source of sulfur dioxide and the greenhouse gas carbon dioxide—is taking a growing share of the electricity market (CEQ, 1991).

## THE DEVELOPING WORLD

Many developing countries are experiencing rapid growth of both population and industry—leading to skyrocketing urban concentrations and corresponding concentrations of pollution. Automobile sales, for example, are rising fastest in the industrializing nations of Asia. Heavy industry, fleeing the higher wages and taxes (and stricter environmental regulations) of the developed world, has shifted much production to the Third World, along with the environmental burden. These countries are only beginning to recognize these problems, and many do not have the resources to respond. Mexico, Brazil, South Korea, and Taiwan are among the few developing countries to have adopted automotive emission controls in the past few years (World Resources Institute, 1990:212).

As a result, the majority of the world's most polluted cities by many measures are in developing countries, where urbanization and industrialization have outpaced air quality controls (Figures 2.1 and

FIGURE 2.1. SULFUR DIOXIDE LEVELS IN SELECTED CITIES.

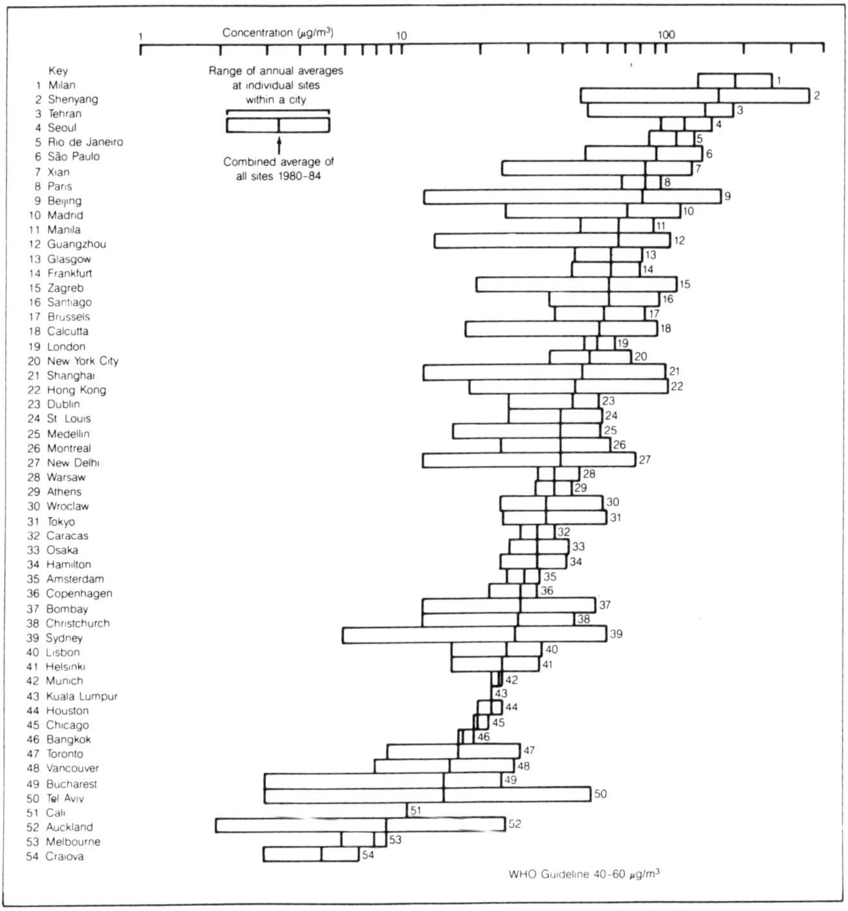

Concentration (µg/m³)   1   10   100

Key
Range of annual averages at individual sites within a city

Combined average of all sites 1980-84

1 Milan
2 Shenyang
3 Tehran
4 Seoul
5 Rio de Janeiro
6 São Paulo
7 Xian
8 Paris
9 Beijing
10 Madrid
11 Manila
12 Guangzhou
13 Glasgow
14 Frankfurt
15 Zagreb
16 Santiago
17 Brussels
18 Calcutta
19 London
20 New York City
21 Shanghai
22 Hong Kong
23 Dublin
24 St Louis
25 Medellin
26 Montreal
27 New Delhi
28 Warsaw
29 Athens
30 Wroclaw
31 Tokyo
32 Caracas
33 Osaka
34 Hamilton
35 Amsterdam
36 Copenhagen
37 Bombay
38 Christchurch
39 Sydney
40 Lisbon
41 Helsinki
42 Munich
43 Kuala Lumpur
44 Houston
45 Chicago
46 Bangkok
47 Toronto
48 Vancouver
49 Bucharest
50 Tel Aviv
51 Cali
52 Auckland
53 Melbourne
54 Craiova

WHO Guideline 40-60 µg/m³

SOURCE: World Health Organization and U.N. Environment Program (1987).

2.2). The U.N. Environment Program's Global Environmental Monitoring System (GEMS) is a network of stations in many countries collecting comparable data on air quality. In 1985, some 85 percent of the GEMS cities in the developing world (but only half of those in developed countries) exceeded the World Health Organization (WHO) standard for airborne particulates. The major cities of Brazil, China, India, Iran, Malaysia, Thailand, and Indonesia experience particulate concentrations two to eight times the WHO standards. It should be noted that much of the particulate matter comes from sources other than fossil fuel combustion (such as dust). Half of the GEMS stations in developing countries exceeded the WHO standards for sulfur dioxide in 1984, compared with only 15 percent of the stations in developed countries (World Resources Institute and IIED, 1988:171). Severe sulfur dioxide pollution, from fossil fuel combustion, plagues the industrial areas of China, Brazil, and Korea (World Resources Institute and IIED, 1988:166–169).

Mexico City is notorious for its smog. The ingredients are supplied by millions of cars, trucks, and buses, along with refineries and chemical plants, all operating without effective emission controls in a physical setting that confines the city's air mass in what amounts to a huge chemical reactor under the hot tropical sun. The lack of adequate sewage treatment and the failure to control heavy metals from industry contribute to the pollution. The new national vehicle emis-

FIGURE 2.2. SUSPENDED PARTICULATE MATTER LEVELS IN SELECTED CITIES.

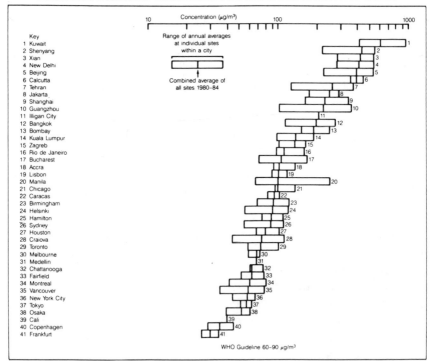

SOURCE: World Health Organization and U.N. Environment Program (1987).

sion standards and the recent closing of the city's biggest refinery will probably help, but the experience of U.S. cities shows how difficult and expensive it can be to regain healthy air once high levels of pollution are reached.

Some regions of Eastern Europe (whose economies have recently entered the Third World) are being forced to deal with the environmental legacy of years of inept central planning that put production goals ahead of health and safety. Inefficient industry, based on high-sulfur coal, has turned areas of Poland, eastern Germany, and Czechoslovakia into some of the most polluted places on earth.

If human beings are to thrive on this planet in the future, we will have to learn a new way of thinking about these problems in all their complexity. Since the atmosphere has no boundaries, nations will need to collaborate with one another—rich and poor, North and South—in solving many of the most intractable air pollution problems, sharing costs as they will share the benefits of cleaner air.

The aspirations of the poorer countries are an especially difficult part of the puzzle. Increasing Third World energy use is a given. China's billion poor people, for example, cannot be expected to forgo the benefits of their huge coal reserves for the sake of environmental concerns of interest largely in rich countries. Nor can millions of new drivers in developing countries be expected to abandon their cars. Meeting these countries' demands for energy services as cleanly and efficiently as possible is a task for the entire world. There is only one atmosphere, and we all share it.

## HOW AIR QUALITY IS REGULATED

Most industrialized countries and many developing countries regulate air quality. The regulations are generally based on ambient air quality standards, which establish maximum concentrations of specific pollutants measured at some distance from the sources. Controls are then imposed on emission sources to keep the ambient concentrations within the specified limits. This approach—working backwards from air quality to emissions—was developed by the U.S. Public Health Service and incorporated in the 1970 Clean Air Act and its 1977 and 1990 amendments. The air pollutants regulated by these standards usually include sulfur dioxide, nitrogen oxides, carbon monoxide, particulates, and often heavy metals such as lead and cadmium. Sampling intervals and measuring techniques vary. Generally each standard has two levels: a "primary" standard to protect human health and a "secondary" standard to protect human welfare and the environment (World Resources Institute, 1990).

## NATIONAL AMBIENT AIR QUALITY STANDARDS

In the United States, the EPA determines and enforces National Ambient Air Quality Standards (NAAQS) under the Clean Air Act and its amendments. These ambient standards cover six so-called criteria pollutants: sulfur dioxide ($SO_2$), nitrogen oxides ($NO_x$), carbon monoxide (CO), particulate matter smaller than 10 $\mu$m ($PM_{10}$), ozone ($O_3$), and lead (Pb). Table 2.1 gives the NAAQSs for these substances. In setting standards, EPA considers only the health and environmental risks of the various pollutants (with some margin of safety to ensure protection of especially susceptible people, such as asthmatics or sufferers from heart disease), not the costs of compliance or the availability of suitable control technology.

Individual states bear the immediate responsibility for regulating emissions to meet the NAAQSs. Each state divides its geographic area into air quality control regions (AQCRs), and EPA regularly publishes data on which AQCRs are in compliance with the primary and secondary standards for all six criteria pollutants. States submit, for EPA approval, state implementation plans (SIPs) detailing their strategies for meeting and maintaining the standards, with specific emission standards for specific types of sources. If a state's SIP is inadequate, EPA may draw up its own plan for the state.

## NEW SOURCE PERFORMANCE STANDARDS

In addition to the ambient air quality standards, EPA imposes specific emission standards on new sources of emissions or existing sources that undergo major modifications. These New Source Performance Standards (NSPS) are often more stringent than those applied by the states to existing sources. The object is to prevent states with relatively clean air from trying to attract new industry by offering lenient air quality standards.

TABLE 2.1. NATIONAL AMBIENT AIR QUALITY STANDARDS (NAAQS)

| POLLUTANT | PRIMARY STANDARD (WELFARE-RELATED) | | SECONDARY STANDARD (WELFARE-RELATED) | |
|---|---|---|---|---|
| | Averaging time | Standard level concentration[a] | Averaging time | Standard level concentration |
| PM$_{10}$ | Annual arithmetic mean[b] | 50 μg/m³ | | Same as primary |
| | 24-hour[b] | 150 μg/m³ | | Same as primary |
| SO$_2$ | Annual arithmetic mean | (0.03 ppm) 80 μg/m³ | 3-hour[c] | 1300 μg/m³ (0.50 ppm) |
| | 24-hour[c] | (0.14 ppm) 365 μg/m³ | | |
| CO | 8-hour[c] | 9 ppm (10 mg/m³) | | No secondary standard |
| | 1-hour[c] | 35 ppm (40 mg/m³) | | No secondary standard |
| NO$_2$ | Annual arithmetic mean | 0.053 ppm (100 μg/m³) | | Same as primary |
| O$_3$ | Maximum daily 1-hour average[d] | 0.12 ppm (235 μg/m³) | | Same as primary |
| Pb | Maximum quarterly average | 1.5 μg/m³ | | Same as primary |

[a] Parenthetical values are approximately equivalent concentrations.

[b] Total suspended particulates (TSP) was the indicator pollutant for the original particulate matter (PM) standards. This standard has been replaced with the new PM$_{10}$ standard and is no longer in effect. New PM standards were promulgated in 1987, using PM$_{10}$ (particles less than 10 μm in diameter) as the new indicator pollutant. The annual standard is attained when the expected annual arithmetic mean concentration is less than or equal to 50 μg/m³; the 24-hour standard is attained when the expected number of days per calendar year above 150 μg/m³ is equal to or less than 1; as determined in accordance with Appendix K of the PM NAAQS.

[c] Not to be exceeded more than once per year.

[d] The standard is attained when the expected number of days per calendar year with maximum hourly average concentrations above 0.12 ppm is equal to or less than 1, as determined in accordance with Appendix H of the ozone NAAQS.

SOURCE: EPA (1991).

## PREVENTION OF SIGNIFICANT DETERIORATION

The 1977 amendments to the Clean Air Act added measures to ensure that regions with relatively clean air do not get dirtier. These measures require EPA to classify all such AQCRs as Class I, II, or III (with Class I the most pristine). The amendments specify the maximum allowable increases in $SO_2$ and particulate concentrations for each class. They also require EPA to develop similar standards for other pollutants. In these areas, new major sources of emissions (100 or more tons per year of any pollutant regulated by the act) must receive special permits, before construction, based on reviews of their expected emissions. They are required to use the "best available control technology" to limit emissions.

## TOXIC AIR POLLUTANTS

Toxic substances are chemical, physical, or biological agents that interfere with life processes and may endanger human health. Airborne toxic substances (mainly substances that cause cancer, genetic damage, or birth defects) are controlled at the national level directly by EPA emission standards (in contrast to the so-called criteria pollutants, for which ambient air quality standards have been adopted). The Clean Air Act amendments of 1977 left the decision on which toxic substances to control up to EPA. By 1990, EPA had established emission standards for only seven toxic air pollutants (asbestos, beryllium, mercury, vinyl chloride, arsenic, radionuclides, and benzene). In the 1990 amendments, Congress—rather than leaving the regulatory decisions up to EPA—listed 189 hazardous pollutants and required EPA to review the list periodically and set emission standards for industry that require the use of "maximum achievable control technology." Controls are to be phased in over a 10-year period (EPA, 1990). The president's Council on Environmental Quality (1991) estimates that industrial emissions of these substances will be reduced by 75 to 90 percent as a result.

California has adopted a two-phase risk-assessment/risk-management approach to toxic substance control that goes beyond the national effort. In the first phase, a toxics "Hot Spots" Act establishes a formal inventory of toxic emissions and a risk quantification program. Currently the statewide list of substances identified for evaluation as toxic air contaminants (TAC) contains 232 substances. Through a statewide process of scientific evaluation, 14 of these substances have thus far been identified as TACs. The program's second phase requires development and implementation of a "best available" control strategy for each identified TAC. To date, control measures have been adopted for benzene, chromium, asbestos, ethylene oxide, and dioxins; these measures affect over 2000 industrial facilities in the state. According to the state Air Resources Board, the cancer risk from chromium emissions (from plating shops) has been reduced by more than 99 percent; the risk from ethylene oxide (from sterilizers and fumigators) has been reduced even more.

## THE PROBLEM OF NONATTAINMENT

By 1977, it had become apparent that many AQCRs had failed to meet the federal ambient air quality standards for one or more pollutants. The law as written prohibited any new major sources of emissions in such areas, thus closing off most industrial growth. The 1977 amendments, among other things, established new, more distant deadlines for compliance. The 1977 amendments also included enforcement procedures for these "nonattainment areas." All new major sources of emissions were subject to review before construction and were required to meet the "lowest achievable emission rate" (in essence, the lowest emissions achieved by any facility of the same type).

By 1990, nearly every large urban area in the country still failed to meet one or more of the NAAQSs. The most widespread problem was ozone, with ninety-six nonattainment areas in 1989, but carbon monoxide (forty-one areas) and particulate matter (seventy-three areas) were also problems. The EPA announced that more than 100 million Americans live in cities that are out of attainment with the public health standards for ozone (EPA, 1990:2).

In the 1990 Clean Air Act amendments, Congress gave the states more time to meet the air quality standards—20 years in the case of ozone in Los Angeles. At the same time, it imposed a variety of new federal requirements for nonattainment areas, including improved inspection and maintenance programs for automotive emission controls, tighter and more explicit emission standards for stationary emission sources, permit requirements for new sources of emissions, and—depending on the severity of the problem—such additional measures as gasoline-vapor recovery systems at service stations, additional monitoring of emissions, and clean-fuel requirements for fleets of vehicles. The 1990 amendments also impose much higher penalties for violations, give EPA the power to subpoena compliance data, increase civil and administrative penalties, and allow private citizens to bring suit against violators.

## CRITERIA POLLUTANTS

As explained earlier, the Clean Air Act requires EPA to set ambient standards for six criteria pollutants: sulfur dioxide ($SO_2$), nitrogen oxides ($NO_x$), carbon monoxide (CO), particulate matter smaller than 10 μm ($PM_{10}$), ozone ($O_3$), and lead (Pb). These standards include safety margins to protect especially susceptible people.

### Ozone

Photochemical smog, a chemical by-product of society's heavy dependence on the automobile, is a complex mixture of hydrocarbons and other gaseous and particulate substances formed from chemical reactions that take place mostly in sunlight. Its hallmark is high concentrations of ozone ($O_3$), a chemically reactive three-atom molecule of oxygen. Ozone is formed in the atmosphere over urban areas from volatile organic compounds (VOC), such as petroleum and solvent vapors, and nitrogen oxides ($NO_x$) with energy provided by sunlight. (Automobile exhausts contain both VOCs and $NO_x$ in large amounts.) Smog is a typically urban problem, but neighboring rural areas—Shenandoah National Park, for example—may feel its effects. Photochemical smog is appearing even in the tropics, where the periodic burning of savannah grasses releases ozone precursors under the intense sunlight of those regions.

Ozone itself is a serious health and environmental hazard. At concentrations common in many cities, some healthy adults and children experience coughing, painful breathing, and temporary reduction of lung function after an hour or two of exercise. The long-term effects are uncertain, but many scientists believe that, even at concentrations below those set by air quality standards, ozone exposure can cause progressive lung damage, especially in people who work or play outdoors during the smog season (U.S. Congress, 1989). Ozone also damages crops and other vegetation.

Ozone concentrations have been measured for about a century. Preindustrial concentrations are thought to have been about 0.01 ppm; today levels ten times that high are common in urban areas. The National Ambient Air Quality Standard for ozone is 0.12 ppm (1-hour average, not to be exceeded more than one day per year).

The formation of ozone is chemically complex and full of scientific uncertainties that have frustrated control. Because of the link with VOCs, and because $NO_x$ emissions have proved difficult to reduce, efforts to reduce ozone concentrations have focused on cutting VOC emissions. Tailpipe emissions of unburned fuel have been cut by 90 percent since the 1970s through the use of catalytic converters. In areas that fail to meet the NAAQS for ozone, moreover, the siting of new sources of VOC emissions is severely limited, and many areas are being forced to adopt tighter and tighter controls on existing emission sources. (The 1990 Clean Air Act amendments, as explained earlier, bring to bear a range of new restrictions on VOC sources.)

Nationally, emissions of VOCs have been reduced by 20 percent in the 1980s (Table 2.2) and $NO_x$ emissions by just under 5 percent. Yet

TABLE 2.2. NATIONAL VOC EMISSION ESTIMATES
(MILLIONS OF METRIC TONS PER YEAR): 1980–1989

| Source | 1980 | 1981 | 1982 | 1983 | 1984 | 1985 | 1986 | 1987 | 1988 | 1989 |
|---|---|---|---|---|---|---|---|---|---|---|
| Transportation | 9.0 | 8.9 | 8.3 | 8.2 | 8.1 | 7.6 | 7.2 | 7.1 | 6.9 | 6.4 |
| Fuel combustion | 0.9 | 0.9 | 1.0 | 1.0 | 1.0 | 0.9 | 0.9 | 0.9 | 0.9 | 0.9 |
| Industrial processes | 9.2 | 8.3 | 7.5 | 7.9 | 8.8 | 8.5 | 8.1 | 8.3 | 8.1 | 8.1 |
| Solid waste | 0.6 | 0.6 | 0.6 | 0.6 | 0.6 | 0.6 | 0.6 | 0.6 | 0.6 | 0.6 |
| Miscellaneous | 2.9 | 2.5 | 2.2 | 2.7 | 2.7 | 2.6 | 2.3 | 2.5 | 2.9 | 2.5 |
| *Total* | 22.7 | 21.3 | 19.5 | 20.3 | 21.1 | 20.2 | 19.1 | 19.4 | 19.5 | 18.5 |

*Note:* The sums of subcategories may not equal the total due to rounding.

SOURCE: EPA (1991).

ozone concentrations have remained in the unhealthy range in cities throughout the United States. Nearly every major urban area in the United States exceeds the standard at least once a year; in Southern California concentrations of several times the ozone NAAQS are common. Despite draconian limits on new hydrocarbon emission sources in smoggy areas, nearly 67 million Americans live in counties that exceed the federal NAAQS for ozone.

Ironically, the nation's success in removing lead from the atmosphere has dealt ozone control a setback. Beginning in the mid-1970s, gasoline refiners—required to phase out lead-based octane-boosters—substituted cheaper but highly volatile butane. Lead concentrations in the air fell by 90 percent. But as gasoline volatility rose, more of each tankful evaporated into the atmosphere, increasing the formation of ozone. In 1989 EPA—to head off independent action by several states—reduced the permissible volatility of summer gasoline and announced that the move would cut hydrocarbon emissions by 13 percent at a cost of a half cent per gallon (Weisskopf, 1989).

A recent study by the U.S. Congress's Office of Technology Assessment (1989) ranked VOC reduction measures by cost per ton of emissions abated. Reducing gasoline volatility, the study found, is most cost-effective ($120 to $740 per annual ton abated). But this measure alone will not bring the ozone nonattainment areas into attainment. The most costly measure, the study found, is converting fleet vehicles to methanol (methyl alcohol)—one of the "clean fuels" whose use is required by the 1990 Clean Air Act in nonattainment areas—at $8700 to $51,000 per annual ton abated. (It is worth noting that alcohol fuels, although cleaner burning with regard to currently controlled pollutants, are known to emit other pollutants, such as formaldehyde, which are themselves toxic and implicated in smog formation.) The average cost of the measures studied was $30,000 per annual ton. Since 100 or so ozone nonattainment areas will be required to reduce emission in total by several million tons, the cost of VOC abatement under the 1990 amendments is likely to be in the billions of dollars.

Some air quality control regions may need further restrictions on $NO_x$ emissions to meet the ozone standard. ($NO_x$ emission reductions can actually increase ozone formation where the ratio of VOC to $NO_x$ is low, as it is in most urban areas.) According to the Office of

Technology Assessment study (1989), too little is known about the pollutant interactions in most regions to justify generalizations about the costs and benefits of such ozone control measures. The 1990 Clean Air Act amendments impose new limits on $NO_x$ emissions nationally, under their acid rain provisions, and also offer scope for emission reductions, where appropriate, to help meet the ambient ozone standard (EPA, 1990).

Ozone is produced by the reaction of atomic oxygen (O) and molecular oxygen ($O_2$) in the presence of another molecule such as nitrogen ($N_2$) or oxygen ($O_2$). (Note for chemists: The other molecule is necessary to conserve momentum in the reaction.)

$$O + O_2 + M \rightarrow O_3 + M \tag{1}$$

The atomic oxygen comes from the breakup of $O_2$ by ultraviolet (high-energy) light at high altitudes or, at low altitudes, the breakup of nitrogen dioxide ($NO_2$) into nitric oxide (NO) and atomic oxygen (O) by less energetic photons:

$$NO_2 + photon \rightarrow NO + O \tag{2}$$

The NO reacts with ozone to regenerate $NO_2$:

$$NO + O_3 \rightarrow NO_2 + O_2 \tag{3}$$

Through these reactions ozone is continuously formed and destroyed, and a steady-state concentration of ozone results.

The presence of hydrocarbons or other VOCs in the air—characteristic of urban areas—short-circuits this cycle. As Equation (3) shows, a molecule of ozone is consumed in regenerating each nitrogen oxide molecule. Organic peroxyl radicals ($RO_2$), formed by oxidation of hydrocarbons, can regenerate $NO_2$ from NO without removing ozone. As a result, ozone concentrations generally rise where hydrocarbon vapors are present in the air (Seinfeld, 1989:eqs. 5–8). The oxidation of hydrocarbons, in turn, is largely due to reactions with hydroxyl radicals (OH), produced by photochemical dissociation of ozone, carbonyl compounds, and nitrous acid. The rates at which these reactions take place depend on temperature, sunlight, and the presence of other chemicals in the air.

But reductions of VOC emissions do not always bring down ambient ozone concentrations. When the ratio of VOCs to $NO_x$ in the air is high (as it often is in rural areas), there is a shortage of NO for peroxyl radicals to oxidize into $NO_2$, so that reducing hydrocarbon concentrations has little effect on ozone formation. In this case, reducing ozone formation requires cuts in $NO_x$ levels. Reducing $NO_x$ levels in other cases may actually increase ozone formation. When VOC/$NO_x$ ratios are low, ozone formation is limited by the amount of organic substances available. This is because the $NO–O_3$ reaction in Equation (3) is more effective than the peroxyl radical oxidation of NO to $NO_2$ and because the $NO_2–OH$ reaction (forming nitric acid) competes with organics for OH, inhibiting the production of peroxyl radicals. Biogenic hydrocarbons, mainly from trees, play a significant but poorly known role. They are approximately as reactive as other hydrocarbons in the air. Again, at high VOC/$NO_x$ ratios—in rural areas, for example—they are net destroyers of ozone.

Thus efforts to reduce ozone levels must take into account the details of the local air chemistry, including the VOC/$NO_x$ ratios, daily sunlight and temperature cycles, and the mixing of air by convection. There is no simple solution. Many scientists argue that $NO_x$ control should be emphasized more in programs to reduce ozone concentrations. The Clean Air Act amendments of 1990 offer exemptions from $NO_x$ standards for areas in which further reductions would increase ozone formation; the amendments also allow states to develop plans for coordinated reductions of VOCs and $NO_x$ to meet the ozone standard.

PARTICULATES

Particles in the air come from a variety of sources, natural and human-made. They include fly ash, sea salt, dust, metals, liquid droplets, and soot. They are emitted by factories, power plants, vehicles, fires, windblown dust, and the like, or they are formed in the atmosphere by condensation or chemical transformation of emitted gases, including sulfur and nitrogen oxides and VOCs. Acted on by the other substances in the air and by sunlight, particles are transformed into a variety of nitrate, sulfate, and carbonaceous solid suspensions called aerosols or particulates.

Particulates have a variety of health and environmental effects that matches their chemical diversity. The nitrate and sulfate particulates are themselves acidic and are deposited downwind, with or without precipitation, where they may cause serious harm to plant life and human structures. The visibility problems in areas such as Los Angeles are due mainly to airborne particles with diameters less than 2 μm (Seinfeld, 1989). Particulates may also be human health hazards, since they are breathed deep into the lungs and many have toxic or mutagenic components. A recent epidemiological study found that nonaccidental death rates rise and fall in near lockstep with daily levels of particulates—even at very low levels, well below federal limits (Raloff, 1991).

Until 1987, EPA regulations limited concentrations of "total suspended particulates" (that is, particulates of all sizes). Since that time standards have dealt only with particles less than 10 μm in diameter ($PM_{10}$), because only these are small enough to reach the lower regions of the respiratory tract. The EPA estimates that emissions of total suspended particulates fell by 15 percent in the 1980s to 7.2 million metric tons (Table 2.3). $PM_{10}$ emissions in 1989 were 5.9 million metric tons. These estimates, however, do not include "fugitive emissions" from such sources as unpaved roads, construction sites, and tilling of fields; fugitive $PM_{10}$ emissions totaled nearly 48 million metric tons in 1985 (EPA, 1991:3.9).

The average concentration of total suspended particulates at EPA

TABLE 2.3. NATIONAL TOTAL SUSPENDED PARTICULATE EMISSION ESTIMATES (MILLIONS OF METRIC TONS PER YEAR): 1980–1989

| Source | 1980 | 1981 | 1982 | 1983 | 1984 | 1985 | 1986 | 1987 | 1988 | 1989 |
|---|---|---|---|---|---|---|---|---|---|---|
| Transportation | 1.3 | 1.3 | 1.3 | 1.3 | 1.3 | 1.4 | 1.4 | 1.4 | 1.5 | 1.5 |
| Fuel combustion | 2.4 | 2.3 | 2.2 | 2.0 | 2.1 | 1.8 | 1.8 | 1.8 | 1.7 | 1.8 |
| Industrial processes | 3.3 | 3.0 | 2.6 | 2.4 | 2.8 | 2.8 | 2.5 | 2.5 | 2.7 | 2.7 |
| Solid waste | 0.4 | 0.4 | 0.3 | 0.3 | 0.3 | 0.3 | 0.3 | 0.3 | 0.3 | 0.3 |
| Miscellaneous | 1.1 | 0.9 | 0.7 | 1.1 | 0.9 | 1.1 | 0.9 | 1.0 | 1.3 | 1.0 |
| Total | 8.5 | 8.0 | 7.1 | 7.1 | 7.4 | 7.3 | 6.8 | 7.0 | 7.5 | 7.2 |

Note: The sums of subcategories may not equal the total due to rounding.

SOURCE: EPA (1991).

monitoring sites fell 20 percent between 1978 and 1989 (EPA, 1991:3.3). Still, as of 1990, seventy-three areas, with 27.4 million inhabitants, failed to attain the $PM_{10}$ air quality standards (50 µg/m³, annual arithmetic mean, or 150 µg/m³, 24-hour average, not to be exceeded more than once per year). The 1990 amendments to the Clean Air Act set an attainment deadline of 1994 and require EPA to prescribe control technology for those areas (EPA, 1990).

### CARBON MONOXIDE

Carbon monoxide (CO), an odorless, colorless gas, is an important pollutant in urban air, where it arises mostly from the incomplete combustion of automobile fuels. The human health risk from inhaled CO arises from the fact that it enters the bloodstream and disrupts the delivery of oxygen to the body's tissues by combining with hemo-globin, the molecule that normally carries oxygen. At very high concentrations (such as those that occur with unvented combustion in enclosed spaces) CO is a deadly poison. At lower concentrations, like those in some cities, it can lead to angina, blurred vision, and impaired physical and mental coordination. The NAAQS for carbon monoxide is 9 ppm (8-hour average, not to be exceeded more than once per year) or 35 ppm (1-hour average, not to be exceeded more than once per year).

Total emissions of CO declined by 23 percent in the 1980s (Table 2.4). Motor vehicles account for about two-thirds of the nation's CO

TABLE 2.4. NATIONAL CARBON MONOXIDE EMISSION ESTIMATES
(MILLIONS OF METRIC TONS PER YEAR): 1980–1989

| Source | 1980 | 1981 | 1982 | 1983 | 1984 | 1985 | 1986 | 1987 | 1988 | 1989 |
|---|---|---|---|---|---|---|---|---|---|---|
| Transportation | 56.1 | 55.4 | 52.9 | 52.4 | 50.6 | 47.9 | 44.6 | 43.3 | 41.2 | 40.0 |
| Fuel combustion | 7.4 | 7.7 | 8.2 | 8.2 | 8.3 | 7.5 | 7.5 | 7.6 | 7.6 | 7.8 |
| Industrial processes | 6.3 | 5.9 | 4.3 | 4.3 | 4.7 | 4.4 | 4.2 | 4.3 | 4.6 | 4.6 |
| Solid waste | 2.2 | 2.1 | 2.0 | 1.9 | 1.9 | 1.9 | 1.8 | 1.8 | 1.7 | 1.7 |
| Miscellaneous | 7.6 | 6.4 | 4.9 | 7.7 | 6.3 | 7.9 | 5.9 | 7.2 | 9.9 | 6.7 |
| Total | 79.6 | 77.4 | 72.4 | 74.5 | 71.8 | 69.6 | 64.0 | 64.2 | 65.0 | 60.9 |

Note: The sums of subcategories may not equal the total due to rounding.

SOURCE: EPA (1991).

emissions (aside from natural sources such as decaying vegetation), and it is significant that vehicle emissions tend to take place near people. Emissions from vehicles fell by a third, despite a 39 percent increase in the number of vehicle-miles driven, owing to the retirement of older cars unequipped with catalytic converters. The average concentration of CO at EPA monitoring stations fell roughly proportionately to the decline in emissions.

Still, CO remains a problem in many urban areas. EPA found that forty-one areas, with a total population of 33.6 million, failed to meet the NAAQS for carbon monoxide in 1988–1989, down from forty-four areas a year earlier (EPA, 1991). In some of these areas, special efforts have been made to reach attainment. Denver, for example, exceeded the CO standard on 29 days in the winter of 1986–1987; tighter inspection and maintenance requirements for automotive emission controls and the use of oxygenated motor fuels brought the total down to 3 days by the winter of 1989–1990 (EPA, 1991:3.18).

The 1990 Clean Air Act amendments set CO attainment deadlines of 1995 for "moderate" nonattainment areas and the year 2000 for "serious" areas, with specific requirements for annual reductions of emissions. They also require varied control measures, including transportation plans to reduce traffic, clean fuels for fleets of vehicles, and emission controls for stationary sources (EPA, 1990).

SULFUR DIOXIDE

Sulfur dioxide ($SO_2$) is a corrosive gas that is hazardous to human health and harmful to the natural environment. $SO_2$ is emitted worldwide by natural processes, such as volcanoes and sea spray, and by human activities, notably combustion of sulfur-containing fuels (mainly coal and fuel oil) and smelting of nonferrous metal ores, oil refining, and pulp and paper manufacture. Electricity generation from fossil fuels is the main source of $SO_2$ emissions in the United States and many other industrialized countries. As coal is generally much higher in sulfur than oil, its combustion therefore yields more sulfur dioxide. The main health effects of sulfur dioxide are breathing difficulty, lung damage, and aggravation of existing respiratory and cardiovascular disease. $SO_2$ also damages the foliage of plants and is a precursor of acid precipitation (see Chapter 3).

Global $SO_2$ emissions from human sources have risen steeply for more than a century, paralleling the growth of heavy, energy-intensive industry. While many industrialized countries have cut their $SO_2$ emissions dramatically in the past 20 years, the worldwide increase appears to be unabated. Much heavy, sulfur-emitting industry has shifted to developing countries without effective emission regulations (World Resources Institute and IIED, 1988:164).

In the United States, $SO_2$ emissions fell by about 10 percent in the 1980s and by 1989 stood at an estimated 21.1 million metric tons (Table 2.5). This decline has several main sources. To meet the requirements of the Clean Air Act, utilities in the 1970s began installing flue-gas scrubbers on new coal-fired electric power plants and turned to lower-sulfur fuels. Emission controls were also added to smelters and sulfuric acid factories, and several large smelters—unable to meet emission standards economically—were closed. Finally, industrial plants turned from coal to lower-sulfur fuels such as oil and natural gas (EPA, 1991). These measures brought average ambient $SO_2$ concentrations down by 24 percent in the 1980s. The NAAQS for sulfur dioxide is 80 $\mu g/m^3$ (annual arithmetic mean) or 365 $\mu g/m^3$ (24-hour average, not to be exceeded more than once per year). A secondary, "welfare-related," standard of 1300 $\mu g/m^3$ (3-hour average, not to be exceeded more than once per year) supplements the health-related primary standard. Nonetheless, to reduce acid precipitation, Congress passed the 1990 Clean Air Act amend-

TABLE 2.5. NATIONAL SULFUR OXIDE EMISSION ESTIMATES
(MILLIONS OF METRIC TONS PER YEAR): 1980–1989

| Source | 1980 | 1981 | 1982 | 1983 | 1984 | 1985 | 1986 | 1987 | 1988 | 1989 |
|---|---|---|---|---|---|---|---|---|---|---|
| Transportation | 0.9 | 0.9 | 0.8 | 0.8 | 0.8 | 0.9 | 0.9 | 0.9 | 0.9 | 1.0 |
| Fuel combustion | 18.7 | 17.8 | 17.3 | 16.7 | 17.4 | 17.0 | 16.9 | 16.6 | 16.6 | 16.8 |
| Industrial processes | 3.8 | 3.9 | 3.3 | 3.3 | 3.3 | 3.2 | 3.2 | 3.2 | 3.4 | 3.3 |
| Solid waste | 0.0 | 0.0 | 0.0 | 0.0 | 0.0 | 0.0 | 0.0 | 0.0 | 0.0 | 0.0 |
| Miscellaneous | 0.0 | 0.0 | 0.0 | 0.0 | 0.0 | 0.0 | 0.0 | 0.0 | 0.0 | 0.0 |
| Total | 23.4 | 22.6 | 21.4 | 20.7 | 21.5 | 21.1 | 20.9 | 20.7 | 20.9 | 21.1 |

Note: The sums of subcategories may not equal the total due to rounding.

SOURCE: EPA (1991).

ments requiring further reductions of about 8 million tons of $SO_2$ annually by the year 2000. Emission-trading provisions, described earlier, are intended to make these reductions as cost-effective as possible.

## NITROGEN OXIDES

Nitric oxide (NO) is emitted by several natural and human activities and is quickly converted to nitrogen dioxide ($NO_2$) in the atmosphere. The nitrogen oxides are known collectively as $NO_x$. Wherever combustion takes place at high enough temperatures for oxygen and nitrogen (contained in air) to react, $NO_x$ is produced. $NO_x$ is also formed from nitrogen in the fossil fuels themselves (especially coal). In the United States, the two main sources of $NO_x$ are stationary-source fuel combustion (56 percent in 1989) and vehicles (40 percent). Transportation—cars, trucks, and buses—is the largest source worldwide. $NO_2$ can irritate lungs and lower resistance to respiratory infections such as influenza. A key ingredient in ozone formation, it also forms acids that can damage structures and natural systems.

Controlling $NO_x$ emissions is more challenging than controlling sulfur dioxide. $SO_2$ emissions come overwhelmingly from large facilities, such as power plants, that are relatively easy (although expensive) to identify and control. $NO_x$ sources (most of which are motor vehicles) are smaller, more mobile, and much more numerous and varied. The United States and many other countries have phased in catalytic converters and engine design modifications to reduce $NO_x$ emissions from gasoline-powered vehicles over the past decade or so, and some are beginning to apply controls to diesel trucks and buses (World Resources Institute and IIED, 1988:167). Yet average annual $NO_x$ emissions have fallen by only about 5 percent since 1980 (Table 2.6). In fact, $NO_x$ emissions in 1989 were 8 percent higher than in 1970 (EPA, 1991).

The relevant ambient air quality standard applies to concentrations of nitrogen dioxide ($NO_2$). Like $NO_x$ emissions, average $NO_2$ concentrations at EPA's monitoring sites have been nearly constant throughout the 1980s. Los Angeles is the only urban area that has recorded violations of the annual $NO_2$ concentration standard (100 $\mu g/m^3$, annual arithmetic mean) in the 1980s.

TABLE 2.6. NATIONAL NITROGEN OXIDE EMISSION ESTIMATES
(MILLIONS OF METRIC TONS PER YEAR): 1980–1989

| Source | 1980 | 1981 | 1982 | 1983 | 1984 | 1985 | 1986 | 1987 | 1988 | 1989 |
|---|---|---|---|---|---|---|---|---|---|---|
| Transportation | 9.8 | 10.0 | 9.4 | 8.9 | 8.8 | 8.9 | 8.3 | 8.1 | 8.1 | 7.9 |
| Fuel combustion | 10.1 | 10.0 | 9.8 | 9.6 | 10.2 | 10.2 | 10.0 | 10.5 | 10.9 | 11.1 |
| Industrial processes | 0.7 | 0.6 | 0.5 | 0.5 | 0.6 | 0.6 | 0.6 | 0.6 | 0.6 | 0.6 |
| Solid waste | 0.1 | 0.1 | 0.1 | 0.1 | 0.1 | 0.1 | 0.1 | 0.1 | 0.1 | 0.1 |
| Miscellaneous | 0.2 | 0.2 | 0.1 | 0.2 | 0.2 | 0.2 | 0.2 | 0.2 | 0.3 | 0.2 |
| *Total* | 20.9 | 20.9 | 20.0 | 19.3 | 19.8 | 19.9 | 19.1 | 19.4 | 20.0 | 19.9 |

*Note:* The sums of subcategories may not equal the total due to rounding.

SOURCE: EPA (1991).

The 1990 Clean Air Act amendments tighten emission standards, for both vehicles and other combustion facilities, with the aim of reducing acid precipitation. The amendments also include provisions for reducing $NO_x$ emissions to control ozone pollution in ozone nonattainment areas.

LEAD

Lead exposure may cause neurological damage, most notably in children and fetuses, whose nervous systems are still developing. The results can be mental retardation, learning disabilities, or hyperactivity. As many as 20 percent of preschool children in the United States have excessive levels of lead in their blood (Grove, 1987:525). At higher doses lead can damage kidneys and degrade the immune system. Between 75 and 95 percent of the lead inhaled or ingested stays in the body for life, accumulating in the bones and other tissues. Lead is also toxic to microorganisms that decompose organic matter. Because lead is not removed from natural ecosystems to any significant degree by natural processes, these effects are essentially permanent.

Most of the world's lead pollution comes from the use of lead-based gasoline additives to increase octane ratings. Since the 1970s many countries, recognizing the health risks of lead, have taken steps to phase out these lead-based additives. Japan began to reduce the lead content of gasoline in 1970; lead-free gasoline was introduced there in 1975; nearly all gasoline sold in Japan today is lead-free. The European Community reduced the allowable lead content of gasoline from 0.4 g/liter in 1978 to 0.013 g/liter in 1989. The United Kingdom, slower to move, introduced lead-free gasoline only in 1985, but a gradual phasing out of lead in gasoline reduced its emissions by more than two-thirds between 1975 and 1986 (World Resources Institute and IIED, 1988).

In the United States, lead emissions fell by 90 percent in the 1980s, mainly from the gradual phasing out of leaded gasoline (Table 2.7). (Leaded gasoline also damages catalytic converters, which are used to reduce tailpipe emissions of unburned hydrocarbons, carbon monoxide, and nitrogen oxides.) Emissions from lead smelters and battery factories have also been substantially reduced. The average concentrations of airborne lead at EPA monitoring stations fell by 87 percent in the 1980s (EPA, 1991). The NAAQS for lead is 1.5 $\mu g/m^3$ (maximum quarterly average).

The downside of the lead story is a dramatic increase in hydrocarbon vapor emissions. Refiners, in phasing out lead-based octane boosters, phased in exceedingly volatile alternatives, notably butane, as substitutes. These volatile compounds, in evaporating from fuel tanks, tankers, and barges, have aggravated the problem of urban ozone formation. To combat this side effect of lead control, EPA in 1989 issued regulations reducing the volatility of summer gasoline.

TABLE 2.7. NATIONAL LEAD EMISSION ESTIMATES
(MILLIONS OF METRIC TONS PER YEAR): 1980–1989

| Source | 1980 | 1981 | 1982 | 1983 | 1984 | 1985 | 1986 | 1987 | 1988 | 1989 |
|---|---|---|---|---|---|---|---|---|---|---|
| Transportation | 59.4 | 46.9 | 46.9 | 40.8 | 34.7 | 15.5 | 3.5 | 3.0 | 2.6 | 2.2 |
| Fuel combustion | 3.9 | 2.8 | 1.7 | 0.6 | 0.5 | 0.5 | 0.5 | 0.5 | 0.5 | 0.5 |
| Industrial processes | 3.6 | 3.0 | 2.7 | 2.4 | 2.3 | 2.3 | 1.9 | 1.9 | 2.0 | 2.3 |
| Solid waste | 3.7 | 3.7 | 3.1 | 2.6 | 2.6 | 2.6 | 2.6 | 2.6 | 2.5 | 2.3 |
| Miscellaneous | 0.0 | 0.0 | 0.0 | 0.0 | 0.0 | 0.0 | 0.0 | 0.0 | 0.0 | 0.0 |
| Total | 70.6 | 56.4 | 54.4 | 46.4 | 40.1 | 20.9 | 8.4 | 8.0 | 7.6 | 7.2 |

*Note:* The sums of subcategories may not equal the total due to rounding.

SOURCE: EPA (1991).

## AIRBORNE TOXIC SUBSTANCES

A number of substances of varying degrees of toxicity (mainly carcinogens, mutagens, and toxins causing birth defects) are released into the air by industry with little regulation. Only in the past few years, under provisions of the 1986 Superfund amendments, has the EPA been required to inventory the major sources of these releases by manufacturers. The 1988 inventory (the latest available) estimated that 2.4 billion pounds of 302 different toxic substances were emitted into the air in that year by U.S. manufacturers handling large amounts of these substances (CEQ, 1991:87). Emissions from smaller sources have not been estimated.

Toxic substances were first placed in the Clean Air Act in the 1977 amendments. But EPA had brought only seven toxic air pollutants under control by the time of passage of the 1990 Clean Air Act amendments, so these measures specifically mandate regulation, for the first time, of 189 different toxics. Existing sources must meet emission standards based on the most effective control technologies available in the affected industries. New sources must match the emission controls of the best-controlled similar existing source. These regulations are to be phased in over 10 years. Special regulations will help limit accidental releases. Studies of the risks of toxics will determine whether additional regulation is needed.

## INDOOR AIR POLLUTION

Energy use in buildings produces a variety of air pollutants, including carbon monoxide, carbon dioxide, smoke from stoves and fireplaces, and various gaseous oxides of nitrogen and sulfur from furnaces, as well as stray natural gas and heating oil vapors. The radioactive gas radon, present in small quantities in natural gas, is emitted by gas-burning appliances. Especially in well-insulated, tightly sealed, energy-efficient homes, these pollutants and others—such as cigarette smoke, formaldehyde from plywood and glues, radon from the surrounding soil, and asbestos—can build up to significant levels. Highly energy-efficient buildings may be so tightly sealed that they experience a complete air change only once every 2 to 10 hours, compared to once every hour or half hour for older, leaky buildings (Shurcliff, 1986).

While many of these pollutants can be eliminated at the source— for example, by not smoking or not using wood stoves or unvented gas heaters—others are so much a part of modern life (or, in some cases, the natural world) as to be nearly inescapable. Ventilation is the key. Exhaust fans are used to change the air in many energy-efficient buildings. Often air-to-air heat exchangers are used to re-capture the heat that would otherwise go out with the indoor air.

A more troubling, and far more widespread, problem of indoor air pollution is the use of so-called traditional fuels in the Third World. Smoke from wood, crop residues, and dung contains a variety of carcinogens, mutagens, and other toxic substances in the form of easily respirable particles. Especially among those who spend long hours cooking over fires in enclosed spaces, health risks are great (although inadequately studied).

## SOUTHERN CALIFORNIA: OUR URBAN FUTURE?

Southern California is a revealing case study because the Los Angeles area, with its sunny climate, air-trapping bowl-shaped site, millions of cars, and concentrated industry, has long been a byword for polluted air. In response, the state of California has pioneered the development of air quality control measures, which are being emu-lated in other parts of the country. Although California's tailpipe emission standards for VOCs, $NO_x$, and CO from vehicles have been stricter than national standards for years, the Los Angeles area's ozone levels have nonetheless remained far above the 0.12-ppm ozone NAAQS, with peak levels of 0.33 ppm in 1989 (EPA, 1991:4.21). Moreover, the Los Angeles basin is far from meeting CO and particulate standards. Its citizens breathe air rated "unhealthful" for more than 200 days each year, according to the EPA Pollutant Standard Index. (In the 1990 Clean Air Act amendments, Southern California has the dubious distinction of its own ozone pollution ranking: "severe.")

The region has been under increasing pressure in the past few years to improve the situation before EPA imposes its own plan to remedy the region's failure to meet air quality standards. With emissions of cars and other major sources of volatile organic compounds already regulated so tightly, it was necessary to go after smaller and smaller sources. In 1989 the South Coast Air Quality Management District announced a sweeping long-term plan to fight pollution by declaring war on some of Californians'—and Americans'—most cherished customs.

The plan begins with an attack on small but easy-to-control pollution sources. Paints and solvents would be reformulated to eliminate oil-based preparations. Use of barbecue lighter fluid and gasoline-powered lawn mowers would be prohibited. Limits would be placed on the number of cars per family. Parking fees would be raised. The plan even includes bans on fast-food drive-through windows and on bias-ply tires (which reduce fuel efficiency and wear faster than radial tires, emitting organic particles to the atmosphere).

By the late 1990s, all diesel buses, 40 percent of cars, and 70 percent of freight vehicles would be replaced with new models using clean fuels such as natural gas, methanol, or electricity. Emissions from industry and consumer products would be halved. The year 2007 would see further emission reductions, including a complete ban on gasoline-powered automobiles. By 2010 the area would be in full compliance with the NAAQSs for ozone, carbon monoxide, nitrogen oxides, and particulate matter. These and other smog-control measures would change the lives of all 12 million Southern Californians profoundly (and many millions more by 2010). Public transport would be encouraged by a variety of measures (higher parking fees, for example). Workers would be given incentives to live closer to their jobs. The cult of the automobile would be quelled or at least diminished.

As might be expected, debate has swirled around this plan. One industry and labor group set the cost at $2200 a year per household and predicted the loss of 52,500 jobs (Mathews, 1989). In contrast, the Air Quality District itself estimated annual costs of $200 per resident (offset by $600 in benefits) and the creation of 80,000 jobs. A recent independent study estimates costs at $13 billion annually ($2700 per household), with benefits of $3 to $4 billion (fewer premature deaths, less lung disease, less damage to buildings and materials from bad air), but acknowledges important uncertainties in estimates of both costs and benefits (Krupnick and Portney, 1991). This study, in turn, has been challenged (Lipfert and Morris, 1991; Friedman, 1991; Lents, 1991; Miller, 1991; Chapman, 1991).

The technology to meet the plan's ambitious goals has also been questioned. Automobile manufacturers regard the requirement that $NO_x$ emission standards (already the strictest in the nation) be halved as especially problematic (Weisskopf, 1989). The pollution abatement benefits of the methanol fuels thought likely to meet much of the requirement for clean-fueled vehicles have been questioned, too. While methanol itself is much less photochemically reactive than gasoline or diesel fuel, formaldehyde emitted by methanol-fueled engines is very reactive, and emissions of these vehicles have yet to be thoroughly tested (Seinfeld, 1989). Moreover, the use of methanol fuels will not have much impact on ozone formation in places where the VOC/$NO_x$ ratio is high (as in areas downwind of major urban areas) because ozone formation in this case is limited by $NO_x$ concentrations. Further, the cost of switching to methanol-fueled vehicles will be very high per ton of VOC emissions abated (U.S. Congress, 1989).

These economic and technical uncertainties will be narrowed in the years to come. Doubtless, there will be changes in the plan. What is not in doubt is that the future health of millions of Americans may hinge on similar changes in lifestyle. Many of the plan's measures are identical to those that later emerged in the 1990 amendments to the Clean Air Act. The *New York Times* called the plan "a new vision of the American urban future" (Reinhold, 1989). As if to underscore the point, twelve northeastern states and the District of Columbia are considering adopting similar ozone-control regulations (Wald, 1991).

## OPTIONS

The 1990 amendments to the Clean Air Act are expected to impose costs of about $25 billion a year on the American economy by the end of the century—in addition to the more than $30 billion or so that the nation already spends each year on air pollution control

(Gutfeld and Rosewicz, 1990; CEQ, 1991). That, economists say, is how it should be: The prices of goods and services should reflect their true total costs, including environmental impacts and other "externalities." The costs of energy services—home heating and cooling, transportation, industrial process heat, and so on—will tend to rise in the United States to reflect the costs imposed by the new clean air legislation. Offsetting these increases at least partially will be the energy savings brought about by more energy-efficient technologies, which are under development in every area (see Part II).

In the short run, the distribution of costs and prices will not perfectly reflect the environmental externalities. Specific pollution-control technologies required by state or federal regulators are unlikely to display a perfect balance of cost and environmental benefit. This is why the 1990 amendments included emission-trading provisions in the sulfur dioxide emission reduction program: the trading of emission allowances in principle should distribute costs and benefits more efficiently.

In the long run, utilities will take into account the environmental protection costs of fossil-fueled power plants as they make their decisions on new generating units, on retiring old ones, and on investments in increased efficiency of energy use. Other industries will make similar calculations as they consider their own investments in new plant. Regulators, also, will be able to make adjustments in their control-technology requirements in the interest of economic efficiency.

On the question of transportation, U.S. consumers still receive mixed signals regarding environmental costs. On the one hand, they are being informed that their automobiles are a significant source of nitrogen oxides, smog-forming organic compounds, and greenhouse gases (mainly carbon dioxide) that may be affecting global climate. On the other, they are aware that U.S. gasoline prices are the lowest ever in relation to the total costs of owning and operating an automobile, so that they have little incentive to purchase more energy-efficient cars. Clearly regulators must play a larger role—by imposing taxes on fuels (as many European countries have done) that reflect the environmental costs of driving, by legislatively mandating higher automobile and truck fuel efficiencies, or by a combination of both approaches.

The economics of air pollution will never be simple, but consumers are likely to see clearer price signals in the next few years. At the top of their list of priorities should be energy efficiency. At home, this means more efficient furnaces, air conditioners, windows, lights, and houses too. More efficient automobiles and cars burning cleaner fuels will see substantial markets for the first time, as will electric cars as soon as the requisite new battery technology becomes competitive. In industry, the new pollution costs will make nonfossil options more attractive. Utility planners, aided by favorable regulation, will consider seriously a range of renewables, including various forms of solar generation and the next generation of nuclear-power technologies, as well as increased emphasis on energy efficiency—the most cost-effective of all options to reduce pollution.

All the measures to control air pollution will affect measures to control concentrations of the greenhouse gases. The National Academy of Sciences and expert groups worldwide have called for actions, on grounds of halting global climate change, that would also have the effect of reducing overall air pollutant emissions from fossil fuels. These actions—carbon taxes, for example—would further enhance the attractiveness of nonfossil options.

In the end, human society will learn the arts of pollution avoidance. Pollution abatement is simply too costly. It is fitting that innovative technology—the source of much of our past and present environmental pollution—will also be the source of these new arts.

## REFERENCES

Chapman, Duane. 1991. Letter. *Science* 253 (9 Aug.):607–608.

Council on Environmental Quality (CEQ). 1991. 21st Annual Report, Executive Office of the President. Washington: Government Printing Office.

Davis, Ged R. 1990. Energy for planet earth. *Scientific American* 263(3): 55–62.

Friedman, Robert M. 1991. Letter. *Science* 253 (9 Aug.):607–608.

Fulkerson, William, Roddie R. Judkins, and Manoj K. Sanghvi. 1990. Energy from fossil fuels. *Scientific American* 263(3):129–135.

Graedel, Thomas E., and Paul J. Crutzen. 1989. The changing atmosphere. *Scientific American* 261(3):58–68.

Grove, Noel. 1987. Air: An atmosphere of uncertainty. *National Geographic* 171(4):502–537.

Gutfeld, Rose, and Barbara Rosewicz. 1990. Clean-air accord is reached in Congress that may cost industry $25 billion a year. *Wall Street Journal,* 23 Oct., p. A2.

Krupnick, Alan J., and Paul R. Portney. 1991. Controlling urban air pollution: A benefit-cost assessment. *Science* 252 (26 Apr.):522–528.

Lents, James M. 1991. Letter. *Science* 253 (9 Aug.):607–608.

Levine, Mark, et al. 1990. *Energy Efficiency, Developing Nations, and Eastern Europe.* Washington: International Institute for Energy Conservation.

Lipfert, Frederick W., and Samuel C. Morris. 1991. Letter. *Science* 253 (9 Aug.):607–608.

Mathews, Jay. 1989. California plans war on smog; lives of 12 million people could be altered. *Washington Post,* 13 March, p. A1.

Miller, Bernard. 1991. Letter. *Science* 253 (9 Aug.):607–608.

Raloff, J. 1991. Dust to dust: A particularly lethal legacy. *Science News* 139 (6 Apr.):212.

Reinhold, Robert. 1989. Southern California takes steps to curb its urban air pollution. *New York Times,* 18 March, p. 1.

Rosewicz, Barbara. 1990. Price tag is producing groans already. *Wall Street Journal,* 29 Oct., p. A7.

Sathaye, Jayante, Andre Ghirardi, and Lee Schipper. 1987. Energy demand in developing countries: A sectoral analysis of recent trends. *Annual Review of Energy,* vol. 11. Palo Alto, Calif.: Annual Reviews.

Schwartz, Stephen E. 1989. Acid deposition: Unraveling a regional phenomenon. *Science* 243 (10 Feb.):753–763.

Seinfeld, John H. 1989. Urban air pollution: State of the science. *Science* 243 (10 Feb.):745–752.

Shurcliff, William A. 1986. Superinsulated houses. *Annual Review of Energy,* vol. 11. Palo Alto, Calif.: Annual Reviews.

U.S. Bureau of the Census. 1990. *Statistical Abstract of the United States.* Washington: Government Printing Office.

U.S. Congress. Office of Technology Assessment. 1989. *Catching Our Breath: Next Steps for Reducing Urban Ozone.* Washington: Government Printing Office.

U.S. Environmental Protection Agency (EPA). Office of Air and Radiation. 1990. *The Clean Air Act Amendments of 1990: Detailed Summary of Titles.* Washington: Government Printing Office.

———. Office of Air Quality Planning and Standards. 1991. *National Air Quality and Emissions Trend Report, 1989.* EPA-450/4-91-003. Washington: Government Printing Office.

Wald, Matthew. 1991. 12 states consider smog curb. *New York Times,* 5 July, p. D1.

Weisskopf, Michael. 1989. EPA sets new gasoline rules to cut smog during summer. *Washington Post,* 11 March.

Whitten, Gary Z., and Michael W. Gery. 1986. The interaction of photochemical processes in the stratosphere and troposphere. In *Effects of Changes in Stratospheric Ozone and the Global Climate*, vol. 2: *Stratospheric Ozone*. Washington: Government Printing Office.

World Health Organization and U.N. Environment Program. 1987. *Global Pollution and Health*. London: Yale University Press.

World Resources Institute. 1990. *World Resources: 1990–91*. New York and Oxford: Oxford University Press.

World Resources Institute and International Institute for Environment and Development (IIED). 1988. *World Resources: 1988–89*. New York: Basic Books.

CHAPTER 3

# ACID RAIN

*John Harte*

Acids produced by the combustion of fossil fuels and the smelting of
nonferrous ores can be transported long distances through the atmosphere and deposited on earth on ecosystems that are exceedingly
vulnerable to damage from excessive acidity (Oden, 1976; Likens
et al., 1979). This is the threat to man and the biosphere discussed
here—a threat commonly referred to as the "acid rain" problem. To
understand the source of the problem and the extent of damage that
has occurred, or is likely to occur, we must explore phenomena on
many scales—from the submicroscopic world of molecules and ions
to the planet earth and, in between, cells, organisms, lakes, hillside
slopes, and forests. Midway between the molecular and the planetary realms is man, the final stop in our journey.

This chapter originally appeared in *The Cassandra Conference: Resources and the Human Predicament,* edited by Paul R. Ehrlich and John P. Holdren (Texas A&M University Press, 1988).
Reprinted with permission.

## A MOLECULAR PERSPECTIVE

Chemical processes, acting molecule by molecule, govern the forma-tion of environmental acidity and ultimately shape its fate. The acidity of precipitation, lake water, or a patch of soil is determined by the concentration, $[H^+]$, of hydrogen ions in that medium. The most commonly used unit of acidity is pH, defined as the negative log-arithm (to the base 10) of $[H^+]$. A low pH corresponds to high acidity, and a unit decrease in pH corresponds to a tenfold increase in $[H^+]$. A pH of 7 is the neutral point—water with such a pH contains as much acid as it does base. Figure 3.1 shows the pH scale, with some representative values marked on it to provide a reference frame for what follows. Particularly noteworthy is that in large regions of Europe and eastern North America the annual mean pH of rain and snow is below 4.5.

The two major acids in acid rain are sulfuric acid, $H_2SO_4$, and nitric acid, $HNO_3$. These acids are formed in the atmosphere from sulfur dioxide ($SO_2$) and oxides of nitrogen ($NO_x$), the gaseous products of fossil fuel combustion and nonferrous-ore smelting. The hydrogen ions in these acids typically originate from the hydrogen in cloud water or in atmospheric water vapor. The chemistry of the produc-tion process is only partly understood at present; it appears that a variety of mechanisms can cause acids to form and that the dominant chemical reactions depend on the location and weather conditions as well as on the chemical composition of the local atmosphere. Since acid formation from the gaseous precursors is an oxidation process, certain oxidants, such as the hydroxyl radical (HO) and hydrogen peroxide ($H_2O_2$) that occur as trace constituents in the atmosphere, play a major role in the formation process. Sunlight, soot, and trace metals may also expedite the process of acid formation under certain circumstances (NRC, 1981; Novakov et al., 1974). Some of the substances (such as $H_2O_2$, soot, and trace metals) that convert the precursor gases to acids can result from combustions, just as do the precursor gases themselves.

In chemical reactions where several chemical species combine to form some reaction products, some of the reactants may be so scarce and others so abundant that most of the abundant molecules are

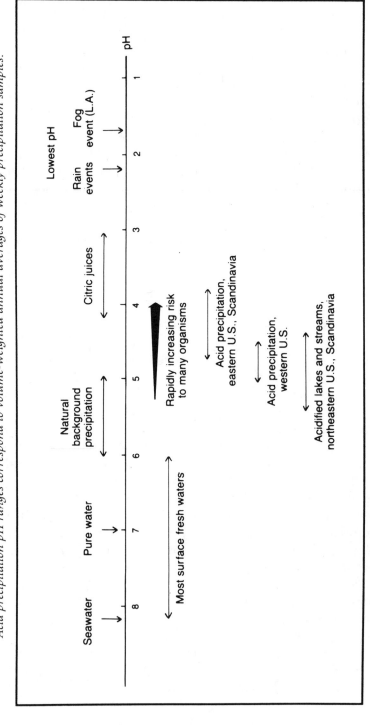

FIGURE 3.1. THE pH SCALE WITH SOME REPRESENTATIVE VALUES AND RANGES.
*Acid precipitation pH ranges correspond to volume-weighted annual averages of weekly precipitation samples.*

unlikely to participate in the reaction. The scarcer reactants are called "limiting" chemicals, and the abundant ones are called "nonlimiting" chemicals. A decrease (brought about by pollution control, for example) in the concentration of a nonlimiting reactant will not have as large an effect on the reaction rate as would a decrease in the concentration of a limiting reactant. This clearly can be an important consideration in the formulation of strategies for acid-rain reduction.

We turn now from sources of acidity in the atmosphere to the fate of acidity when it is deposited in an ecosystem in the form of rain, snow, fog droplets, or even dry particles. Society's main concern is with the biological consequences of acid deposition. But before we can predict the magnitude of the biological damage that acid rain causes, we must first estimate the increase in acidity of a volume of lake water or soil when a specified quantity of acid is added to it. How is this done?

If the substance added to the water or soil were, say, common table salt, then calculation of the increase in salt concentration would be straightforward: the quantity of salt dissolved in the water or added to the soil divided by the volume of water or weight of the soil would give the increase in concentration of the substance. For acidity, however, the situation is more complex. Unlike salt molecules, hydrogen ions can readily be removed or produced in soil and water by internal chemical processes. For example, most natural waters and soils contain alkaline chemicals that annihilate (or neutralize) acidity, removing $H^+$ ions by chemically converting them into $H_2O$. Alkalinity is thus one component of an ecosystem's defense against acid precipitation (Stumm and Morgan, 1981). Unfortunately, the neutralizing process is reciprocal: acidity annihilates alkalinity. Because alkalinity is produced through processes (weathering reactions) that proceed at a rate much lower than the rate at which it is used up (through acid precipitation), it is essentially a finite resource and can be exhausted.

Figure 3.2 illustrates graphically what happens when acid is added to a medium containing alkalinity. When the alkalinity of a water volume has been totally exhausted, we say that the water has "acidified." In lake and stream water the most prevalent form of alkalinity is the bicarbonate ion, $HCO_3^-$, which reacts with $H^+$ as follows: $H^+ + HCO_3^- \rightarrow H_2O + CO_2$. The evolved $CO_2$ can escape harmlessly to the atmosphere in this neutralizing process.

FIGURE 3.2. EFFECT ON pH (RIGHT AXIS) AND ALKALINITY (LEFT AXIS) OF
ADDING ACID TO A CLOSED AQUEOUS SYSTEM.

*As more acid is added, the alkalinity steadily declines, whereas pH declines only gradually until the alkalinity is used up: at that point pH dips sharply. Every mole of $H^+$ added to the system is neutralized by a mole of alkalinity until the alkalinity is used up in the process. The alkalinity present initially is thus a measure of the lake's buffering capacity, or ability to withstand acid input without suffering a large drop in pH.*

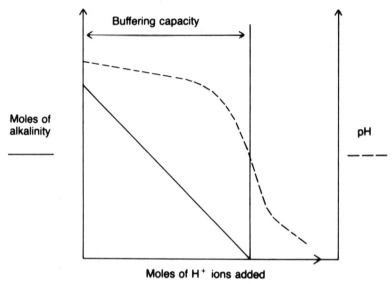

Another acid-neutralizing process that takes place in lakes, streams, and soils is the exchange of $H^+$ ions with nonacidic positive ions, such as $Ca^{2+}$, $K^+$, and $Na^+$, that are found in the solid particles that constitute soils and lake-bottom or stream-bed sediments. This process is called "cation exchange." Yet another set of processes is mineral weathering reactions. Weathering reactions initiated by sulfuric or nitric acid reduce the level of those acids in the environment. Weathering reactions initiated by the natural carbonic acid that results from the dissolution of atmospheric $CO_2$ in precipitation water can create alkalinity, which subsequently can neutralize acid precipitation.

Another reaction that neutralizes acid precipitation occurs as acidic runoff water percolates through soil containing aluminum, lead, cadmium, mercury, and other metals. In a cation exchange process called "metals mobilization," ionic forms of these metals may be exchanged for hydrogen ions. This leads to a decreased level of acidity and an increased level of dissolved metals in the runoff water. As discussed in more detail below, these toxic metals are more biologically accessible in runoff than when they are soil-bound, and under some circumstances they pose a threat to life that can exceed that produced by the unneutralized acid runoff (Cronan and Schofield, 1979).

With the fundamental molecular processes relevant to acid deposition described, albeit briefly, we are now ready to journey over seventeen orders of magnitude of size to the global realm, where we will examine the impact of anthropogenic acidity on the earth.

## ACIDIFICATION ON A GLOBAL SCALE

In what ways, if any, is acidic deposition changing our planet as a whole? There are two types of planetary changes to consider. First, there are direct chemical changes, which may be of global proportions. In particular, acid deposition may be altering significantly the global $H^+$ or alkalinity budgets and, in a related fashion, may be accelerating global weathering processes or the rate of mobilization of trace metals. Second, the biological consequences of acid deposition may be of global proportions.

Consider, first, the global anthropogenic rate of emission to the atmosphere of $SO_2$ and $NO_x$, the two acid-forming pollutants of concern. In 1980 approximately $85 \times 10^9$ kg of sulfur in the form of $SO_2$ and $35 \times 10^9$ kg of nitrogen in the form of $NO_x$ were emitted to the atmosphere as a result of fossil fuel burning and metal smelting. In contrast, the natural background rates of emission to the atmosphere of these gases are about $100 \times 10^9$ kg per year of sulfur in the form of $SO_2$ and a highly uncertain range of between 10 and $100 \times 10^9$ kg per year of nitrogen in the form of $NO_x$ (NRC, 1981). (An additional $50 \times 10^9$ kg per year of sulfur in the form of hydrogen sulfide and sulfate are emitted to the atmosphere from natural sources as well.) Thus the anthropogenic emissions are a sizable contributor to the total rate at which these gases are loaded into the atmosphere.

Of greater concern to us is knowing (1) what the globally averaged pH of precipitation would be in the absence of the pollutants $SO_2$ and $NO_x$ and (2) how these pollutants alter that pH value. The background emissions of $SO_2$ and $NO_x$ have very little influence on the average pH of precipitation because they are very inefficiently converted to sulfuric and nitric acid. Were all the background $SO_2$ and $NO_x$ emissions converted to these strong acids and the acidifying influence of natural sulfates emitted to the atmosphere were included as well, then the average background pH of rain and snow would be about 4.5. In fact, very small percentages of the background sulfur and nitrogen are converted to strong acids, and the best estimates suggest that the pH of precipitation would be between 5 and 6 in the absence of pollution (Charlson and Rodhe, 1982). The slightly acidic value for this background pH stems largely from the weak carbonic acid, $H_2CO_3$, which is formed as the carbon dioxide in the atmosphere dissolves in cloud water or falling raindrops, from sulfurous acid ($H_2SO_3$) formed from background $SO_2$ dissolved in precipitation water, and from natural background sulfates. In some areas of the world, however, the natural background pH of precipitation can be 6 or even higher because of the presence of alkaline materials, such as limestone dust, in the atmosphere. The Great Basin of the western United States (Harte et al., 1985) and the Tibetan Plateau in central Asia (Harte, 1983) are examples of such regions. In other areas intense volcanic activity or biological processes cause sulfuric acid to form naturally in the atmosphere. Downwind of such sites the natural background pH of precipitation can be slightly below 5. A value of 5.6 is widely considered to be a good working estimate of the global average background pH of precipitation.

In contrast to the background emissions, anthropogenic emissions of $SO_2$ and $NO_x$ are rather efficiently converted to sulfuric and nitric acid in the atmosphere. There is now fairly convincing evidence that the conversion process is expedited by the presence of certain oxidizing agents that are found in highest concentration in polluted atmospheres (NRC, 1981). Thus the chemical processes in the atmosphere that are responsible for the conversion of these gases to acidic form take place more rapidly in the polluted air in which anthropogenic emissions are found. Were all the anthropogenic $SO_2$ and $NO_x$ converted to sulfuric and nitric acid, then the rate of $H^+$ deposition to the earth's surface would be about $7 \times 10^{12}$ moles per year of $H^+$. This can be compared to the background flux of $H^+$ to earth's surface, which is about $1.5 \times 10^{12}$ moles per year of $H^+$. If this anthropogenic flux of $H^+$ were uniformly mixed in the $5 \times 10^{14}$ cubic meters per year of global precipitation, then the average pH of precipitation would be about 4.9. Of course, the anthropogenic acidity is not uniformly dispersed around the earth; indeed, on about 1 percent of the earth's surface annually averaged pH's in the low 4s are observed. Additional acidity can fall to the earth's surface in dry form; subsequent wetting of dry acid particles by rain or snow can acidify surface waters or soils in the same manner that acid precipitation does. All told, at least half of the amount of acid that could potentially be formed from anthropogenic emissions (the $7 \times 10^{12}$ moles per year of $H^+$) is actually formed in the atmosphere and descends to the surface in either wet or dry form. The remaining fraction of the sulfur and nitrogen oxides forms non-acidic or only weakly acidic chemical species. The best available evidence is that the fluxes of wet and dry acidity are roughly equal on a globally averaged basis.

We can also compare the global rate of sulfuric and nitric acid formation from pollution with the rate that these acids are produced naturally by various other processes, such as nitrification and photosynthesis. Moreover, we can also compare the rates at which alkalinity is produced globally and trace metals such as aluminum are mobilized with the rate that anthropogenic acidity uses up alkalinity and mobilizes aluminum. Table 3.1 presents a global comparison of natural and anthropogenic global acid-forming and acid-neutralizing processes.

It should be clear from the table that on a global basis anthropogenic acidity dominates precipitation chemistry but that, in turn, the chemical effects of anthropogenic acidity are dominated on a global basis by chemical processes occurring in terrestrial and aquatic habitats. The oceans exemplify this global resilience to acid deposition; the effect of many centuries of anthropogenic acidity deposited in the oceans would have an undetectable influence on seawater pH. By way of contrast, the effect of raising the $CO_2$ level of the atmosphere from the preindustrial value of about 275 ppm to a new future

TABLE 3.1. GLOBAL FLUXES OF H$^+$, STOCKS OF ALKALINITY, AND FLOWS OF ALKALINITY

| *Global Fluxes of H$^+$ (10$^{12}$ equivalents per year)[a]* | |
|---|---|
| Precipitation flux from atmosphere to surface in form of carbonic acid, assuming preindustrial value of 275 ppm for atmospheric $CO_2$ concentration | 1.04 |
| Same as above but assuming present $CO_2$ concentration of 340 ppm | 1.16 |
| Precipitation acidity from natural background emissions of sulfur | 0.3–1.0 |
| Gross production of H$^+$ in global nitrification process | 400 |
| Gross uptake of H$^+$ in biological nitrogen assimilation | 400 |
| Acidity produced from anthropogenic $SO_2$ and $NO_x$ emissions (approximately half is deposited to earth in wet and half in dry form) | 4–5 |
| *Global Stocks of Alkalinity (10$^{15}$ equivalents)* | |
| Cation exchange capacity in top 1 m of soil | 30–50 |
| Alkalinity of surface fresh waters | 0.05–0.1 |
| Alkalinity of seawater | 2700 |
| *Global Flows of Alkalinity (10$^{12}$ equivalents per year)* | |
| Production by weathering processes | 1–10 |

[a] The last four H$^+$ fluxes are associated with alkalinity fluxes. A decrease in alkalinity corresponds to an increase in H$^+$ and vice versa. On an equivalents basis, the magnitudes of these fluxes are equal. The first two processes, involving $CO_2$ dissolution, do not involve any flow of alkalinity.

equilibrium value of 400 ppm as a result of fossil fuel burning would be to lower the pH of seawater by about 0.15 pH units, which is a measurable amount.

The purely and directly chemical comparisons expressed in the table should not be construed to mean that anthropogenic acidity will have no global-scale chemical consequences. Widespread chemical changes may arise through the disruption of biological processes, which can act as points of leverage. For example, the process of biological nitrogen fixation is sensitive to the pH of the medium— soil or water—that surrounds the nitrogen-fixing organisms. Conceivably the rate of global nitrogen fixation could be diminished significantly by acidic deposition, and, in turn, this diminution could affect the global nitrogen cycle. Far more likely, however, is the possibility that such damage will occur on a regional scale long before it shows up on a global one. For instance, in habitats where nitrogen fixation plays a critical role in the nitrogen cycle, acidic deposition may cause major chemical changes.

Widespread forest dieback (discussed in more detail below) caused by acid deposition could have major global climatic repercussions. One reason is that such dieback would increase the concentration of atmospheric $CO_2$ and thereby contribute to the $CO_2$ greenhouse effect. It would also increase the albedo of the earth's surface, thereby decreasing the amount of solar radiation absorbed at the earth's surface. And it would affect the global transpiration rate, which in turn would alter the amount of water vapor (the major greenhouse gas) and the amount of cloud cover in the atmosphere. The climatic consequences of these changes would undoubtedly be more severe at the regional level, but the possibility of global-scale climate disruption cannot be ruled out.

Having journeyed from the molecular to the global domain and thereby viewed the acid-rain problem from the submicroscopic perspective of sources and sinks and then from a holistic perspective, we now go to where the real action is—in organisms, ecosystems, and, finally, society.

## DAMAGE TO CELLS AND ORGANISMS

Most organisms live in an environment whose chemical and physical characteristics are conducive to the organisms' well-being. That is not a fortuitous circumstance, of course, for in the course of both evolution and succession organisms adapt to, and to some extent create, their environment. If the chemical properties of their environment are altered too much from those that they are used to, then biological damage is likely to result. For example, typical organisms thrive best at a particular value of the pH of the medium—be it soil or water—in which they live. Many fish, particularly those of the family Salmonidae, which includes trout and salmon, will not reproduce successfully when the pH of their surrounding water drops below approximately 5.5. Most adult salmonids do poorly and eventually succumb to the stress of pH's below approximately 5 (Beamish, 1976; Glass et al., 1982).

Although fish kills resulting from acid rain have seized the head-lines, biological damage is by no means limited to these organisms. Salamanders and other amphibia are particularly sensitive to acidic conditions. One of the first victims of lake acidification in the Adiron-dacks was the spotted salamander, *Ambystoma maculatum* (Pough, 1976). Ironically, in areas of the United States where acid deposition is occurring and surface-water acidification is likely to occur in the future, yet another member of the genus *Ambystoma* may have the honor of presaging more severe damage to come. The area is the western slope of the Colorado Rockies, where very low alkalinity lakes and ponds are found and where precipitation with a mean annually averaged pH of 4.7 to 5 has been observed over the past 5 years (Harte et al., 1985). During snowmelt in late spring and early summer at a study site in the Rockies, the pH in some of the low-alkalinity surface waters has been observed to drop to values in the neighborhood of 5. This drop in pH occurs during the most acid-sensitive stage (egg development) in the life cycle of the tiger sala-mander, *A. tigrinum*, which inhabits small subalpine lakes at the site. If acid deposition continues there, these salamander populations could be the sacrificial "canaries in the mine," providing early warn-ing of widespread damage from acid snow and rain that is prevalent throughout the higher-elevation, low-alkalinity regions of the west-ern United States.

Microscopic life is also at risk from acidification. Many species of phytoplankton have a pH optimum in what is called the circum-neutral range—6 to 8—and exhibit low reproduction and high mor-tality at pH's either below or above that range. Both field and laboratory studies confirm that in acidified lakes and streams entire populations of some phytoplankton species have been killed (Mar-morek, 1983; Stokes, 1983; Yan and Stokes, 1978; Schindler and Turner, 1982; Tonnessen, 1983). This is not just of esoteric ecological interest, for the phytoplankton are the base of the aquatic food chains. If chemical and physical conditions permit, sunlight and nutrients are captured by phytoplankton, which are then eaten by zooplankton, crustacea, fish, and aquatic birds. In any particular lake or stream some species of phytoplankton may actually thrive under more acidic conditions, but generally such species will not provide a base for the same food chains that the original community of species did. Many species of zooplankton are also adversely affected by acidification. For example, members of the genus *Daphnia* are quite sensitive to water pH, with a tolerance threshold pH of about 5 (Raddum, 1980; Yan and Strus, 1980).

Many nitrogen-fixing organisms (organisms capable of converting molecular nitrogen into ammonia, a nutrient form of nitrogen) are believed to be particularly sensitive to acidification. Although more experimentation is needed here to quantify dose-response relations, it is quite likely that legumes and certain species of blue-green algae and lichen are adversely affected when the pH of their surroundings drops significantly below its accustomed value.

The damages described above are associated directly with low pH values of the media surrounding the organisms. Although the detailed physiological cause of the damage is not known in many cases, limited information available suggests that excessive $H^+$ concentrations interfere with ion transport, particularly in body organs, such as fish gills, that communicate directly with the surrounding media. Another direct biological effect of acidic deposition is associated with the corrosive action of acid on many substances. Plant leaf surfaces often have a waxy layer that affords the plant protection against fungal and other kinds of invasion and disease. Corrosion of this layer by acid deposition has been suggested as a mechanism to explain at least part of the damage observed on trees growing in regions such as New England and Germany that receive intense acidic deposition. For a review of this and other mechanisms of acid-induced tree damage, see Postel (1984).

Other, more indirect biological effects of acidification have also been identified. As acid runoff flows through soils, the process of metal mobilization reduces the acidity of the water at the same time that it increases the availability to organisms of trace metals such as aluminum, lead, mercury, and cadmium. Unfortunately, for most organisms this tradeoff appears to be an unfavorable one—the increase in trace-metal availability at toxic concentrations can do more damage than would the original acidity. When aluminum concentrations in lake and stream waters approach 100 ppb, toxic effects on fish can be detected (Baker and Schofield, 1982). Normally, unacidified waters have aluminum concentrations sufficiently below this value that the metal does not pose a threat to aquatic life. In acidified lakes in many areas of Scandinavia and eastern North America, lake-water concentrations of aluminum exceed 200 ppb;

the fish mortalities that have been observed there are now believed to be due to a combination of the direct effects of low pH and the indirect effect of acid-mobilized toxic metals.

Acid runoff in soil can also leach from the root zones of trees such cations as calcium and potassium, which are essential to plant growth. The contrast between this process and toxic metal mobilization is worth highlighting. The toxic metals were formerly bound in soil particles and were biologically unavailable; acidification rendered them soluble and thereby accessible for plant or animal uptake. In contrast, the essential cations were formerly found in bound form from which uptake could proceed at a pace adequate for normal plant needs. When they become soluble as a result of leaching by acidic runoff, they flow out of the root zone (indeed, out of entire watersheds) and are no longer available to meet plant requirements.

Tree damage from acidification illustrates the potential multiple hazards associated with acidic deposition (Johnson and Siccama, 1983). A number of mechanisms by which acidic precipitation could damage trees have been proposed. As mentioned above, one such mechanism is direct damage to the waxy protective layer found on needles and leaves by the corrosive action of acids, which, among other things, may increase the susceptibility of the plant to fungal attack. Acidic precipitation may also interfere with the various mechanisms by which trees obtain essential nutrients such as nitrogen and phosphorus by killing the microorganisms that live in soil or are symbiotically associated with the root structure and that process organic forms of nutrients into inorganic forms that can be used by vegetation. Yet another proposed mechanism is a leaching process in which acid runoff percolating through soil dissolves essential cations and carries them away from the root zone. A fourth possible mechanism involves the toxic metals mobilized by acid runoff; they can damage trees either by direct toxic action or by physical interference with plant functions. As an example of a physical action, it has been suggested by Bernhard Ulrich, a soil scientist in Germany, that aluminum might clog the fine root hairs of trees, thereby interfering with their ability to take in sufficient water; droughtlike symptoms thus might show up even in years with plenty of precipitation. A fifth hypothesis is that the nitrate in nitric acid acts as a growth stimulant, causing trees to extend their growing period into the cold season, which leads to damage to sensitive new growth from freezing. The facts are not yet in place from which to make a convincing assignment of cause to effect. It is possible that some combination of stresses is at work, perhaps different combinations in different forests.

This brief overview of the kinds of biological damage that acid deposition can induce may give the impression that every organism suffers when the surroundings acidify. There is, however, evidence that a number of organisms can tolerate, and indeed thrive in, levels of acidity in their surroundings that exceed those normally encountered. Unfortunately, this prospect should not be greeted with approval or even equanimity, for in many instances these organisms would not be considered desirable, at least by our species. Among fish populations the least interesting from a sportfishing perspective (for example, suckers as compared with trout) survive the best at low pH. We can even speculate about the possibility of certain disease-associated organisms thriving in acidic waters. Giardia is a protozoan that is associated with a severe gastrointestinal disease that afflicts hikers and mountain climbers. It is contracted by drinking seemingly clean, pure mountain-stream waters in which the microscopic organism lives. This protozoan spends part of its lifetime in the mammalian gut and part free-living in natural waters. Consistent with available information is the possibility that the pH optimum of this organism lies below the normal circumneutral values found in natural waters and that acidified lakes and streams provide a more congenial home for it.

## ECOSYSTEMS

Over vast regions of Europe and North America precipitation occurs with a pH substantially below the estimated natural background value of 5.6. In the northeastern United States, southeastern Canada, Scandinavia, northern and central Europe, and parts of China the annually averaged pH of precipitation is between 4 and 4.5. In many areas of the western United States, and undoubtedly in areas that have not yet been investigated, precipitation pH averages in the range of 4.5 to 5. Wherever precipitation with a pH significantly below background value occurs on ecosystems with low buffering capacity, chemical and biological impacts are likely to occur eventually.

Freshwater lakes sitting atop granitic bedrock and thus having low buffering capacity provided the first evidence that acidic precipitation could damage whole ecosystems. Such lakes are like the lymph nodes in the human body—they exhibit the first signs of environmental disease and thus serve as early-warning indicators of eventual wider damage. Scandinavian lakes exhibited the first signs of trouble about two decades ago, and in the subsequent years a pattern of wholesale ecological devastation was documented (Almer et al., 1978; Henriksen, 1980). First, the alkalinity of these lakes totally disappeared, and pH values sank below 5 in many of them. As a result the lake-water concentrations of trace toxic metals increased. Trout populations died out, the composition of the planktonic and microbial populations shifted from those formerly habituated to a circumneutral range of pH's to those that could tolerate the increased lake-water acidity. Successful reproductive activity of many amphibia dwelling in the lakes ceased. Although it is not true that *all* life has been extinguished in such lakes, the phrase "dead lakes" has been widely used in the popular media to refer to them. At the present time it is estimated that in Norway and Sweden at least 20,000 lakes of a total of 90,000 investigated exhibit these symptoms and are considered to be "dead" or "dying." Because the organisms that can adapt to acidified lake waters are rarely the sort that are of interest to sports fishers and because the acidified lakes cannot fulfill the ecological role formerly played by the unacidified lakes, these lakes are, indeed, dead functionally, even though not literally.

Stimulated by the early findings in Scandinavia, scientists in eastern North America (where precipitation acidity was similar to that in Scandinavia) began to look systematically at lakes with low buffering capacity to see whether the Scandinavian phenomenon was repeated there. In the Adirondacks and subsequently in parts of Canada compelling evidence for a similar pattern of ecological damage was soon obtained. Presently some 400 lakes in the Adirondacks are considered to be "dead." Good reviews and summaries of these developments are contained in Cowling (1982) and Glass et al. (1982).

The evidence indicating a causal linkage between surface-water acidification and acidic deposition in Scandinavia and eastern North America is of two forms: (1) direct or positive and (2) indirect, resulting from the failure of scientists to develop convincing alternative causal mechanisms to account for the observed patterns of lake acidification. Three pieces of positive evidence stand out. First, a clear correlation exists between geographic areas where precipitation is particularly acidic and areas where low-alkalinity lakes became acidic. Second, elevated sulfate levels are observed in acidified lakes, as expected when the acidifying agent is sulfuric acid in precipitation. Third, an intensive study of lake acidification in the Adirondacks demonstrated that observations of varying degrees of acidification found in lakes in close proximity to one another are consistent with the hypothesis that acid precipitation is the primary cause of lake acidification in that area.

This third argument is worth expansion. Initially, a puzzling feature of the investigations in eastern North America emerged. In the Adirondacks there were situations in which two lakes, which historically were believed to have similar chemical characteristics and were receiving similar amounts of acidic precipitation, exhibited differing responses to that precipitation. One lake might acidify and its pH drop to the mid-4s, while the pH of its neighbor might remain at, say, 6. Such observations were initially interpreted by some scientists to indicate that acidic precipitation could not be the cause of lake acidification. Further research revealed, however, that in such instances soil depth differed in the watersheds surrounding the lakes. Lakes situated in basins with substantial soil accumulation responded more slowly to acidic precipitation (EPRI, 1981). The reason is obvious: the soils surrounding a lake provide buffering agents that offer the lake some protection against acidification—the deeper the soils, the greater the availability of such protection and the longer it takes for the lake to acidify.

While the positive evidence to date that links lake acidification to acidic deposition is compelling, at least two major alternative causative factors have been suggested. First, it has been claimed that acidification could result from altered land-use practices. The absence of correlation between locations of acidified lakes and locations in which land-use alteration has occurred suggests that this explanation does not generally apply. Second, it has been suggested that lakes become acidic naturally because of biological processes and therefore that the role of acidic precipitation is uncertain. It is true that there are naturally occurring acidic lakes, the so-called brown-water lakes, in which organic acids produce low-pH waters. However, the acidified lakes of concern in Scandinavia and eastern North America are clear-water lakes. In contrast to the brown-water lakes, these lakes are biologically very unproductive. Havas et al. (1984) give a particularly insightful critique of these and other proposed alternative causative factors.

In the western United States, where acidic deposition is also occurring but is not as intense as that in Europe and eastern North America, the risk of future lake acidification is a real one (Roth et al., 1985). In that region there are numerous sites, particularly at the higher elevations in the Rocky Mountains, the Sierra Nevada, and the Cascades, where acid deposition is occurring on highly sensitive watersheds. As yet no biological damage from acid deposition has been documented in the West, but evidence for acid-induced chemical change has been reported. In particular, the first stage of acidification—a reduction in surface-water alkalinity—was detected in the West in a survey of stream-water chemistry conducted by the U.S. Geological Survey over a 10- to 15-year period (Smith and Alexander, 1983).

Available data are insufficient to predict the rate at which alkalinity levels will decline in the future in lake and stream waters of the West. But a number of factors suggest that many ecosystems in the high mountains of the western United States may be even more sensitive to acidic precipitation than were those in the northeast. Most noticeably, there is the extremely low alkalinity found in the soils and surface waters of the Sierra Nevada of California, the Cascades of Washington, and parts of the Rocky Mountains (all sites where precipitation is considerably more acidic than background). However, these alkalinity concentrations are probably no lower than those in Scandinavia and the Adirondacks before the onset of acidic deposition there. Five factors distinguish these western sites:

1. A large fraction of precipitation falls as snow, which means that the pulse of acid during spring melt observed in Scandinavia and the Adirondacks may pose a particularly great threat in the western mountains.
2. It is likely that dry deposition of acids contributes relatively more in the West than in Scandinavia or eastern North America (because of the presence of long dry periods in much of the West), which means that a comparison of wet precipitation pH values will give an underestimate of the total acid loading rate in the West.
3. The soils in the western mountains are extremely thin, and the slopes are steep in comparison to those in the East or in Scandinavia, which means that available acid-neutralizing capacity

will be exhausted faster for a given pH, or at a comparable rate for a higher pH.

4. High concentrations of potentially toxic acid-mobilizable trace metals occur in the western surface rocks and soils, which pose a biological threat.

5. A short growing season characterizes the high mountains in the West, which means that ecosystem recovery following a late-spring acid pulse will be retarded.

All these factors should dispel any illusions one might harbor that western ecosystems are not at risk simply because precipitation acidity is lower in the West than in areas of the world where severe damage has already occurred.

Acidic precipitation will induce a drop in lake-water pH only after the buffering capacities of the lake water, the lake-bottom sediments, and the soils that surround the lake and interact with snowmelt and rain runoff are largely exhausted. With soil-buffering capacity depleted by acidic runoff, one would expect that damage to terrestrial vegetation might also occur, and yet until only a few years ago no evidence for such damage existed. Then in 1982 scientists in West Germany discovered that major damage to trees was occurring in approximately 8 percent of the forested area in that country. A year later the damaged area was up to 34 percent, and in 1984 it was 55 percent (Postel, 1984).

Forest damage has also been documented in the United States in the past few years, particularly but not limited to stands of red spruce in New England (Johnson and Siccama, 1983). Many possible causes of this widespread forest damage have been proposed. Acid precipitation is one; others include oxidative damage induced by ozone (which is produced in the lower atmosphere mainly by chemical reactions that require the presence of $NO_x$—the same oxides of nitrogen that can also form nitric acid), the direct effects of $NO_x$ and $SO_2$, and outbreaks of forest pests. While pollutant levels have generally been rising or have remained steadily high in regions with forest damage, the suggestion that drought or pest damage is the underlying cause is weakened by the absence of information correlating the outbreak of recent forest damage with such stresses.

Forest damage may spread to huge regions of the planet in the future, but it is not the only potential regional ecological hazard of acid precipitation. The discussion of other hazards given below is necessarily speculative in view of the sparse information available, but it gives some idea of the potential regional risks of acidic precipitation.

Damage to agricultural crops has been investigated. The evidence at hand, while scanty, suggests that in most regions agriculturally exploited soils are not at great risk. Most such soils are naturally well buffered. In addition, soil amendments used in modern agriculture generally augment natural soil-buffering capacity. Moreover, most crops are annuals, and their growing period is probably too short to allow acid precipitation to cause extensive damage. If damage to crops occurs, it most likely is by direct corrosive action to the plants. For this reason orchard crops may be most at risk, over the long run, but information is lacking here.

In certain habitats slow soil-building and nutrifying processes are vital to long-term ecological integrity. For instance, in subalpine and alpine ecosystems lichens and other organisms play an important role in replacing nutrients lost through leaching processes that wash nutrients downhill and out of the ecosystem. Nitrogen fixation is not the major pathway by which most green plants obtain their nitrogen, but it is nevertheless a critical process in certain ecosystems. In some nitrogen-limited lakes and coastal waters in the arctic and subarctic, for example, nitrogen fixation carried on by alders and lichens on the surrounding land is an important source of this essential nutrient. In some tropical forests where soil reserves of nitrogen are particularly low, nitrogen-fixing trees constitute a sizable fraction (up to one-half in some areas) of the canopy individuals. Because nitrogen-fixation activity by any organism and by most lichens (whether or not of the nitrogen-fixing kind) is likely to be quite sensitive to acidic deposition, major long-term impacts could occur in such areas. Such damage is likely to be slow in developing and to require a very carefully designed monitoring strategy to detect it in its early stages.

## THE HUMAN DIMENSION

Acid deposition can degrade the quality of human life in three ways. First, it can damage ecosystems and thereby deprive us of a priceless source of beauty, joy, and inspiration. Second, it can damage ecosystems and thereby degrade the quality of material benefits that human society derives from healthy ecosystems; these benefits include the maintenance of clean air and water, the moderation of climate, the production of food and fiber, and the maintenance of a genetic "library." Third, it can directly lead to damaging effects on human health.

A nation that poisons its forests and lakes poisons an aesthetic and spiritual well of its people. There is a pathetic sort of positive feedback lurking here, for as children grow up in a world that affords meager opportunity to enjoy wilderness and natural ecosystems, then later, as adults, their incentive to save what little is left for future generations will be less.

The existence of a major confluence of interest between the material well-being of people and the health of ecosystems has been discussed by many authors in a variety of contexts (Harte and Socolow, 1971; Holdren and Ehrlich, 1974). For acid-precipitation damage the examples abound. Recreational opportunities are lost when lakes and streams no longer support viable fish populations; major sources of revenue and jobs dry up when commercial forests and fisheries become less productive; water supplies become more erratic when the integrity of forested watershed slopes is eroded. We cannot now predict the extent to which these linkages between acid deposition and loss of ecosystem services to society will become manifest. Many of the mechanisms by which the damage might occur are identified. It is the inherent unpredictability of ecosystem behavior and uncertainty about the degree to which society will regulate sources of acidity that force us to view the future with uncertainty in this regard.

Some potential direct effects of acid deposition on human health have likewise been identified, but, again, large uncertainties exist in our ability to predict how severe these will be. Extremely acidic rain and fog have been observed in some heavily polluted urban areas. In a small city on the Tibetan Plateau of China, a rainstorm with a pH of 2.25 was observed (Harte, 1983), while Los Angeles fogs with pH as low as 1.7 have been observed (Munger et al., 1983). Toxic metals mobilized by acidic water can enter human water supplies and produce chronic toxicity. These toxic metals could come from runoff/soil interactions and then enter well water in areas with low carbonate buffering capacity, or they could be mobilized from plumbing pipes, solder joints, and aluminum water containers. Another pathway through which toxic metals can damage our health is food-chain biomagnification. Recently an excess of aluminum has been found in the damaged brain tissue of victims of Alzheimer's disease and of amyotrophic lateral sclerosis. It is too soon to say whether aluminum is a cause of these afflictions and, if so, whether acid deposition is a significant contributing factor (Maugh, 1984).

The works of man are at risk as well. Evidence of pollution damage to buildings, statuary, metals, and paints exposed to urban air abounds. Less clear is the precise role that acid deposition plays in the ensemble of corrosive pollutants found in urban air.

## WHAT SHOULD SOCIETY DO?

There is compelling evidence in hand that acid deposition has severely altered the chemistry of streams and lakes and destroyed populations in large areas of Europe and eastern North America. Equally compelling evidence exists that the problem is spreading; for instance, there are regions of the western United States that are at risk. And the possibility that acid deposition is partly or entirely responsible for extensive forest dieback in Europe and eastern North America cannot be dismissed. In the face of the scientific uncertainties a number of political postures are possible. We can assume that we have seen the worst and that no new damage will occur; a case could then be made that regulatory action is not needed. Or we can take a worst possible scenario and approach the problem under the assumption that massive forest dieback, devastating human health effects, plummeting productivity of major commercial fisheries, and severe disruption of water supplies will occur in the near future if drastic regulatory action is not taken immediately. The probability that either of these scenarios is correct is small indeed. Much more likely is the possibility that the story of the recent past will continue to unfold, with new forms of damage catching us by surprise and with already identified damage continuing to spread in geographic extent and magnitude. A prudent society should assume no less.

The rapidity with which symptoms of forest damage have come to light is awesome to contemplate. Controlling acidic precipitation, in contrast, is a time-consuming process. Not only does it take time to draft and pass legislation to regulate emissions of acid-precursor gases, but it will take time to implement controls, once enacted. Is it necessary to wait until clear evidence is in hand detailing quantitatively the amounts of acidic deposition in individual sensitive areas that can be attributed to each source of $NO_x$ or $SO_2$ before seeking regulatory action? It may be argued that it is wasteful of money to regulate a source of acid precursors when compelling data showing the precise degree of culpability of that source are not in hand. However, the symptoms of forest damage develop so rapidly and the magnitude of the damage could be so great that a prudent society ought not make regulatory action contingent on understanding every last nuance of the problem. The risk of underreaction, resulting in massive forest damage as well as continuing aquatic impacts and possible health effects, appears to overwhelm the risk of overreaction and the expense of unnecessary preventive action. To be specific, suppose that legislation is "overreactive" and results in the regulation of individual pollution sources that are not primarily responsible for acid deposition on sensitive sites and that forest damage is not in the least influenced by acid deposition. Has society been more damaged than if no action had been taken and ecosystem damage accelerates? The following consideration suggests that the answer is no.

Acid precipitation is not the only environmental problem caused by the emissions under discussion, and thus there are other reasons to control emissions of $NO_x$ and $SO_2$ independent of the acid-precipitation phenomenon. $SO_2$ in combination with other pollutants, such as fine particulates from coal and petroleum combustion, is implicated in human health damage. Tropospheric ozone, which affects adversely both people and vegetation, is created by reactions involving atmospheric nitrogen oxides and hydrocarbons, also produced in fossil fuel combustion. Therefore, appropriate control strategies that limit the production of the acid precursors would be of benefit, even if it should turn out that the cause of the forest damage is not primarily acidic precipitation or that some of the regulated sources are not responsible for the acid deposition falling at sites where lake acidification is a major problem. Should it turn out that acid deposition does damage forests, however, then the nations that regulate these emissions will have one more reason to be glad they do.

Recent political events in the United States are not encouraging. Legislative attempts to control major sources of acid deposition have been derailed, and so the risks and damages continue. Earl Cook, with his usual insight, perceived the underlying problem in 1976: "It is a paradox that an affluent society, possessing more resources for study and experimentation than a poor society, appears to be prevented by that very affluence from making decisions that require any sacrifice" (Cook, 1976:379).

## REFERENCES

Almer, B., W. Dickson, C. Ekstrom, and E. Hornstrom. 1978. Sulfur pollution and the aquatic environment. In J. O. Nriagu, ed., *Sulfur in the Environment*. New York: Wiley.

Baker, J. P., and C. L. Schofield. 1982. Aluminum toxicity to fish in acidic waters. *Water, Air, and Soil Pollution* 18:289–309.

Beamish, R. J. 1976. Acidification of lakes in Canada by acid precipitation and the resulting effects on fishes. *Water, Air, and Soil Pollution* 6:501–514.

Charlson, R. J., and H. Rodhe. 1982. Factors controlling the acidity of natural rainwater. *Nature* 295:683–685.

Cook, E. 1976. *Man, Energy, Society*. San Francisco: Freeman.

Cowling, E. 1982. Acid precipitation in historical context. *Environmental Science and Technology* 16(2):110A–23A.

Cronan, C. S., and C. L. Schofield. 1979. Aluminum leaching response to acid precipitation: Effects on high-elevation watersheds in the Northeast. *Science* 204(20):304–306.

Ehrlich, P. R., A. H. Ehrlich, and J. P. Holdren. 1977. *Ecoscience: Population, Resources, Environment*. San Francisco: Freeman.

Electric Power Research Institute (EPRI). 1981. *The Integrated Lake-water Acidification Study (ILWAS)*. Electric Power Research Institute Report EA-1825. Palo Alto, Calif.: EPRI.

Glass, N. R., D. E. Arnold, J. N. Galloway, G. R. Hendrey, J. J. Lee, W. W. McFee, S. A. Norton, C. F. Powers, D. L. Rambo, and C. L. Schofield. 1982. Effects of acid precipitation. *Environmental Science and Technology* 16(3):162A–169A.

Harte, J. 1983. An investigation of acid precipitation in Qinghai Province, China. *Atmospheric Environment* 17:403–408.

_____. 1985. *Consider a Spherical Cow: A Course in Environmental Problem Solving.* Los Altos, Calif.: Kaufmann.

Harte, J., and R. H. Socolow. 1971. *Patient Earth.* New York: Holt, Rinehart & Winston.

_____, G. Lockett, R. Schneider, C. Blanchard, and H. Michaels. 1985. Acid precipitation and surface water vulnerability on the western slope of the high Colorado Rockies. *Water, Air, and Soil Pollution* 25:313–320.

Havas, M., T. C. Hutchinson, and G. E. Likens. 1984. Red herrings in acid rain research. *Environmental Science and Technology* 18(6):176A–186A.

Henriksen, A. 1980. Acidification of freshwaters—A large-scale titration. In D. Drablos and A. Tollan, eds., *Ecological Impact of Acid Precipitation: Proceedings of an International Conference.* Sandefjord, Norway: SNSF Project.

Holdren, J. P., and P. R. Ehrlich. 1974. Human population and the global environment. *American Scientist* 62:282–292.

Johannessen, M., and A. Henriksen. 1978. Chemistry of snow meltwater: Changes in concentration during melting. *Water Resources Research* 14(4):615–619.

Johnson, A. H., and T. G. Siccama. 1983. Acid deposition and forest decline. *Environmental Science and Technology* 17(7):294a–305a.

Lawson, D. R., and J. G. Wendt. 1982. *Acid Deposition in California.* Society of Automotive Engineers Technical Paper no. 821246.

Likens, G., E. Wright, R. F. Galloway, and T. J. Butler. 1979. Acid rain. *Scientific American* 241:43–51.

Marmorek, D. R. 1983. Changes in the behavior and structure of plankton systems of intermediate pH. In J. Teasley, ed., *Proceedings of a Symposium on Acid Precipitation.* Ann Arbor, Mich.: Ann Arbor Science Publishers.

Maugh, T. H. 1984. Acid rain's effects on people assessed. *Science* 226:1408–1410.

Munger, J. W., D. J. Jacob, J. M. Waldman, and M. R. Hoffman. 1983. Fogwater chemistry in an urban environment. *Journal of Geophysical Research* 88:5109–5121.

National Research Council (NRC). 1981. *Atmosphere-Biosphere Interactions: Toward a Better Understanding of the Ecological Consequences of Fossil Fuel Combustion.* National Research Council, Committee on the Atmosphere and the Biosphere. Washington: National Academy Press.

Novakov, T., S. Chang, and A. Harker. 1974. Sulfates as pollution particulates: Catalytic formation on carbon (soot). *Science* 186:259–261.

Oden, S. 1976. The acidity problem—An outline in concepts. *Water, Air, and Soil Pollution* 6:137–166.

Postel, S. 1984. *Air Pollution, Acid Rain, and the Future of Forests.* Worldwatch Paper no. 58. Washington: Worldwatch Institute.

Pough, F. H. 1976. Acid precipitation and the embryonic mortality of spotted salamanders, *Ambystoma maculatum. Science* 192:68–70.

Raddum, G. G. 1980. Comparison of benthic invertebrates in lakes with different acidity. In D. Drablos and A. Tollan, eds., *Ecological Impact of Acid Precipitation: Proceedings of an International Conference.* Sandefjord, Norway: SNSF Project.

Roth, P., C. Blanchard, H. Michaels, J. Harte, and M. El-Ashry. 1985. *Acid Deposition in the Western United States: An Assessment.* Washington: World Resources Institute.

Schindler, D. W., and M. A. Turner. 1982. Biological, chemical, and physical responses of lakes to experimental acidification. *Water, Air, and Soil Pollution* 18:259–271.

Schofield, C. L. 1980. Processes limiting fish populations in acidified lakes. In D. S. Shriner, C. R. Richmond, and S. E. Lindberg, eds., *Atmospheric Sulfur Deposition.* Ann Arbor, Mich.: Ann Arbor Science Publishers.

Smith, R. A., and R. B. Alexander. 1983. *Evidence for Acid Precipitation-Induced Trends in Common Ion Concentrations at Bench-Mark Stations.* USGS Circular no. 910. Denver: U.S. Geological Survey.

Stokes, P. M. 1983. pH related changes in attached algal communities of soft water lakes. In J. Teasley, ed., *Proceedings of a Symposium on Acid Precipitation.* Ann Arbor, Mich.: Ann Arbor Science Publishers.

Stumm, W., and J. Morgan. 1981. *Aquatic Chemistry: An Introduction Emphasizing Chemical Equilibria in Natural Waters.* 2nd ed. New York: Wiley.

Tonnessen, K. 1983. Aquatic ecosystem acidification in the Sierra Nevada, California: Potential for chemical and biological changes. In G. Hendrey, ed., *Acid Precipitation: Aquatic Effects.* Ann Arbor, Mich.: Ann Arbor Science Publishers.

Wright, R. F., and A. Henriksen. 1978. Chemistry of small Norwegian lakes, with special reference to acid precipitation. *Limnology and Oceanography* 223:407–498.

Yan, N. D., and P. M. Stokes. 1978. Phytoplankton of an acidic lake and its responses to experimental alterations of pH. *Environmental Conservation* 5(2):93–100.

————, and R. Strus. 1980. Crustacean zooplankton communities of acidic, metal-contaminated lakes near Sudbury, Ontario. *Canadian Journal of Fisheries and Aquatic Science* 37:2282–2293.

CHAPTER 4

# GLOBAL CLIMATE CHANGE

*Stephen H. Schneider*

MANY SCIENTISTS ARE UNEASY about public discussion of critical scientific issues involving significant uncertainties. This is especially true if important policy implications attend speculative scientific subjects. The increasing concentrations of so-called greenhouse gases ($CO_2$, $CH_4$, $N_2O$, chlorofluorocarbons) provide one of the best current examples of a problem in which the public need for reliable scientific knowledge exceeds the ability to provide it. Indeed, potential societal and environmental effects have been assessed and reassessed by national and international groups for decades (Bach et al., 1979; Carbon Dioxide Assessment Committee, 1983; Bolin et al., 1986; Williams, 1978; Climate Research Board, 1978; Farrell, 1987; Seidel and Keyes, 1983; Pearman, 1987; NRC, 1987; IPCC, 1990a, 1990b;

Adapted in part from Stephen H. Schneider, "Climate-Change Scenarios for Greenhouse Increases," presented at Technologies for a Greenhouse-Constrained Society Conference, Oak Ridge, Tennessee, 1991, and "Global Climate Change: Ecosystem Effects" in *Interdisciplinary Science Reviews* (in press). Any opinions, findings, conclusions, or recommendations expressed in this chapter are those of the author and do not necessarily reflect the views of the National Science Foundation.

NAS, 1991). As our understanding of the scope of these effects has improved, the need for regional and temporal details about future climatic conditions has become increasingly urgent.

The desire to predict climatic changes arises from concern over the rapidity and magnitude of these changes and from the need in some cases to plan how to respond to changes far in advance. Society may be affected by rises in sea level, intensification of tropical cyclones, decline in the quantity and quality of freshwater resources, alterations in agricultural productivity, increases in direct threats to health, and impacts on unmanaged ecosystems (Smith and Tirpak, 1988). Yet we cannot easily or precisely determine how the complex climatic system will react to anthropogenic pollutants.

One response to the need for information on future climatic changes has been the analysis of large climatic changes in the geological past (Schneider, 1987; Budyko et al., 1987; Barron and Hecht, 1985; Berger et al., 1984). Although such paleoclimatic metaphors are relevant for estimating future climatic sensitivity to large changes in the radiation that forces climate to change, they are not exact analogies to the rate and character of present greenhouse-gas increases. (See, for example, the discussion in IPCC, 1990a: sec. 5.5.3.) Looking at more recent climatic records—the so-called historical analog method—can also provide insights into climatic behavior and societal vulnerabilities (Lough et al., 1983; Jager and Kellogg, 1983; Pittock and Salinger, 1982). But these methods too are based on climatic cause-and-effect processes that could differ from future greenhouse-gas radiative effects.

Therefore, scientists estimating future climatic changes have focused on large-scale models of the climate—general circulation models (GCMs)—that attempt to represent mathematically the complex physical interactions among the atmosphere, oceans, ice, biota, and land. As these models have evolved, more and more information has become available and more comprehensive simulations have been performed. Nevertheless, the complexities of the real climate system still vastly exceed the comprehensiveness of today's GCMs and the capabilities of today's computers. (See IPCC 1990a and 1990b for a state-of-the-art review.) Many important uncertainties are unlikely to be resolved before some significant climatic changes are felt, and certainly not before we are committed to some long-term environmental and societal effects.

Society is thus faced with a classic example of the need to make decisions based on imperfect information. Some projected climatic effects appear severe, but perhaps they can be mitigated if we know what to expect and if we choose to respond. At the same time, there is a risk of investing resources to prevent a problem that may not appear (or may appear where least expected). The need to know details about the timing and distribution of future climate changes has been stated in many scientific and political forums, and detailed climate impact studies have been commissioned (Kates et al., 1985; Parry, 1985; Senate Committee on Governmental Affairs, 1979; CEQ, 1980; IPCC, 1990b).

The American Association for the Advancement of Science has conducted one such study on climatic variability, climatic change, and U.S. water resources (Waggoner, 1990). In that study, one contribution (Schneider, Gleick, and Mearns, 1990) involved formulating scenarios of plausible climatic changes that might be used by hydrologists and water resource planners for detailed studies of regional hydrological effects. These scenarios are summarized here and updated. Hydrologists simply cannot make informed statements on the water-resource implications of future climatic changes without such scenarios. The same is true for biologists concerned with species movements, habitat transformation, and forest yields.

## FORECASTING CHANGES IN METEOROLOGICAL VARIABLES

To shed some light on these questions, I offer here a set of forecasts on changes in certain key meteorological variables over a range of temporal, spatial, and statistical scales. I believe that carefully qualified, explicit scenarios of plausible future climatic changes are preferable to impact speculations based on implicit or casually formulated forecasts. Therefore I include Table 4.1 to provide impact-assessment specialists with ranges of climate changes that reflect state-of-the-art modeling results. If the table sparks discussion among climatologists, improved and expanded projections may evolve.

My projections are based on analyses of available results and provide what are believed to be plausible estimates of the direction and magnitude of important anthropogenic climatic changes over

TABLE 4.1. RANGES OF CLIMATE CHANGES REFLECTING STATE-OF-THE-ART MODELING RESULTS AND INFERENCES

| Phenomenon | Projection of probable global annual average change[a] | DISTRIBUTION OF CHANGE | | | Significant transients? | CONFIDENCE OF PROJECTION | | Estimated time for research leading to consensus (years) |
|---|---|---|---|---|---|---|---|---|
| | | Regional average | Change in seasonality? | Interannual variability[b] | | Global average | Regional average | |
| Temperature[c] | +2 to +5°C | −3 to +10°C | yes | down? | yes | high | medium | 0 to 10 |
| Sea level | 0 to 80 cm[d] | global[e] | no | ???[g] | yes[d] | high | medium | 5 to 20 |
| Precipitation | +7 to +15% | −20 to +20% | yes | up? | yes | high | low | 10 to 40 |
| Direct solar radiation | −10 to +10% | −30 to +30% | yes | ???[g] | possible | low | low | 10 to 40 |
| Evapotranspiration | +5 to +10% | −10 to +10% | yes | ???[g] | possible | high | low | 10 to 40 |
| Soil moisture | ???[g] | −50 to +50% | yes | ???[g] | yes | ???[g] | medium | 10 to 40 |
| Runoff | increase | −50 to +50% | yes | ???[g] | yes | medium | low | 10 to 40 |
| Severe storms | ???[g] | ???[g] | [f] | ???[g] | yes | ???[g] | ???[g] | 10 to 40 |

a For an "equivalent doubling" of atmospheric $CO_2$ from the preindustrial level.
b Inferences based on preliminary results for the United States from Rind et al. (1990).
c Based on three-dimensional model results. If only trace gas increases were responsible for a twentieth-century warming trend of about 0.5°C, this range should be reduced by perhaps 1°C.
d Assumes only small changes in Greenland or West Antarctic ice sheets in twenty-first century. For equilibrium, hundreds of years would be needed and up to several meters of additional sea level rise could be accompanied by centuries of ice sheet melting from an equilibrium warming of 3°C or more.
e Increases in sea level at approximately the global rate except where local geological activity prevails or changes occur to ocean currents.
f Some suggestions of longer season and increased intensity of tropical cyclones as a result of warmer sea surface temperatures.
g No basis for quantitative or qualitative forecast.

SOURCE: Adapted from Schneider et al. (1990).

the next 50 years or so—a typical estimate for an equivalent doubling of carbon dioxide—together with a simple high, medium, or low level of confidence for each variable. ("Equivalent doubling" means that carbon dioxide together with other trace greenhouse gases has a radiative effect equivalent to doubling the preindustrial value of carbon dioxide from about 280 ppm to 560 ppm.) As another measure of the nature of the uncertainties, I include a rough estimate of the time that may be necessary to achieve a widespread scientific consensus on the direction and magnitude of the changes. In some cases—the magnitude of changes in global annual average temperature and precipitation, for example—such a consensus has virtually been reached. (See NRC, 1987, or IPCC, 1990a, 1990b.) In other cases—such as changes in the extent of cloud cover, patterns of regional precipitation changes evolving over decades, the potential for intensifying storms, and the daily, monthly, or interannual variance of many climatic variables—the large uncertainties surrounding present projections will be reduced only with considerably more research, probably measured in decades (see IPCC, 1990a:fig. 11.4). It is important to emphasize the high degree of uncertainty in climate-change forecasts. This uncertainty may be underrepresented by some who develop or use detailed regional climate-change scenarios for the purpose of studying impacts or advocating policy responses.

Consider, for example, the first row in Table 4.1: temperature change. The global average change of +2 to +5°C is the current range for general circulation models (IPCC, 1990a:chap. 5). This range is typical of that in most national and international assessments (as in Dickinson, 1986) for an equivalent doubling of greenhouse gases, neglecting transient delays. Based on curve fits to current global warming trends, many scientists prefer to revise that range downward somewhat to 1.5 to 4.5°C or perhaps even 1 to 4°C (Wigley and Raper, 1990a). NAS (1991), for example, used a 1 to 5°C range. The neglect of transients means that the range given is based on the assumption that trace-gas concentrations have been elevated over a long enough period for the climate to come into equilibrium with the increased concentration of greenhouse gases. In reality, the large heat capacity of the oceans will delay realization of full equilibrium warming by perhaps many decades (Hoffert et al., 1980; Schneider and Thompson, 1981; Bryan et al., 1982). This implies that at any specific time when we reach an equivalent $CO_2$

doubling (by, say, the year 2030), the actual global temperature increase may be considerably less than the $+2$ to $+5°C$ listed in Table 4.1. However, this "unrealized warming" (Hansen et al., 1985) will eventually occur when the climate system's thermal response catches up to the greenhouse-gas forcing.

On a finer scale, forecasts of regional or watershed changes in temperature, evaporation, and precipitation are most germane to estimating hydrological or other regional consequences of greenhouse warming. But, as Table 4.1 suggests, such regional forecasts are more uncertain. Regional temperature ranges given in the table are much larger than global changes and even allow for some regions of negative change (as in Schlesinger and Mitchell, 1987; IPCC, 1990a, 1990b). Surface temperature increases projected for the higher northern latitudes, for example, are up to several times larger than the projected global average response, at least in equilibrium. Given the importance of regional impact information, other techniques are being developed to evaluate small-scale hydrological effects of large-scale climatic changes (Gleick, 1986). P. H. Gleick has used a regional hydrological model driven by large-scale climate change scenarios from various GCM inputs (Gleick, 1987). Other techniques embed a mesoscale atmospheric model into a limited region of the global-scale GCM (see Giorgi, 1990).

Even more uncertain than regional details, but perhaps most important, are estimates for such measures of climatic variability as the frequency and magnitude of severe storms, enhanced heat waves, or reduced frost probabilities (Parry and Carter, 1985; Mearns et al., 1984, 1990; Wigley, 1985). Theoretical physical reasoning suggests, for example, that hurricane intensities will increase with climatic changes (Emanuel, 1987). Such issues are just now beginning to be considered and evaluated from equilibrium climate-model results and will, of course, have to be studied for realistic transient cases to be of maximum value to impact assessors.

Other uncertainties raised by the transient nature of the actual trace-gas forcing are the emission and removal rates of $CO_2$, $CH_4$, and other greenhouse gases. Figure 4.1 shows three plausible scenarios based on high, medium, and low emission rates (WMO, 1988). These uncertainties have been added to those associated with estimates of climate sensitivity and the delay associated with oceanic heat capacity. In any case, since the earth has apparently not experienced global average temperatures more than 1 to 2°C higher than at

FIGURE 4.1. THREE SCENARIOS FOR GLOBAL TEMPERATURE CHANGE TO THE YEAR 2100.
*These scenarios are derived from combining uncertainties in future trace green-house gas projections with those of modeling the climatic response to those projec-tions. Sustained global temperature changes beyond 2°C (3.6°F) would be unprecedented during the era of human civilization. The middle to upper range represents climatic change at a pace 10 to 100 times faster than long-term natural average rates of change.*

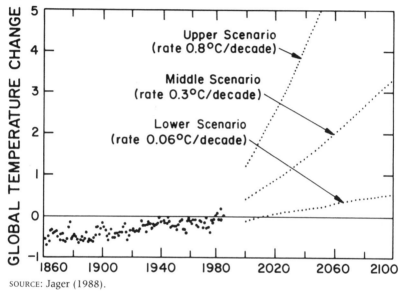

SOURCE: Jager (1988).

present over the past glacial cycle (150,000 years) (Barnola et al., 1987), all but the slowest scenario represents a rapid, large climatic change to which the environment and society will have to adapt.

The principal technical advance needed to build consensus on the reliability of time-evolving regional forecasts of hydrological vari-ables is the development, testing, and verification of coupled atmo-sphere, ocean, and land-surface models for realistic transient scenarios (Thompson and Schneider, 1982). There has been recent progress in the development of such models (Stouffer et al., 1989; Washington and Meehl, 1989), although they reaffirm the conten-tion (Schneider and Thompson, 1981) that reliable regional climatic projections evolving over decades will require coupling of dynamic atmosphere and ocean models driven by realistic scenarios of tran-sient trace-gas increase. Therefore, I suggest in Table 4.1 that it will be a decade or more before the scientific community reaches such a consensus, because it will be at least a decade before high-resolution models of atmospheric, oceanic, biospheric, and hydrospheric subsystems can be run in coupled mode and data can be obtained to validate these simulations. (See also IPCC, 1990a:fig. 11.4.)

Some scientists may object in principle to the approach taken in Table 4.1, arguing that since the confidence levels cited are intuitive, they may be incorrect or some predicted change could even be in the opposite direction of that listed. Indeed, predictions about something as complex as global climate will always be somewhat uncertain. Nevertheless, many policymakers are likely to have the opposite reaction—the information in the table, even if certain, may still not contain enough detail to justify major policy decisions. Policy analysts typically want regional details even finer than those available in the survey article by Schlesinger and Mitchell (1987), a few of which are reproduced here as Figures 4.2 to 4.5. Note especially the large warming in high latitudes (especially over sea ice margins) in winter associated with simulated reductions in snow cover or sea ice. Warming could cause thermal expansion of oceans, melting of ice on land, and possibly increased buildup of snow on high, cold ice sheets. Taken together, these factors contribute to the rise in sea level of zero to about 1 m typically estimated for the twenty-first century (IPCC, 1990a:chap. 9).

Perhaps more important for biological assessment is the prediction of soil moisture, for which one GCM example is given in Figure 4.6. Note the large decrease in summertime soil moisture (especially in mid-latitudes), a result common to many models (as in Rind et al., 1990). The schism between the reticence of some scientists to make any forecasts and the insistence of some policy analysts on high levels of regional detail cannot be resolved simply by having the latter fashion their own implicit scenarios for impact assessment. Rather, I believe, it is better to have knowledgeable scientists put forth "forecasts" based on the best available information.

FIGURE 4.2. GEOGRAPHICAL DISTRIBUTION OF THE SURFACE AIR
TEMPERATURE CHANGE (°C) FOR DJF.

*Distribution is simulated with* (top) *a version of the GFDL GCM,* (middle) *the
GISS GCM, and* (bottom) *the NCAR GCM. Stippling indicates temperature
increases larger than 4°C. For reference to particular versions of the GCMs used, see
Schlesinger and Mitchell (1987).*

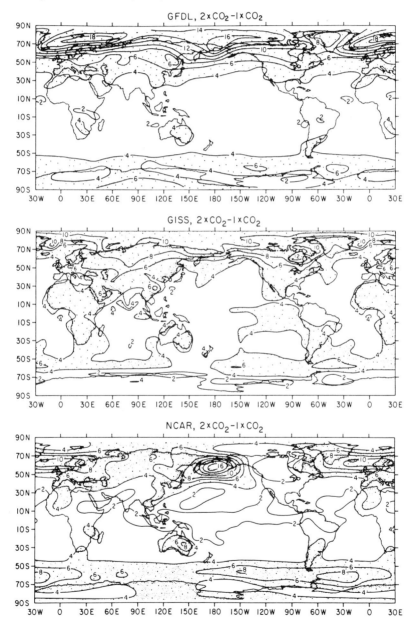

GFDL, $2 \times CO_2 - 1 \times CO_2$

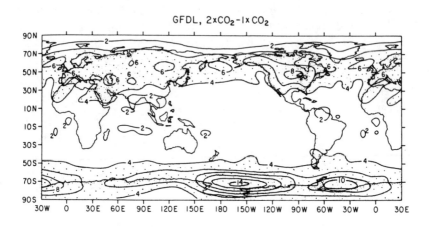

GISS, $2 \times CO_2 - 1 \times CO_2$

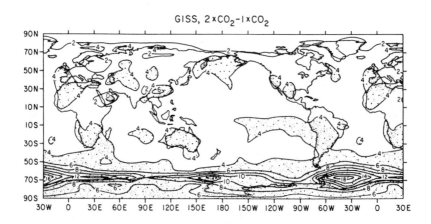

NCAR, $2 \times CO_2 - 1 \times CO_2$

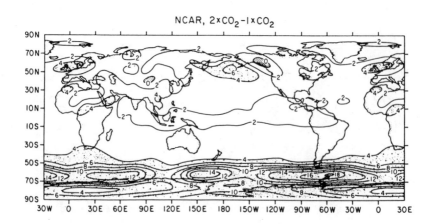

*Distribution is simulated with a version of* (top) *the GFDL GCM,* (middle) *the GISS GCM, and* (bottom) *the NCAR GCM. Stippling indicates a decrease in precipitation rate. For details of which specific GCM versions were chosen, see Schlesinger and Mitchell (1987).*

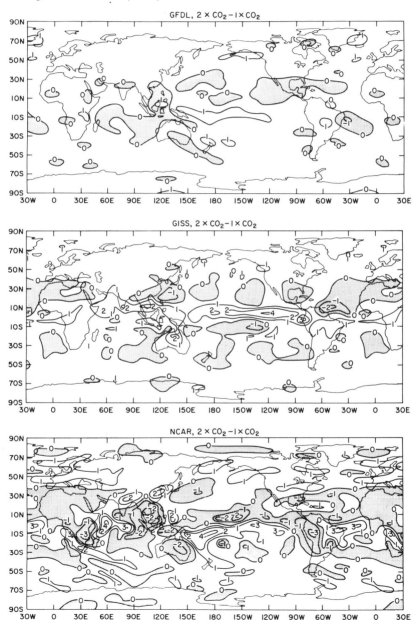

FIGURE 4.5. GEOGRAPHICAL DISTRIBUTION OF THE PRECIPITATION
RATE CHANGE (MM/DAY) FOR JJA.

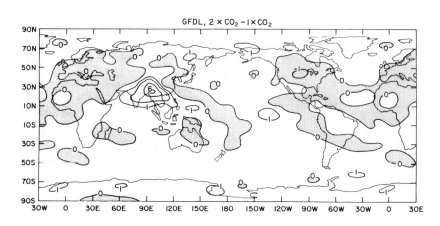

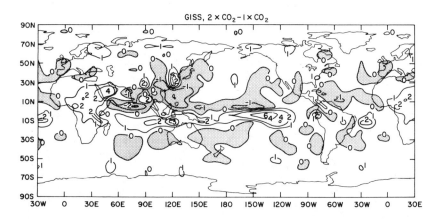

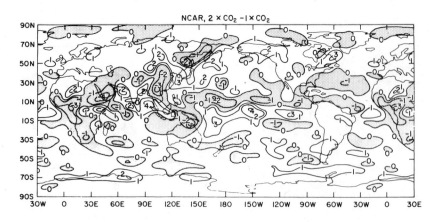

FIGURE 4.6. CO$_2$-INDUCED CHANGE IN SOIL MOISTURE EXPRESSED AS A
PERCENTAGE OF SOIL MOISTURE.

*Simulation obtained from a computer model with quadrupled CO$_2$ compared to a
control run with normal CO$_2$ amounts. Note the nonuniform response of this
ecologically important variable to the uniform change in CO$_2$.*

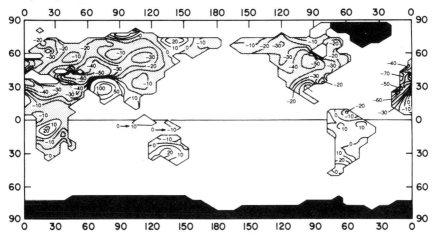

SOURCE: Manabe and Wetherald (1986).

Finally, even the highest-resolution three-dimensional GCMs
suitable for integrations over 50 or more years are not likely to have a
grid much less than 100 km; individual clouds and most biological
field research, for example, are on scales far smaller than that. GCMs
will not, therefore, be able to resolve individual thunderstorms or the
important local and mesoscale effects of hills, coastlines, lakes, vegetation boundaries, and heterogeneous soil. For regions that have
relatively uniform surface characteristics, such as a 1000-km-scale
savannah or a tropical forest with little elevation change, GCM grid-
scale parametric representations of surface albedo, soil type, and
evapotranspiration could be used to estimate local changes. Alterations in climate predicted within a box are likely to apply fairly
uniformly across such nicely behaved, homogeneous areas. But
steep topography or lakes smaller than GCM grids can mediate local
or regional climate. Therefore, even if GCM predictions were accurate at grid scale, they would not necessarily be appropriate to local
conditions.

Large-scale observed climatic anomalies are translated to local

FIGURE 4.7. DISTRIBUTION OF THE RELATIONSHIP BETWEEN LARGE-SCALE
(AREA-AVERAGED) AND LOCAL VARIATIONS OF THE MONTHLY MEAN
SURFACE AIR TEMPERATURE (*ABOVE*) AND PRECIPITATION (*BELOW*).

*Distribution is given by the first empirical orthogonal function determined from 30
years' observational monthly means at forty-nine stations in Oregon in comparison
with the statewide average.*

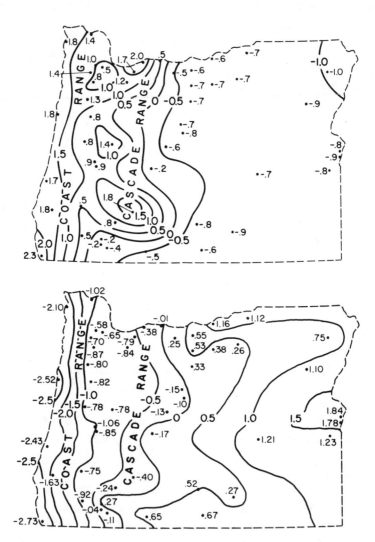

SOURCE: Gates (1985).

variations in Figure 4.7. This analysis (Gates, 1985) of the local climatic variability for the state of Oregon was based on several years of data using a technique known as empirical orthogonal functions. The north–south Cascade Mountains translate a simple change in the frequency or intensity of westerly winds into a characteristic climatic signature of either wetter on the west slope and drier on the east or vice versa. In other words, a general circulation model producing altered westerlies could be applied to the map to determine the effect on a local area. Such a map, constructed from variations of climate observed over several years, seems an ideal way to translate the GCM grid to the local scale or mesoscale. Because empirical data have been used, however, such a relation would be valid only where the causes of recent climatic variations or oscillations carry forward and include the effect of climatic changes forced by trace gases. It is not obvious that the signature of climatic change from increases in trace gases will be the same as past vacillations—many of which could have been internal oscillations within the climate system, not the result of external forcing such as changes in trace gases. Thus other translations of scale must be considered.

One might embed a high-resolution mesoscale model within one box of a GCM, using as boundary conditions for the mesoscale model the wind, temperature, and so forth predicted by the GCM at the grid boundaries (Dickinson et al., 1989). The mesoscale model could then account for local topography, soil type, and vegetation cover and translate GCM forecasts to local topography. Figure 4.8 is an example for the western United States. For such a method to have any reasonable hope of success, however, the GCM must produce accurate climatic statistics for the special grid box. To return to the Oregon case in Figure 4.7, if the climatic average of the GCM's winds in the unperturbed case (the control case) has the wrong westerly component, the climate change will probably be misrepresented in a region where topography amplifies any such error in the wind direction. A prerequisite for this kind of scale transition, therefore, is a sufficiently accurate control climate for the key variables. Only then does it make sense to take the next step: imposing a scenario of trace-gas increase on the GCM to estimate how the local-scale climate might change.

While testing scale transitions in steep topography and other rapidly varying local features, modelers should examine the behavior of their models using grid boxes that are much less "pathological"—

FIGURE 4.8. AVERAGE JANUARY TOTAL PRECIPITATION (CM).

(a) Observations; (b) R15 general circulation model (4.5° latitude × 7.5° longitude); (c) mesoscale model driven by output of R15 model.

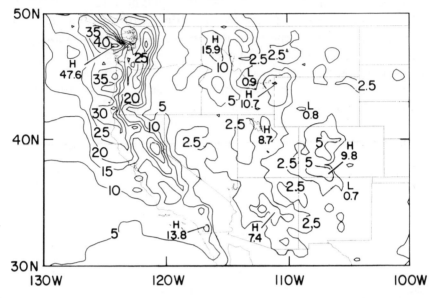

**(a)** Observations

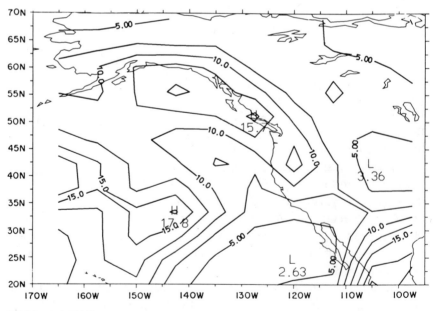

**(b)** Model R15

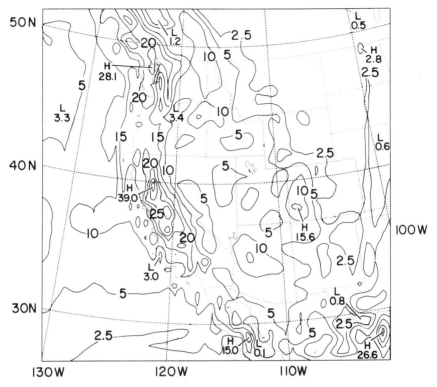

**(c)** Mesoscale model

SOURCE: Giorgi (1990).

that is, examine boxes where local features are relatively homogeneous and where translation of local-to-grid scales should prove a less serious obstacle (as in Mearns et al., 1990).

Uncertainty about parametric representations of feedback mechanisms like clouds or sea ice is one reason why the goal of climate modeling—reliable, verified forecasting of key variables such as temperature and rainfall—is not yet possible. Another source of uncertainty external to the models is human behavior. Forecasting the effect of carbon dioxide on climate, for example, requires knowing how much $CO_2$ or $CH_4$ is going to be emitted (Nordhaus and Yohe, 1983; Edmonds and Reilly, 1984; Ausubel et al., 1988) and how that emission will be distributed or removed by the physical, chemical, and biological processes of the carbon cycle.

What the climate models can do well is analyze the sensitivity of the climate to uncertain or even unpredictable variables. In the case of carbon dioxide, one could construct plausible scenarios of economic, technological, and population growth to project growth of $CO_2$ emission and model the climatic consequences. Such uncertain climatic factors as cloud feedback could be varied over a plausible range. The calculations would indicate which uncertain factors are most important in making the climate sensitive to $CO_2$ increases. One could then concentrate research on these factors. The results would also suggest the range of climatic futures that ecosystems and societies may be forced to adapt to and at what potential rates. How to respond to such information, of course, is a political issue (Schneider, 1989, 1990a).

## VERIFICATION OF CLIMATE MODELS

The most perplexing question about climate models is whether they can be trusted to provide grounds for altering social policies, such as those governing carbon dioxide emissions. How can models so fraught with uncertainties be verified? In fact, there are several methods. Although none is sufficient alone, together they can provide significant, albeit circumstantial, evidence of a model's credibility.

The first verification method is checking the model's ability to simulate today's climate. The seasonal cycle (see Figure 4.9) is one good test because the temperature changes are several times larger, on a hemispheric average, than the temperature change from an ice age to an interglacial period. GCMs map the seasonal cycle well, which suggests they are on the right track as far as fast physics is concerned (cloud feedback, for example). The seasonal test, however, does not indicate how well a model simulates such slow processes as changes in deep-ocean circulation or ice cover, which may have an important effect on the time scales (decade to century) over which $CO_2$ is expected to double.

A second verification technique is isolating individual physical components of the model, such as its parameterizations, and testing them against reality. One can check, for example, whether the model's parameterized cloudiness matches the observed cloudiness

FIGURE 4.9. THE SEASONAL CYCLE AS A TEST OF A MODEL'S VALIDITY.

*A three-dimensional climate model has been used to compute the winter to summer temperature extremes all over the globe (above). The model's performance can be verified against the observed data (below). This verification exercise shows that the model quite impressively reproduces many of the features of the seasonal cycle. These seasonal temperature differences are mostly larger than those occurring between ice ages and interglacials or for any plausible future carbon dioxide change. Although this approach cannot validate models for processes occurring on medium to long time scales (greater than 1 year), they are very encouraging for validating to a rough factor of 2 to 3 such fast-physics parameterizations as clouds.*

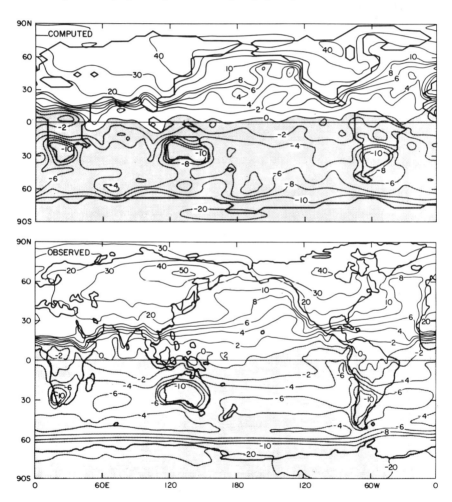

SOURCE: Manabe and Stouffer (1980).

93

of a particular box. But this technique cannot guarantee that the complex interactions of individual model components are properly treated. The model may be good at predicting average cloudiness but bad at representing cloud feedback. In that case, simulation of overall climatic response to increased carbon dioxide, for example, is likely to be inaccurate. A model should reproduce to better than, say, 10 percent accuracy the flow of thermal energy among the atmosphere, surface, and space, which are well-measured quantities. Together these energy flows compose the well-established greenhouse effect on earth and constitute a formidable and necessary test for all models.

A model's performance in simulating these energy flows is an example of physical verification of model components. Some scientists (Ellsaesser, 1984; Lindzen, 1990) have argued that critically important hydrological feedback processes are not well simulated by GCMs. Recent satellite analyses (Figure 4.10) by A. Raval and V. Ramanathan (1989) suggest that models do a credible job of simulating the so-called water vapor greenhouse effect so critical to a model's sensitivity to greenhouse-gas forcing (see Schneider 1990b for a discussion). Ramanathan and Collins (1991), however, have shown that over the region of the earth covered by tropical cumulus clouds (on the order of 10 percent of the surface), the formation of cirrus cloud shields around such cumulus clouds could limit ocean temperature buildups to less than 305° K. If this is true, this negative feedback on surface temperature would limit the ultimate warming of the earth for very large $CO_2$/trace-gas buildup but would have little effect on the Table 4.1 estimates—especially the likelihood of hard-to-predict, significant, regional, time-evolving climatic changes.

A third method for determining overall, long-term simulation skill is the model's ability to reproduce the diverse climates of the ancient earth (COHMAP, 1988; Barron and Hecht, 1985; Imbrie, 1987; Schneider, 1987) or even those of other planets (Kasting et al., 1988). Paleoclimatic simulations of the Mesozoic era, glacial/interglacial cycles, or other extreme past climates help in understanding the coevolution of the earth's climate with living things. As verifications of climate models, however, they are also crucial to estimating the climatic and biological future.

Overall validation of climatic models thus depends on constant

FIGURE 4.10. COMPARISON OF GREENHOUSE EFFECT, HEAT-TRAPPING
PARAMETER (*G*), AND SURFACE TEMPERATURE.

*Data were obtained from three sources: ERBE annual values, obtained by averaging April, July, and October 1985 and January 1986 satellite measurements; 3-D climate-model simulations for a perpetual April simulation (NCAR Community Climate Model); line-by-line radiation-model calculations by A. Arking using $CO_2$, $O_3$, and $CH_4$. The line-by-line model results come close to the CCM and the ERBE values.*

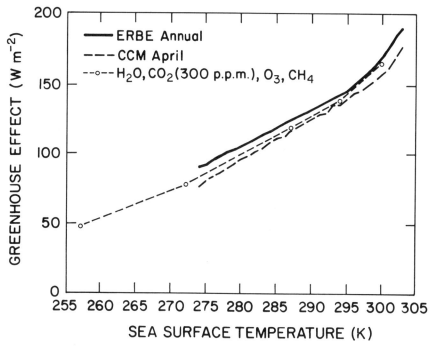

SOURCE: Raval and Ramanathan (1989).

appraisal and reappraisal of performance in the categories noted here. But these are indirect or surrogate validations for greenhouse-gas increases. Also important are direct validations—for example, a model's response to such century-long forcings as the 25 percent increase in carbon dioxide and large increases of other trace greenhouse gases (Ramanathan et al., 1985) since the industrial revolution. Indeed, most climatic models are sensitive enough to predict that a warming of 1°C should have occurred during the past century.

FIGURE 4.11. OBSERVED GLOBAL MEAN TEMPERATURE CHANGES (1861–1989)
COMPARED WITH PREDICTED VALUES FROM A HIGHLY SIMPLIFIED,
UPWELLING-DIFFUSION CLIMATE MODEL.

*The modeled results show how much temperature change would have occurred if the equilibrium sensitivity to $CO_2$ doubling were as labeled on the figure and if greenhouse gas forcing were the only cause of climate change.*

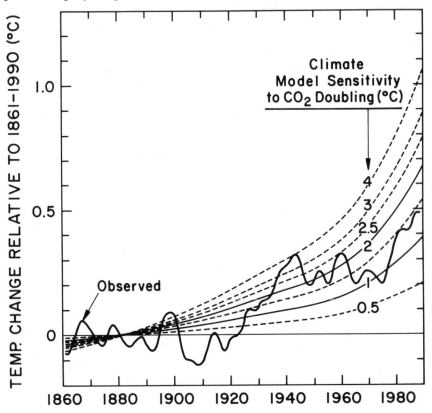

SOURCE: Wigley and Raper (1990a).

However, the precise forecast of the past 100 years also depends on how the model accounts for such factors as changes in the solar constant or volcanic dust (Schneider and Mass, 1975; Gilliland and Schneider, 1984; Hansen et al., 1981). Indeed, as Figure 4.11 shows, the typical prediction of a degree or so of warming is broadly consistent but somewhat larger than observed.

There are several possible explanations (see Schneider, 1989) for the discrepancy:

1. The models are too sensitive to increases in trace greenhouse gases by a rough factor of 2.
2. Modelers have not properly accounted for such competitive external forcings as volcanic dust or changes in solar energy output.
3. Modelers have not accounted for other external forcings such as regional tropospheric aerosols from agricultural, biological, and industrial activity (Wigley and Raper, 1989).
4. Modelers have not properly accounted for internal processes that could lead to stochastic or chaotic behavior (Hasselmann, 1976; Dalfes et al., 1983; Lorenz, 1968), including the possibility of a naturally occurring twentieth-century cooling trend that masked the global warming signal.
5. Modelers have not properly accounted for the large heat capacity of the oceans, which could take up some of the heating of the greenhouse effect and delay (but not ultimately reduce) warming of the lower atmosphere.
6. Both present models and observed climatic trends could be correct, but models are typically run for equivalent doubling of the $CO_2$ concentration, whereas the world has experienced only a quarter of this increase, and nonlinear processes have been properly modeled and produced a sensitivity appropriate for doubling but not for a 25 percent increase.
7. The incomplete and nonuniform network of thermometers has underestimated actual global warming this century.

## ECOSYSTEM RESPONSE TO RAPID CHANGE

One of the Office of Management and Budget officials of the Bush administration said it would be no more difficult to adapt to a few degrees' climate warming than to get off a Boston flight in Washington. This view, of course, is ignorant of ecology. The forests of Virginia cannot call the neighborhood van lines to transplant millions of trees up to Massachusetts. Many species of trees take hundreds, even thousands, of years for substantial migration. Moreover, while individual species may migrate with slowly changing climate, whole habitats usually do not, as species composition is dramatically altered (see COHMAP, 1988). That is indeed what occurred when the ice ages waned about 12,000 years ago and in the next 4000 years were replaced by the relatively warm interglacial period we are now experiencing. At that time a 5°C (9°F) temperature increase accompanied 100-m (330-foot) sea level rises and dramatic habitat changes such as those seen in Figure 4.12. The transition from ice age to interglacial quite literally revamped the ecological face of the earth. The time it took for nature to make this adjustment was on the order of thousands of years, driven by sustained global average temperature changes on the order of 5°C. We are facing the prospect of temperature changes of this magnitude over the course of a century—a rate that most ecologists who study forest ecosystems suggest is simply too fast for the trees, let alone habitats, to keep up. In this connection, ecologist Margaret Davis (1990:108) notes:

> The fossil record shows that most forest trees were able to disperse rapidly enough to keep up with most of the climatic changes that took place in recent millennia. These changes were much more gradual than the climatic changes projected for the future. Even so, there were occasional periods of disequilibrium between plant distributions or abundances, soils, and climate that lasted a century or more. The most rapid dispersal rates known from the fossil record, however, are an order of magnitude too slow to keep up with the temperature rise expected in the coming century.

---

FIGURE 4.12. RELATIVE FRACTION (IN PERCENT) OF FOSSIL POLLEN FROM VARIOUS SPECIES OF TREES.

*These pollen patterns can be correlated to climatic conditions. The maps show, for example, that at the end of the last Pleistocene Ice Age 11,000 years ago much of the northeastern United States was covered with spruce forests, which are found today in central Canada near Hudson Bay. As climatic conditions ameliorated, the spruce "chased" the retreating ice sheet northward into Canada, whereas more temperate-climate trees like oak moved outward from their refuge in the Carolinas into the middle Atlantic states and the Midwest. Around 4000 years ago the distribution of these temperate-climate trees stabilized in what is roughly their present configuration.*

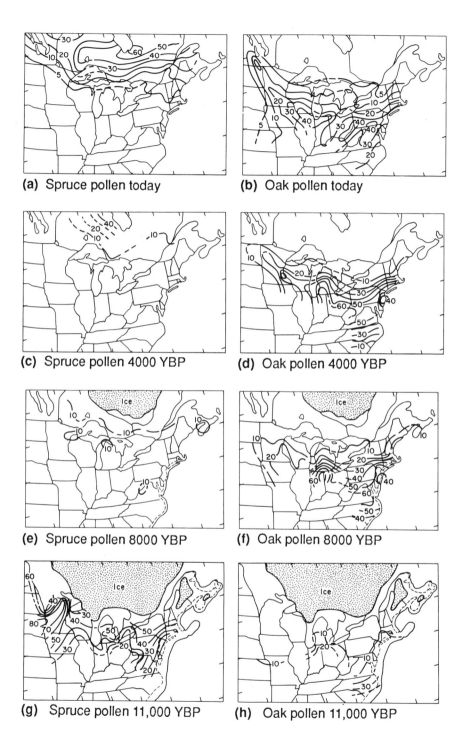

**(a)** Spruce pollen today

**(b)** Oak pollen today

**(c)** Spruce pollen 4000 YBP

**(d)** Oak pollen 4000 YBP

**(e)** Spruce pollen 8000 YBP

**(f)** Oak pollen 8000 YBP

**(g)** Spruce pollen 11,000 YBP

**(h)** Oak pollen 11,000 YBP

SOURCE: Bernabo and Webb (1977).

Figure 4.13 shows the result of calculations (Davis and Zabinski, 1991) suggesting that certain species (sugar maple in this case) could be marooned substantially out of their ranges should the climate in fact change as projected by most modern general climate models. (Goddard Institute for Space Studies is shown in part (*a*) of the figure and Geophysical Fluid Dynamics Laboratory in part (*b*).) Habitats would change unpredictably, and some species would quite literally be stranded in ecologically unsuitable territory. Apart from the climatic effects, $CO_2$ will differentially alter the photosynthetic and water use efficiency of plants. This too will change the fabric of ecosystems.

Trees are not the only potential victims of rapid climate change, for various animals depend on complex habitats for their livelihood,

FIGURE 4.13. PRESENT GEOGRAPHICAL RANGE OF SUGAR MAPLE (HORIZONTAL LINES) AND POTENTIALLY SUITABLE RANGE UNDER DOUBLED $CO_2$ (VERTICAL LINES).

*Cross-hatching indicates the region of overlap. (**a**) Predictions using climate scenario derived from the GISS general circulation model; (**b**) predictions using climate scenario derived from the GFDL model. Gridpoints are sites of climatic data output for each model.*

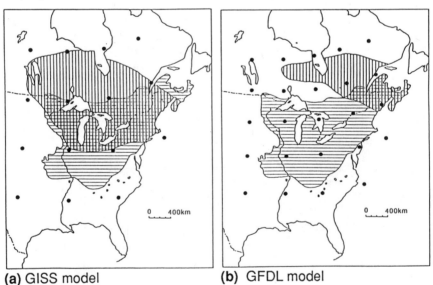

**(a)** GISS model          **(b)** GFDL model

SOURCE: Davis and Zabinski (1991).

even survival. One symbolic case has been described by Botkin et al. (1991): the Kirtland's warbler in northern Michigan, which lives in a narrow area of jack pines that survive on sandy soils in that climate. Ecological models of the growth and decline of jack pine forests suggest that even a small climate change would be enough to devastate this habitat, almost certainly dooming Kirtland's warbler to extinction in 30 to 60 years. Already high rates of extinction could be substantially exacerbated by climate changes occurring more rapidly than ecosystems could adapt (Wilson, 1989, or Peters, 1991) or by having adaptation and migration blocked by highways, urban development, seawalls, agriculture, and so forth. Apart from the issue of extinction is the altered habitat associated with climate change. Ecologist Terry Root (1988a) of the University of Michigan has shown a high degree of statistical correlation between the distribution and abundance of many bird species and large-scale environmental factors such as temperature. Figure 4.14 shows the clear association

FIGURE 4.14. DISTRIBUTION AND ABUNDANCE OF THE
WINTER RANGE OF THE EASTERN PHOEBE.

*The northern boundary lies very close to the −4°C isotherm of January minimum temperature (heavy solid line).*

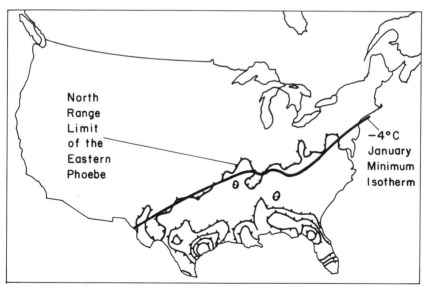

SOURCE: Root (1988a).

between the January mean temperature and the northern range limit of the eastern phoebe. This kind of analysis implies that as climate changes, the physiological tolerance of certain birds would cause them—habitat permitting—to change their ranges as well (Root, 1988b). Since birds can migrate rapidly but vegetation may take centuries to adjust, rapid climatic changes (several degrees Celsius on the scale of decades) could tear apart the fabric of current ecosystems with hard-to-predict consequences for individual species.

## IMPLICATIONS FOR POLICY

Climate change is not necessarily a threat to the existence of every climate-sensitive species. But given the transient nature of the climate change, which nonetheless would be occurring over the course of decades, and the transient nature of forest and other vegetation ecosystems, which could change over decades to centuries, it is quite clear that plausible climatic changes could create substantial disequilibrium within ecosystems. About the only outcome that can be predicted with certainty is major surprises. Substantial change to plant and animal populations, habitat types, and even some extinctions would seem inevitable from most climate-change scenarios. Perhaps the only forecast that seems clearly justifiable is this: The more rapidly the climate changes, the higher the probability of substantial disruption and surprise.

These possibilities have led to a debate as to whether humans need to intervene as "ecological engineers" deliberately replanting or transplanting presently maladapted species in what might be their new habitats of the future. Given our relative ignorance of the full implications of these potential ecological impacts, some conservationists argue that such human intervention (or ecoengineering) would be unethical at best. Others argue that species extinctions that might take place if stranding occurs are even worse risks than habitat alteration that might take place from ecoengineering (Roberts, 1988). All of these wrenching debates are created by the prospect of rapid global warming, which in turn is created by the increasing use of energy and materials demanded by growing populations insisting on using the cheapest available technology to increase their standard of living.

It is unlikely that societies will radically alter these plans. Thus the capacity of ecosystems to respond to rapid change demands urgent study if we are to figure out how to help nature adapt more effectively. In the final analysis, however, slowing down the rate of warming through the application of the "tie-in" principle is the best immediate strategy. Although deciding how to act is always a decision about whether it is appropriate to invest present resources as a hedge against potential change, those actions that have already been proved cost-effective (using and producing energy more efficiently, for example) should be vigorously pursued. All political obstacles slowing the penetration of these cost-effective actions should be removed and incentives created (see NAS, 1991; OTA, 1991). Additional policy actions should include those that not only help to slow the rate of global warming but help to solve other problems as well. Banning CFCs not only reduces about 20 percent of potential global warming forcing, for example, but also eliminates the emission of chemicals that reduce stratospheric ozone. Fuel switching from coal to natural gas not only cuts roughly in half the carbon dioxide released per unit energy produced, but substantially reduces the sulfur dioxide emissions that contribute to acid rain. Energy efficiency and conservation planning not only reduce environmental impacts, but reduce the dependency of countries such as the United States and Japan on imported energy supplies, thereby reducing any balance of payments deficits, military budgets, and even threats to world security.

These tie-in strategies are, I believe, long overdue with or without global warming. And because of the added urgency of slowing down global warming, thereby buying time for us and nature to adapt to whatever changes will take place, such actions clearly seem the most appropriate immediate policy response.

## REFERENCES

Ausubel, J. H., A. Grubler, and N. Nakicenovic. 1988. Carbon dioxide emissions in a methane economy. *Climatic Change* 12:245–264.

Bach, W., J. Pankrath, and J. Williams (eds.). 1979. *Interactions of Energy and Climate.* Dordrecht: Reidel.

Barnola, J. M., D. Raynaud, Y. S. Korotkevich, and C. Lorius. 1987. Vostok ice core provides 160,000-year record of atmospheric $CO_2$. *Nature* 329:408–414.

Barron, J., and A. D. Hecht (eds.). 1985. *Historical and Paleoclimatic Analysis and Modeling.* New York: Wiley.

Berger, A., J. Imbrie, J. Hays, G. Kukla, and B. Saltzman (eds.). 1984. In *Milankovitch and Climate: Parts 1 and 2.* Dordrecht: Reidel.

Bernabo, J. C., and T. Webb III. 1977. Changing patterns in the Holocene pollen record of northeastern North America: A mapped summary. *Quaternary Research* 8:64–96.

Bolin, B., et al. (eds.). 1986. *The Greenhouse Effect, Climatic Change and Ecosystems.* Chichester: Wiley.

Botkin, D. P., D. A. Woodby, and R. A. Nisbet. 1991. Kirtland's warbler habitat: A possible early indicator of climate warming. *Biological Conservation* 56(1):63–78.

Bryan, K., F. Komro, S. Manabe, and M. J. Spelman. 1982. Transient climate response to increasing atmospheric carbon dioxide. *Science* 215:56–58.

Budyko, M. I., A. B. Ronov, and A. L. Yanshin. 1987. *History of the Earth's Atmosphere.* New York: Springer-Verlag.

Carbon Dioxide Assessment Committee. 1983. *Climate Change.* Washington: National Academy Press.

Charlson, R. J., J. Langner, H. Rodhe, C. B. Leovy, and S. G. Warren. 1991. Perturbation of the northern hemisphere radiative balance by backscattering from anthropogenic sulfate aerosols. *Tellus* 91 (in press).

Climate Research Board. 1978. *International Perspectives on the Study of Climate and Society.* Washington: National Academy of Sciences.

COHMAP. 1988. Climatic changes of the last 18,000 years: Observations and model simulations. *Science* 241:1043–1052.

Council on Environmental Quality (CEQ). 1980. *The Global 2000 Report to the President.* Washington: Government Printing Office.

Dalfes, H., S. H. Schneider, and S. L. Thompson. 1983. Numerical experiments with a stochastic climate model. *Journal of Atmospheric Science* 40:1648–1658.

Davis, M. B. 1990. Climatic change and the survival of forest species. In *The Earth in Transition: Patterns and Processes of Biotic Impoverishment*, ed. George M. Woodwell. Cambridge: Cambridge University Press.

Davis, M. B., and C. Zabinski. 1991. Changes in geographical range resulting from greenhouse warming effects on biodiversity in forests. In *Consequences of Greenhouse Warming to Biodiversity*, ed. R. L. Peters and T. Lovejoy. New Haven: Yale University Press.

Dickinson, R. E. 1986. In *Sustainable Development of the Biosphere*, ed. W. C. Clark and R. E. Munn. Cambridge: Cambridge University Press.

Dickinson, R. E., R. M. Errico, F. Giorgi, and G. T. Bates. 1989. A regional climate model for the western United States. *Climatic Change* 15:383–422.

Edmonds, J. A., and J. Reilly. 1984. Global energy in $CO_2$ to the year 2050. *Energy Journal* 4:21.

Ellsaesser, H. W. 1984. The climatic effect of $CO_2$: A different view. *Atmospheric Environment* 18:431–434.

Emanuel, K. A. 1987. The dependence of hurricane intensity on climate. *Nature* 326:483–485.

Farrell, M. P. (ed.). 1987. *Master Index for the Carbon Dioxide Research State-of-the-Art Report Series*. Washington: Department of Energy.

Gates, W. L. 1985. The use of general circulation models in the analysis of the ecosystem impacts of climatic change. *Climatic Change* 7:267–284.

Gilliland, R. L., and S. H. Schneider. 1984. Volcanic, $CO_2$, and solar forcing of northern and southern hemisphere surface air temperatures. *Nature* 310:38–41.

Giorgi, F. 1990. Simulation of regional climate using a limited area model nested in a general circulation model. *Journal of Climate* 3(9):941–963.

Gleick, P. H. 1986. Methods for evaluating the regional hydrologic impacts of global climatic changes. *Journal of Hydrology* 88:97–116.

————. 1987. Regional hydrologic consequences of increases in atmospheric $CO_2$ and other trace gases. *Climatic Change* 10:137–160.

Hansen, J., D. Johnson, A. Lacis, S. Lebedeff, P. Lee, D. Rind, and G. Russell. 1981. Climate impact of increasing atmospheric carbon dioxide. *Science* 213:957–966.

————, G. Russell, A. Lacis, I. Fung, and D. Rind. 1985. Climate response times: Dependence on climate sensitivity and ocean mixing. *Science* 229:857–859.

Hasselmann, K. 1976. Stochastic climate models. Part I: Theory. *Tellus* 28(6):473–485.

Hoffert, M. I., A. J. Callagari, and C. T. Hsieh. 1980. The role of deep sea heat storage in the secular response to climatic forcing. *Journal of Geophysical Research* 85:667.

Imbrie, J. 1987. Abrupt terminations of late Pleistocene ice ages: A simple Milankovitch explanation. In *Abrupt Climatic Change: Evidence and Implications*, ed. W. H. Berger and L. D. Labeyrie. Dordrecht: Reidel.

Intergovernmental Panel on Climate Change (IPCC). 1990a. *Scientific Assessment of Climate Change*. Report prepared for IPCC by Working Group I. Geneva: World Meteorological Organization.

————. 1990b. *Potential Impacts of Climate Change*. Report prepared for IPCC by Working Group II. Geneva: World Meteorological Organization.

Jager, J. 1988. *Developing Policies for Responding to Climatic Change: A Summary of the Discussions and Recommendations* of the Workshops Held in

Villach, 28 September to 2 October 1987. WCIP-1, WMO/TD 225. Geneva: World Meteorological Organization.

Jager, J., and W. W. Kellogg. 1983. Anomalies in temperature and rainfall during warm arctic seasons. *Climatic Change* 5:39—60.

Kasting, J. F., O. B. Toon, and J. B. Pollack. 1988. How climate evolved on the terrestrial planets. *Scientific American* 258:90—97.

Kates, R. H., A. H. Ausebel, and M. Berberian. 1985. *Climate Impact Assessment: SCOPE 27.* New York: Wiley.

Lindzen, R. S. 1990. Some coolness concerning global warming. *Bulletin of the American Meteorological Society* 77:288—299.

Lorenz, E. N. 1968. Climate determinism. *Meteorology Monograph* 8(30): 1—3.

Lough, J. M., T. M. L. Wigley, and J. P. Palutikof. 1983. *Journal of Climate and Applied Meteorology.* 22:1673.

Manabe, S., and R. J. Stouffer. 1980. Sensitivity of a global climate model to an increase in $CO_2$ concentration in the atmosphere. *Journal of Geophysical Research* 85:5529.

————, and R. Wetherald. 1986. Reduction in summer soil wetness induced by an increase in atmospheric carbon dioxide. *Science* 232:626.

Mearns, L. O., R. W. Katz, and S. H. Schneider. 1984. Extreme high temperature events: Changes in their probabilities and changes in mean temperature. *Journal of Climate and Applied Meteorology* 23:1601—1613.

————, S. H. Schneider, S. L. Thompson, and L. R. McDaniel. 1990. Analysis of climate variability in general circulation models: Compared with observations and changes in variability in $2 \times CO_2$. *Geophysical Research* 95:490.

National Academy of Sciences (NAS). 1991. *Policy Implications of Greenhouse Warming.* Washington: National Academy Press.

National Research Council (NRC). 1987. Future carbon dioxide emissions from fossil fuels. *Current Issues in Atmospheric Change.* Washington: National Academy Press.

Nordhaus, W., and G. Yohe. 1983. In *Changing Climate.* Report of the Carbon Dioxide Assessment Committee. Washington: National Academy Press.

Office of Technology Assessment (OTA). 1991. *Changing by Degrees: Steps to Reduce Greenhouse Gases.* Washington: Government Printing Office.

Parry, M. L. 1985. Estimating the sensitivity of natural ecosystems and agriculture to climatic change. *Climatic Change* 7:1—3.

Parry, M. L., and T. R. Carter. 1985. The effect of climatic variations on agricultural risk. *Climatic Change* 7:95—110.

Pearman, G. I. (ed.). 1987. *Greenhouse: Planning for Climate Change.* Leiden: E. J. Brill.

Peters, R. 1991. Climate-change scenarios for impact assessment. In *Proceedings of the World Wildlife Foundation Conference on the Consequences of the Greenhouse Effect for Biological Diversity*. New Haven: Yale University Press.

Pittock, A. B., and J. Salinger. 1982. Towards regional scenarios for a $CO_2$-warmed earth. *Climatic Change* 4:23–40.

Ramanathan, V., and W. Collins. 1991. Thermodynamic regulation of ocean warming by cirrus clouds deduced from observations of the 1987 El Niño. *Nature* 351:27–32.

_____, R. J. Cicerone, H. B. Singh, and J. T. Kiehl. 1985. Trace gas trends and their potential role in climate change. *Journal of Geophysical Research* 90:5547–5566.

Raval, A., and V. Ramanathan. 1989. Observational determination of the greenhouse effect. *Nature* 342:758.

Rind, D., R. Goldberg, J. Hansen, C. Rosenzweig, and R. Ruedy. 1990. Potential evapotranspiration and the likelihood of future drought. *Journal of Geophysical Research* 95(D7):9983–10004.

Roberts, L. 1988. Is there life after climate change? *Science* 242:1010–1012.

Root, T. 1988a. Environmental factors associated with avian distributional boundaries. *Journal of Biogeography* 15:489–505.

_____. 1988b. Energy constraints on avian distributions and abundances. *Ecology* 69:330–339.

Schlesinger, M. E., and J. Mitchell. 1987. Climate model simulations of the equilibrium climatic response to increased carbon dioxide. *Reviews of Geophysics* 25:760–798.

Schneider, S. H. 1987. Climate modeling. *Scientific American* 256(5):72–80.

_____. 1989. The greenhouse effect: Science and policy. *Science* 243:771–781.

_____. 1990a. *Global Warming: Are We Entering the Greenhouse Century?* New York: Vintage Books.

_____. 1990b. The global warming debate heats up! An analysis and perspective. *Bulletin of the American Meteorological Society* 71:1292–1304.

Schneider, S. H., and C. Mass. 1975. Volcanic dust, sunspots, and temperature trends. *Science* 190:741–746.

_____, and S. L. Thompson. 1981. Atmospheric $CO_2$ and climate: Importance of the transient response. *Journal of Geophysical Research* 86:3135–3147.

_____, P. H. Gleick, and L. O. Mearns. 1990. Prospects for climate change. In *Climate and Water: Climatic Change, Climatic Variability, and the Planning and Management of U.S. Water Resources*, ed. P. E. Waggoner. New York: Wiley.

Seidel, S., and D. Keyes. 1983. *Can We Delay a Greenhouse Warming? The Effectiveness and Feasibility of Options to Slow a Build-Up of Carbon Dioxide in the Atmosphere*. Washington: Environmental Protection Agency and Strategic Studies Staff, Office of Policy Analysis, Office of Policy and Resources Management.

Senate Committee on Governmental Affairs. 1979. *Carbon Dioxide Accumulation in the Atmosphere, Synthetic Fuels and Energy Policy*. Washington: Government Printing Office.

Smith, J. B., and D. Tirpak (eds.). 1988. *The Potential Effects of a Global Climate Change on the United States: Draft Report to Congress*. Vol. 2. Washington: Environmental Protection Agency.

Stouffer, R. J., S. Manabe, and K. Bryan. 1989. Interhemispheric asymmetry in climate response to a gradual increase of atmospheric $CO_2$ *Nature* 342:660–662.

Thompson, S. L., and S. H. Schneider. 1982. $CO_2$ and climate: The importance of realistic geography in estimating the transient response. *Science* 217:1031–1033.

Waggoner, P. E. (ed.). 1990. *Climate Change and U.S. Water Resources*. New York: Wiley.

Washington, W. M., and G. A. Meehl. 1989. Climate sensitivity due to increased $CO_2$: Experiments with a coupled atmosphere and ocean general circulation model. *Climate Dynamics* 4:1–38.

Wigley, T. M. L. 1985. Impact of extreme events. *Nature* 316:106–107.

Wigley, T. M. L., and S. C. B. Raper. 1989. Possible climate change due to $SO_2$-derived cloud nuclei. *Nature* 339:365–367.

————. 1990a. Detection of the enhanced greenhouse effect on climate. In J. Jager and H. L. Ferguson, eds., *Climate Change: Science, Impacts, and Policy*. Cambridge: Cambridge University Press.

————.1990b. Natural variability of the climate system and detection of the greenhouse effect. *Nature* 344:324–327.

Williams, J. 1978. *Carbon Dioxide, Climate and Society*. Oxford: Pergamon Press.

Wilson, E. O. 1989. Threats to biodiversity. *Scientific American* 261:108–116.

World Meteorological Organization (WMO). 1988. *Report of the Workshops Held in Villach (28 September–2 October 1987) and Bellagio (9–13 November 1987)*. WCIP-1. Geneva: World Meteorological Organization.

CHAPTER 5

# OIL SPILLS

*Mary Hope Katsouros*

THE WORLD ANNUALLY CONSUMES about 2.9 billion tons of oil,[1] of which 1.7 billion tons are transported by sea. Countries with limited petroleum resources have little recourse but to import. Tankers and pipelines permit oil to be transported over land and seas. About 600 million tons of oil and refined products pass through U.S. waters each year (NRC, 1991a). Some of this oil—roughly 3 million tons annually, according to a 1985 report (NRC, 1985)—enters the ocean. Not all oil in the sea arises from human activity; natural sources such as underwater seeps discharge small amounts of oil for millions of years. Human beings, however, are responsible for increased, often highly concentrated, discharges of oil into the sea from a variety of sources: tanker accidents, routine ship operations, atmospheric fallout, offshore drilling, and municipal and industrial wastes and runoff. Events such as the 1989 *Exxon Valdez* spill in Alaska's Prince William Sound and the 1991 intentional release of oil into the Persian Gulf have focused public awareness on the environmental effects of oil in the sea. This chapter reviews sources of oil in the sea, the effects of oil, and strategies for reducing oil discharges.


## SOURCES

Petroleum, or oil, is a naturally occurring complex mixture of hydrocarbons and small amounts of other organic compounds. It is derived over millions of years from organic materials exposed to great heat and pressure beneath the earth's surface. Many of the compounds contained in petroleum are also produced by grass and forest fires and by the combustion of fossil fuels. For this reason, it is difficult to determine whether hydrocarbons in the marine environment come from oil spillage and seepage or from other sources, except in special cases such as tanker spills. Therefore, most estimates of discharges of oil into the sea are uncertain.

The most recent comprehensive estimates are those of a 1985 National Research Council (NRC) report, *Oil in the Sea*, based on a 1981 workshop of marine experts. *Oil in the Sea* divides oil inputs into five categories: natural sources, marine transportation, offshore production, atmospheric fallout, and municipal and industrial wastes and runoff. For each category, the NRC has issued a "best estimate" and a predicted range of total discharge (see Table 5.1). Together these sources were discharging an estimated 3.2 million tons of oil into the sea each year at the time of the study. The U.S. Coast Guard (1990a) has issued an updated estimate on discharges from marine transportation. Many of these estimates are highly uncertain, however, and give only the approximate magnitude of oil discharge from the various sources, not precise values.


### NATURAL SOURCES

Underwater seeps discharge 200,000 tons of petroleum to the sea each year, within a factor of 10 (NRC, 1985). Estimating underwater seepage rates is difficult. Most seeps lie hidden, scattered across the world's vast continental shelves. Seeps flow intermittently, so that

TABLE 5.1. SOURCES OF PETROLEUM HYDROCARBONS IN THE MARINE ENVIRONMENT
(MILLIONS OF TONS ANNUALLY)

| Source | Probable range | Best estimate |
|---|---|---|
| Natural sources | 0.025–2.5 | 0.25 |
| Atmospheric pollution | 0.05–0.5 | 0.3 |
| Marine transportation | 1.00–2.60 | 1.45 |
| Offshore production | 0.04–0.06 | 0.05 |
| Municipal and industrial wastes and runoff | 0.585–3.21 | 1.18 |
| Total | 1.7–8.8 | 3.2 |

seepage in a given year may vary substantially from the average flow over geological time. Underwater seepage rates must therefore be extrapolated from seeps on land and from known marine seeps, taking into account the geology of the various continental shelves. The NRC estimates that at most 100 trillion tons of oil are available for offshore seepage. Seepage of 2 million tons annually would deplete this supply within 50 million years—a short time compared to the age of the earth. The NRC considers this flow rate an upper bound. Tar from seeps often washes up on shore, as in Santa Barbara, California, where some people mistake it for residue from offshore oil operations. Rivers annually carry to the sea an additional 50,000 tons of oil that have eroded from sedimentary rocks.

MARINE TRANSPORTATION

The NRC estimated oil inputs from marine transportation to be 1.45 million tons per year in 1981. As Table 5.2 shows, most of this oil comes not from tanker accidents but from routine operations. The table also indicates that the amount of oil entering the sea as a result of both ship operations and accidental spills has fallen by about 60 percent since the late 1970s, when data for the 1985 NRC report were compiled. Based on more recent data and on new regulations governing ships, the U.S. Coast Guard (1990a) estimates that ships now empty into the sea about 570,000 tons of oil annually. Although the number of tankers has not quite doubled since 1939, their average size has increased to the extent that they now move more than fifteen times the amount of oil. (The 1956 Arab–Israeli War and the

TABLE 5.2. OIL DISCHARGED INTO THE MARINE ENVIRONMENT FROM MARITIME
TRANSPORT (MILLIONS OF TONS ANNUALLY)

| Source | 1990 | 1981/85 |
|---|---|---|
| Bilge and fuel oil | 0.25 | 0.28 |
| Tanker operations | 0.16 | 0.71 |
| Accidental spillage | | |
|    Tanker accidents | 0.11 | 0.39 |
|    Nontanker accidents | 0.01 | 0.02 |
| Marine terminal operations | 0.03 | 0.02 |
| Drydocking and scrapping of ships | 0.01 | 0.03[a] |
| *Total* | 0.57 | 1.45 |

[a] Does not include scrapping of ships.

simultaneous closing of the Suez Canal affected the traditional navigational routes, necessitating the generation of supertankers.)

The 135-member International Maritime Organization (IMO) has recommended improvements in tanker procedures. The 1973 International Convention for the Prevention of Pollution from Ships as modified by the protocol of 1978—MARPOL 73/78 for short—has been responsible for the reduction of oil entering the seas. MARPOL 73/78 came into force in 1983, superseding the International Convention for the Prevention of Pollution of Sea by Oil—better known as OILPOL 54/62/69. As of 1990, MARPOL 73/78 had been signed by fifty-seven nations, under which 85 percent of the world's merchant fleet was flagged (U.S. Coast Guard, 1990a). Widespread adherence to MARPOL 73/78 has helped reduce oil discharge due to tanker operations from 710,000 tons per year in 1981 to an estimated 160,000 tons per year in 1990.

MARPOL 73/78 contains several measures that reduce the amount of oil dumped into the sea in ballasting, tank washing, and other routine operations. The world's 6400 oil tankers are usually loaded with oil for only one leg of each round-trip voyage (*Lloyd's Register of Shipping*, 1990). On the return trip, they must carry ballast water for stability. If carried in cargo tanks, this ballast water mixes with oil residues. Ballast water emptied into the ocean has long been a source of oil pollution. MARPOL 73/78 reduced such pollution by requiring all tankers over 20,000 deadweight tons to have segregated ballast tanks, dedicated clean ballast tanks, and a crude-oil washing system, depending on the tanker's age and size.[2] Segregated ballast tanks always carry water, never oil, and thus do not pollute when emptied. Dedicated clean ballast tanks are cargo tanks that are cleaned after each loaded voyage and only then filled with ballast water. A crude-oil washing system uses high-pressure jets of crude oil to clean cargo tanks. MARPOL 73/78 also required tankers to monitor the amount of oil they dump at sea. Discharge is limited to 1/15,000 of cargo capacity for existing tankers and 1/30,000 of cargo capacity for new tankers.

MARPOL 73/78 requires all ships (not just tankers) over 400 gross tons to carry slop tanks for machinery space bilge oil, fuel oil sludge, and oily ballast water. These slop tanks are supposed to be emptied at reception facilities on shore. Many nontankers, however, have slop tanks too small to hold accumulated sludge for entire voyages (NRC, 1985:64), and shore-based reception facilities are often inadequate. An estimated 10 percent of oil sludge from tankers and 25 percent from nontankers is still dumped directly into the sea (U.S. Coast Guard, 1990a:21). As shown in Table 5.2, bilge and fuel oil discharges did not decline much during the 1980s.

Oil tankers accidentally spilled an estimated 390,000 tons of oil per year in 1981, compared to 110,000 in 1990. As shown in Table 5.3, the volume of accidental spillage varies widely and unpredictably from year to year, both in U.S. waters and worldwide, so that an estimate only makes sense as an average over several years of spills. The 1981 estimate was inflated by exceptionally heavy spillage in the late 1970s. The 1989 estimate is an average over the years 1981–1989, during which accidental spills were on the whole less frequent and less severe. According to the *Oil Intelligence Report* (21 February 1991, p. 1), 1983 was a markedly bad year; about one-third of that year's spills are attributed to the Iran–Iraq war. The same publication estimates 1990 spillage at 31.75 million gallons, or about 100,000 tons—just below the average for the 1980s.

In 1988, some 5000 to 6000 spills involving oil or other toxic substances occurred along U.S. coasts and other navigable waters. Twelve involved spills of more than 100,000 gallons and ten involved spills ranging from 10,000 to 100,000 gallons. In the late 1970s, there were 13,000 spills a year. The record is improving, but the problem remains serious.

In 1981, accidental tanker spills accounted for an estimated 28 percent of transport-related spillage and 12 percent of all oil

TABLE 5.3. OIL SPILLED (IN TONS) IN TANKER ACCIDENTS: 1976–1989

| Year | U.S. waters[a] | Worldwide[b] |
|------|------|------|
| 1976 | 32,000 | 190,000 |
| 1977 | 400 | 160,000 |
| 1978 | 4,700 | 290,000 |
| 1979 | 11,000 | 580,000 |
| 1980 | 10,000 | 130,000 |
| 1981 | 1,700 | 53,000 |
| 1982 | 4,300 | 8,500 |
| 1983 | 400 | 360,000 |
| 1984 | 16,000 | 21,000 |
| 1985 | 2,300 | 81,000 |
| 1986 | 4,900 | 5,200 |
| 1987 | 2,600 | 14,000 |
| 1988 | 5,400 | 130,000 |
| 1989 | 40,000 | 210,000 |

[a] National Research Council (1991a).
[b] *Lloyd's Register of Shipping* (1990).

TABLE 5.4. MAJOR OIL SPILLS FROM TANKERS: 1960 AND 1965–1989

| Rank | Tanker | Size of spill (tons) | Year | Site of spill |
|------|------|------|------|------|
| 1 | Atlantic Empress | 257,000 | 1979 | West Indies |
| 2 | Castillo de Bellver | 239,000 | 1983 | South Africa |
| 3 | Amoco Cadiz | 221,000 | 1978 | France, Atlantic |
| 4 | Odyssey | 132,000 | 1988 | Mid Atlantic |
| 5 | Torrey Canyon | 124,000 | 1967 | UK, Channel |
| 6 | Sea Star | 123,000 | 1972 | Gulf of Oman |
| 7 | Hawaiian Patriot | 101,000 | 1977 | Hawaiian Islands |
| 8 | Independenta | 95,000 | 1979 | Turkey, Bosphorus |
| 9 | Urquiola | 91,000 | 1976 | Spain, North Coast |
| 10 | Irene's Serenade | 82,000 | 1988 | Greece |
| 11 | Khark 5 | 76,000 | 1989 | Morocco, Med. |
| 12 | Nova | 68,000 | 1985 | Gulf of Iran |
| 13 | Wafra | 62,000 | 1971 | South Africa |
| 14 | Epic Colocotronis | 58,000 | 1975 | West Indies |
| 15 | Sinclair Petrolore | 57,000 | 1960 | Brazil |
| 25 | Burmah Agate | 41,000 | 1979 | USA, Gulf of Mexico |
| 28 | Exxon Valdez | 36,000 | 1989 | USA, Alaska |
| 29 | Corinthos | 36,000 | 1975 | USA, Delaware R. |
| 37 | Argo Merchant | 28,000 | 1976 | USA, East Coast |
| 41 | Texaco Oklahoma | 29,000 | 1971 | USA, East Coast |
| 50 | Ocean Eagle | 21,000 | 1968 | USA, Puerto Rico |

SOURCE: International Tanker Owners Pollution Federation.

discharged to the sea. In 1990, accidents accounted for 19 percent of transport-related spills and only about 5 percent of all oil in the sea, assuming that oil input from other sources remained at its 1981 level. During the 1980s, 1.5 gallons of oil spilled in U.S. waters and 6.7 gallons spilled worldwide for every 100,000 gallons transported (NRC, 1991a:35). Although this ratio has decreased since the 1970s, public reluctance to accept this level of accidental spillage has increased.

Table 5.4 lists the fifteen largest tanker spills between 1960 and 1989, as well as U.S. spills in the top fifty. The top twenty-four occurred outside U.S. waters and were generally not highly publicized, except when oil came ashore and caused substantial environmental damage. The *Torrey Canyon* spill of 1967 and the *Amoco Cadiz* spill of 1978, which oiled hundreds of miles of shoreline in Cornwall and Brittany, respectively, are among the best-known spills. In March 1967, the *Torrey Canyon*, a Liberian-flag tanker, grounded 5 miles off the Cornish coast in the United Kingdom, releasing 860,000 barrels of crude oil and becoming the first spill to attract public and scientific attention. The greatest flow of oil from a tanker in U.S. waters is believed to have come from the *Burmah Agate*—not the *Exxon Valdez*, as is widely reported. The *Burmah Agate* collided with the freighter *Mimosa* and caught fire near the Texas coast in 1979. Since most of the spilled oil burned, and much of the remainder dispersed offshore, the *Burmah Agate* caused far less shoreline damage than the *Exxon Valdez*.

A few major accidents usually account for most of the oil spilled in a given year. For example, three-fourths of the oil spilled worldwide in 1978 flowed from the *Amoco Cadiz*. In the United States, large spills (defined as spills of more than 30 tons, or about 10,000 gallons) produce about 95 percent of all spillage but comprise less than 3 percent of all spill events (NRC, 1991a:27). Most tanker accidents do not result in pollution. Out of 9276 tanker accidents assessed in *Lloyd's Register of Shipping* (1990), only 6 percent definitely resulted in oil spillage (NRC, 1991a:25). In another 18 percent of accidents, it was not known whether any pollution occurred.

Table 5.4 does not list nontanker incidents such as the 1991 Persian Gulf oil spill. When Iraq opened the valves of Kuwait's major oil-loading terminals in January 1991, an estimated 6 million barrels of crude oil flowed into the Persian Gulf (Efron, 1991). That is three times as much oil as any tanker has spilled. The old record for spillage belonged to Ixtoc I, an exploratory well in the Gulf of Mexico that spilled an estimated 3.5 million barrels of oil over 10 months in 1979 and 1980. Since the Persian Gulf is shallow, enclosed, and full of fragile habitats, many observers believe the 1991 Gulf oil spill will prove to be the most damaging that has yet occurred. In the NRC classification scheme, the Gulf spill was technically a marine terminal spill. In 1990, marine terminal operations accounted for an estimated 30,000 tons of oil in the sea; in 1991, such "operations" spilled about 800,000 tons. Iraq also blew up most of Kuwait's 1250 oil wells, of which about 600 caught fire, emitting thousands of tons of soot and sulfur dioxide into the atmosphere (*Golub's Oil Pollution Bulletin*, 10 May 1991, pp. 1, 3, 4). According to Kuwaiti oil industry officials, these fires consumed up to 6 million barrels of oil a day— four times Kuwait's prewar output and about 10 percent of world-wide oil consumption (Horgan, 1991). Much unburned oil fell into the Persian Gulf. In fact, oil rained down as far away as Riyadh, Saudi Arabia, 350 miles from Kuwait, and black snow was reported in the Himalayas (Efron, 1991). Moreover, lakes of oil from out-of-control wells accumulated behind dikes in Kuwaiti oil fields and threatened to spill into the Persian Gulf if not removed in time. According to *Golub's Oil Pollution Bulletin* (10 May 1991, p. 1), one lake contained an estimated 5 million barrels of oil.

### Offshore Oil Production

The NRC estimates that between 40,000 and 70,000 tons of petroleum hydrocarbons enter the marine environment from offshore petroleum production. Of these totals, major spills from platforms contribute between 30,000 and 50,000 tons, minor spills between 3000 and 4000 tons, and operational discharges between 7000 and 11,000 tons. These estimates for the release of petroleum into the marine environment are lower than previous estimates because better data are available for operations.

### Atmospheric Sources

The NRC estimates that between 50,000 and 500,000 tons of petroleum hydrocarbons enter the sea each year from the atmosphere

(NRC, 1985). Most of this oil is emitted or evaporated into the atmosphere by vehicle exhausts and industrial operations, especially oil industry operations. It reaches the ocean primarily by attaching to atmospheric particles and falling out in rain. Precise estimates of discharge amounts are not yet possible. Not only do the many constituent compounds of petroleum undergo complex reactions in the atmosphere, but some of these compounds are also produced by marine phytoplankton and photoplankton. Moreover, few rain samples collected from the open ocean have been analyzed for their petroleum hydrocarbon content. Oil discharges from the atmosphere are greatest near coastlines, downwind of industrial areas.

## MUNICIPAL AND INDUSTRIAL WASTES AND RUNOFF

Runoff from land is probably the largest source of oil in the sea, accounting for an estimated 1.2 million tons per year in 1981. Municipal wastewater alone delivered an estimated 750,000 tons of oil, including 190,000 tons from the United States. The 1985 NRC report assumes that other nations discharge oil in wastewater in the same proportion to their total oil consumption as the United States (with an adjustment for wastewater treatment in developed countries). Primary and secondary treatment removed an estimated 38 percent of oil from wastewater in the United States and 30 percent in Western Europe, Eastern Europe, and the former Soviet Union. These estimates are open to debate, however, given recent revelations about the extent of pollution in the former Warsaw Pact nations. Accurate global data on wastewater discharge and treatment are not available. Oil refineries discharged an estimated 100,000 tons of oil in industrial wastes, and other industries accounted for an additional 200,000 tons. Data on coastal effluent discharges are also scarce. Another 120,000 tons of oil entered the sea in urban runoff. Given the large rise in global population during the 1980s, especially near coastlines, and given the absence of adequate waste treatment in much of the world, oil discharge from wastes and runoffs probably rose during the last decade. The United Nations has predicted that by the year 2000, some 75 percent of the world's population will live within 60 km of the coast (UN, 1985). Unless preventive measures are taken, chronic spillage of oil into coastal waters will probably continue to increase.

## FATES

Once in the ocean, oil may undergo several physical and chemical changes that determine its ultimate environmental effects (NRC, 1985:270ff.). These processes are shown in Figure 5.1. Oil in the sea immediately starts to spread; wind, waves, currents, and the viscosity and surface tension of the oil influence the speed and direction of spreading. Oil typically divides into "thin" slicks, less than one-millionth of a meter thick, and "thick" slicks several millimeters or even centimeters thick. Thin slicks usually comprise the majority of a spill's area, whereas most of the oil volume is in the thick slicks. In quiet waters, thick slicks are usually surrounded and spread more slowly than thin slicks. When wind is present, slicks tend to drift at about 3 or 4 percent of the wind speed, with the thicker part of a slick drifting faster and accumulating at the slick's leading edge. Spreading increases the surface area of oil exposed to air, light, and water and thereby speeds up other processes that determine the oil's fate. Spreading also makes containment and recovery of oil more difficult.

Within a few hours, the light and medium-weight components of the oil begin to evaporate. Hydrocarbons composed of twelve carbon atoms or fewer usually evaporate within a few days; those with up to twenty carbon atoms evaporate within a few weeks. Many of the lighter compounds in crude oil—for example, benzene and toluene—are among the most toxic. For this reason, the oil remaining after evaporation is usually less toxic than freshly spilled oil. The remaining oil is also denser—though still less dense than seawater—and forms thick sludge and tar balls. Evaporation typically removes between 20 and 40 percent of the oil in a spill (NRC, 1985). An estimated 30 percent of the oil spilled by the *Amoco Cadiz* evaporated (Gundlach et al., 1983:122–129), as did perhaps half of the Persian Gulf spill (Efron, 1991), which occurred in exceptionally warm water. Evaporation rates depend on oil composition, spill surface area, water and air temperature, and wind speed. Lighter refined products such as gasoline and kerosene evaporate more quickly than crude oil. Evaporation from a thick, cold slick in calm weather may be much slower than evaporation from a thin, warm slick in stormy weather.

FIGURE 5.1. SCHEMATIC OF PHYSICAL, CHEMICAL, AND BIOLOGICAL PROCESSES.

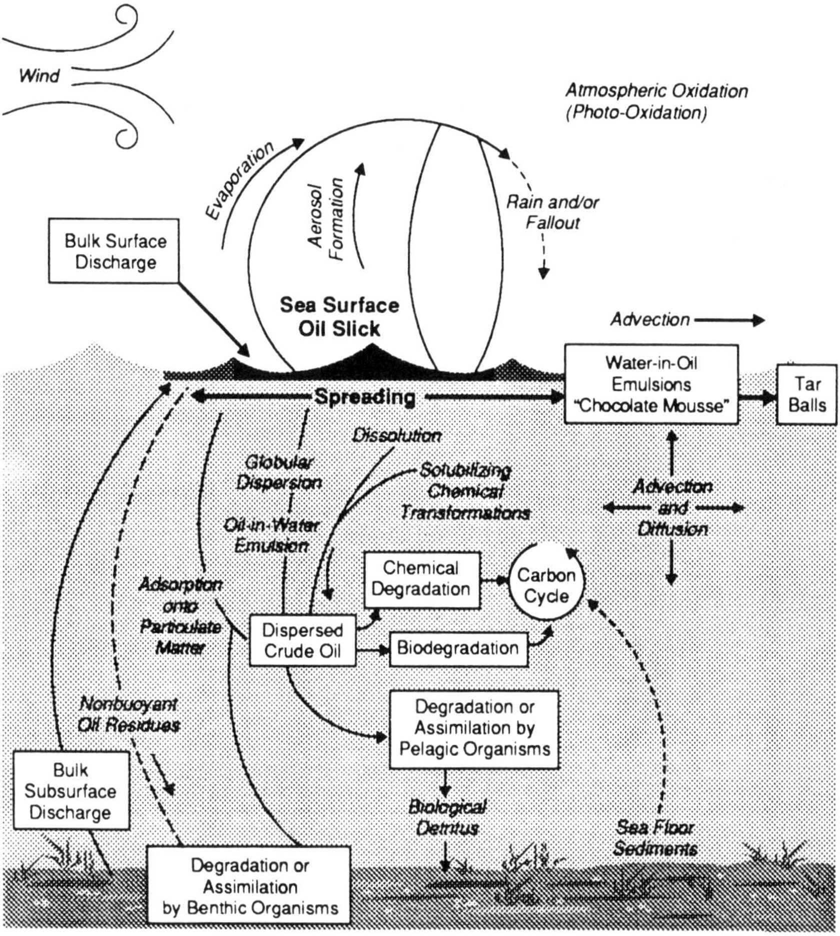

Oil may also dissolve into the water beneath the oil slick. The lighter hydrocarbon compounds tend to dissolve most quickly. Only a small fraction of the volume of a spill usually dissolves, usually in the first few hours after a spill; most of the soluble compounds evaporate instead. The dissolved fraction is highly toxic, however, and may threaten organisms in the water (NRC, 1985:277). Little dissolution occurs after the first few hours of a spill. Another process, dispersion, carries oil into the water in the form of small oil droplets or oil-in-water emulsions. Dispersion proceeds most quickly in the first few days of a spill, before the oil thickens, and is accelerated by high winds and waves. Like dissolution, dispersion may expose subsurface life to toxic oil. Dispersion is generally preferable, however, to the oiling of beaches by an intact oil slick.

After a few hours or a few days of weathering, the remaining oil often forms a sticky, viscous water-in-oil mixture called mousse. Mousse is difficult to pump or absorb from the water surface. In time it is degraded by microorganisms, but first it may sink to the bottom or wash up on shore, often killing wildlife. Weathered oil may also form lumps of tar that degrade much more slowly than mousse. Tar lumps that sink to the bottom or reach shore may remain there for several years.

In the presence of oxygen, sunlight may transform petroleum hydrocarbons into other compounds such as carbon dioxide, organic acids, alcohols, ketones, and phenols (NRC, 1985:282). This process, known as photooxidation, can decompose a thin slick in a few days. The oxidized products are usually more soluble than the original compounds, but they are sometimes more toxic as well. These products are eventually oxidized to carbon dioxide and water. In the first few hours after a spill, before the oil has spread, oil may also oxidize by burning. Spills caused by fires and explosions have occasionally burned themselves out.

About ninety known species of bacteria and fungi are able to degrade oil, both in the water and on land (Mielke, 1990b:14). Before a spill, an average liter of seawater contains only about ten hydrocarbon-decomposing bacteria. After a spill, these bacteria gradually multiply, eventually reaching densities of up to 50 million per liter (Mielke, 1990b:15). Their numbers may be limited by the presence of toxic compounds in the oil or by a lack of nitrate and phosphate nutrients. The warmer the environment, the faster these bacteria remove oil. Microbes removed an estimated 10,000 tons of *Amoco Cadiz* oil at sea (Gundlach et al., 1983:122). Assisted by humans, microbes have also removed much of the oil from beaches in Prince William Sound since the *Exxon Valdez* spill (Pritchard et al., 1991).

Oil that does not evaporate, dissolve, disperse, photooxidize, burn, or biodegrade while in the water generally does the most damage.

This oil, often in the form of mousse or tar lumps, may cling to small particles, sink, and lodge in bottom sediments, or it may wash ashore. It may also be ingested by animals in the water, then excreted to the seafloor. This oil often persists for months or years. Oil in rocky areas with powerful waves and tides, such as Prince William Sound, tends to be removed most quickly. Oil in sheltered areas such as marshes and lagoons remains the longest.

## EFFECTS

Where oil has had an effect, subsequent monitoring has shown biological recovery. Hydrocarbons from seeps and pyrolytic sources are part of the long-term evolution of the oceans, and results of observations made to date indicate that most living organisms can coexist with hydrocarbons when concentrations are very low and oil is weathered. Although accidental oil spills may have devastating short-term effects on coastal environments, the ecosystems generally recover quickly after the pollution source is removed. Natural processes remove nearly all traces of most oil spills in 10 years or less.

The effects of continuous exposure to low levels of oil are less well known. Potentially toxic levels of oil exist in many industrialized coastal areas, but the effects of chronic oil are difficult to isolate from the effects of other contaminants. The NRC concluded in 1985 that "there has been no evident irrevocable damage to marine resources on a broad oceanic scale, by either chronic inputs or occasional major oil spills" (NRC, 1985:6). The NRC cautions, however, that the effects of oil on many specific organisms and environments are poorly understood—for example, few studies have assessed oil's effects on tropical and open-ocean organisms, and macro-algae and larval fish.

The impact of an oil spill on marine life depends on many factors other than the spill's size. Spills usually do the most damage in coastal areas, for example, which support denser, more diverse populations than does the open ocean. Coastal areas are home to many shallow-water bottom-dwelling organisms, including the eggs and larvae of many species, that are especially sensitive to oil. Confined bays and estuaries usually fare worse than areas from which oil can escape easily out to sea. Furthermore, oil persists longer in sheltered, confined areas than in areas exposed to the energy of waves and tides. Weather and ocean conditions such as temperature and sea height determine how much of a spill evaporates and disperses and how much remains intact until it reaches shore. Winds and currents often determine whether a spill dissipates with minor effects at sea or washes ashore. Prevailing winds, for instance, continuously drove oil from the *Amoco Cadiz* ashore for 4 weeks (NRC, 1985:376).

A spill's biological impact also depends on the type and condition of the oil. Fresh oil, for instance, is usually more harmful than weathered oil. The heavy tarry residues left behind after weathering are less toxic to life than the lighter compounds that evaporate and dissolve soon after a spill. Refined products such as fuel oil, which generally contain high concentrations of toxic compounds, often do more ecological damage than crude oil. A relatively small spill of No. 2 fuel oil, the 630-ton *Florida* spill of 1969, killed a large number of small fishes, bottom-dwelling invertebrates, and marsh organisms near Falmouth, Massachusetts (NRC, 1985:551). Refined products tend to dissipate more quickly from sediments than heavier crude oils. Even refined products, however, may remain for many years in sheltered environments. Oil from the *Florida* spill persisted in marshes for over a decade.

The time of year may also determine a spill's impact. Oil is more lethal, for example, when seabirds are feeding or nesting, seals are breeding or nursing, or newly hatched fish are swimming out to sea. The *Exxon Valdez* spill occurred in early spring, a particularly bad time. During the 2 months after the spill, millions of salmon swam to sea; thousands of birds migrated; and most species of birds, mammals, fish, and marine invertebrates reproduced. Seasonal currents and weather patterns may also play a role. The January 1969 well blowout near Santa Barbara, California, for example, occurred during a period of heavy storms. Floodwaters brought large amounts of sediment into coastal waters; much of the spilled oil attached to sediment and sank rather than washing onshore (Mielke, 1990b:7). Bottom-dwelling life would have fared better and intertidal life worse if the spill had occurred at a different time of year.

Oil's effects vary from species to species. Sea mammals, for example, are highly vulnerable. The *Exxon Valdez* spill killed an estimated 3500 to 5500 sea otters and 200 harbor seals (EPA, 1991). Since oil-soaked fur loses its insulating capacity, many sea otters froze to death. Sea otters also died from ingesting oil and perhaps also from inhaling toxic hydrocarbon compounds. Abnormally large numbers of sea otters between the ages of two and eight died in oiled areas more than a year after the spill. Seal populations, already declining before the spill, dropped 35 percent in oiled areas, compared to 13 percent in unoiled areas. Oiled seals behaved lethargically; some seal carcasses were found to have brain lesions. Cetaceans seem to be less prone to injury, perhaps because oil does not adhere easily to their smooth skin (NRC, 1985:427). Some 22 out of 182 killer whales in Prince William Sound, however, are believed to have died during the 2 years after the spill (EPA, 1991).

Seabirds, which forage at the sea surface, are also vulnerable. Oiled feathers lose their buoyancy and their ability to insulate, resulting in death by drowning or freezing. Birds may also ingest lethal amounts of oil while preening. An estimated 300 to 600 bald eagles, 1500 to 3000 pigeon guillemots, and 300,000 murres died following the *Exxon Valdez* spill.

## RESPONSES

Most large oil spills in U.S. waters have caused little damage, thanks in many cases to favorable conditions rather than effective spill response. The *Burmah Agate* lost 41,000 tons of oil off the Gulf Coast in 1979, but most of it burned; 28,000 tons of oil spilled from the *Argo Merchant* off New England in 1976, but most of it dissipated out at sea. In 1984, oil from the 8900-ton *Alvenus* spill washed up repeatedly on Texas beaches, but beach cleanup was fairly effective and environmental damage minor and short-lived. The *Exxon Valdez* spill, however, demonstrated U.S. unpreparedness for a major spill and the limits of oil spill response technology. As the Office of Technology Assessment (OTA) recently reported, spill-fighting technology has not advanced much in the last decade and has rarely led to recovery of more than 15 percent of the oil from large spills (U.S. Congress, OTA, 1990). More rapid reaction by trained, well-equipped response teams could, however, increase recovery rates and reduce damage to shorelines.

The primary response method is mechanical containment and recovery. Mechanical response technologies fall into two categories: containment booms and oil recovery devices. Booms are long, lightweight floating barriers that can concentrate oil in thick layers and direct it toward recovery devices. Booms can also deflect oil from environmentally sensitive areas such as fish hatcheries or desalination plants. In calm water, booms usually succeed in containing oil. In waves more than 5 feet high, however, oil can splash over the top of the boom, making containment impossible. Recovery devices, or skimmers, pump oil from the surface or absorb it with belts, ropes, or disks. Most skimmers work best on thick slicks in calm water. When winds exceed 20 knots or currents exceed 1 knot, most recovery equipment is ineffective. The OTA estimates that mechanical recovery of more than 30 percent of an oil spill is unlikely—even in good weather with trained, well-equipped personnel. Rapid spreading and fragmentation of oil make containment and recovery difficult after the first few days.

Dispersants—chemical agents that reduce surface tension and thereby disperse oil into the underlying water—have an uneven record in fighting oil spills. In 1967, following the *Torrey Canyon* spill, 10,000 tons of dispersants were applied to crude oil stranded on rocky beaches in Cornwall, England. The dispersants proved more toxic than the oil, killing animals and algae in proportion to the dispersant dose and delaying ecological recovery (NRC, 1989:318). Modern dispersants are less toxic, but on many spills they have been ineffective. In 1979, however, dispersants successfully protected Mexican beaches from the Ixtoc I oil spill. No oil came ashore until Hurricane Henri grounded the planes that were dropping dispersants (NRC, 1989). Field studies showed that oil was dispersed from the surface into the upper 3 m of the water, where it quickly became diluted and lost its toxicity. Limited biological studies revealed no adverse effects on commercially valuable shrimp or other species (NRC, 1989:320). Dispersants tend to work best when applied to oil of low viscosity in warm water. Wind and waves help mix the dispersants into the slick and carry droplets away.

A 1989 NRC report, *Using Oil Spill Dispersants on the Sea*, recommends that dispersants be considered as a first response option in treating spills. Dispersion of oil at sea, the report says, generally reduces the impact of oil on shoreline habitats. Moreover, dispersed oil is no more toxic than untreated oil. Since oil dispersibility decreases quickly with weathering, the report recommends that spill contingency plans allow for rapid response, including field testing of dispersants immediately after spills. The report also advises against use of dispersants along shorelines and in shallow water, where dispersed oil can reach bottom sediments. More research is needed, the report says, to improve the effectiveness of dispersants and minimize the toxic effects of dispersed oil.

Burning, though not commonly practiced, can be an effective response to some spills. Oil slicks will ignite if they are fresh and at least 3 mm thick (U.S. Congress, OTA, 1990:22–23). Fire containment booms can keep the oil thick and isolate it from nearby ships. Although burning keeps slicks away from shorelines, it releases sooty black smoke and combustion products to the air. These combustion products may be toxic, but no more toxic than those released by evaporating oil. Burning is usually inadvisable near a stricken ship, especially if the ship still contains oil. The *Exxon Valdez*, for example, spilled only one-fifth of its cargo into Prince William Sound. Most of the rest was successfully offloaded. Burning the entire cargo might have been less costly in the end than letting much of an 11-million-gallon slick reach shore, but that was not apparent in the first few hours after the spill.

Bioremediation—the use of microbes to degrade and oxidize hydrocarbon molecules—is potentially an inexpensive and environmentally safe response technique. Unlike containment and recovery, dispersal, and burning, bioremediation can be effective long after a spill. The *Exxon Valdez* spill gave proponents of bioremediation their first chance to test this technique on a large scale. Following successful field tests, researchers treated 50 miles of beaches in Prince William Sound with fertilizers containing nitrates and phosphates (Pritchard et al., 1991). Oil on beaches treated with fertilizer degraded two to three times faster than oil on control beaches, supporting the thesis that the populations of oil-degrading bacteria had been limited by a lack of nutrients. Although bioremediation has not yet proved effective in the open ocean, efforts to cultivate or genetically engineer more efficient oil-degrading microbes hold promise (U.S. Congress, OTA, 1990:23).

After oil washes on shore, the only treatment other than bioremediation is laborious mechanical cleanup. Much of Exxon's $2 billion cleanup effort following the *Valdez* spill consisted of hosing down beaches and scrubbing rocks. According to a 1991 report of the National Oceanic and Atmospheric Administration (NOAA, 1991), some of Exxon's cleanup methods did more harm than good. Exxon used high-pressure jets of hot water to flush oil from beaches back into the water, where it was contained by booms and pumped into storage tanks. The hot water killed many organisms, such as rockweed and mussels, that had survived the initial oiling. Other organisms were buried and smothered under sand and mud. Furthermore, much of the oil collected in tidal pools, where it killed sensitive organisms such as hard-shelled clams and crustaceans. NOAA found that untreated beaches were recovering from the spill much faster. One NOAA official said that the decision to treat beaches was driven less by science than by public pressure for strong action following the initial failure to contain the spill (Lancaster, 1991).

The central theme of oil spill response is timely action. Once oil has spread and turned to a viscous mousse, it may be impossible to control. Sometimes there is little to do but hope for winds blowing offshore. Better preparedness and faster response could have reduced the effects of the Persian Gulf spill and fires, which were not—or need not have been—unexpected. As early as September 1990, Saddam Hussein threatened to blow up Kuwait's oil wells if U.S.-led forces tried to expel him. Kuwaiti officials, however, failed to buy enough firefighting and well-capping equipment to respond promptly (*Golub's Oil Pollution Bulletin*, 10 May 1991, pp. 1, 3, 4). Allied forces also failed to prepare adequately to fight a large spill. Several months after the war, a lack of funds from the cash-short Saudi government and poor coordination between MEPA and Saudi ARAMCO continued to slow the cleanup (*Golub's Oil Pollution Bulletin*, 26 April 1910, p. 10).

## PREVENTION

Tankers now spill an average of 9000 tons of oil each year in U.S. waters—groundings are the leading cause of spills—and U.S. tanker traffic continues to increase as oil imports rise (NRC, 1991a). It has long been argued that equipping tankers with double hulls would reduce spillage. A recent report of the Tanker Advisory Center (1990) found that in twenty-seven reported groundings of U.S.-flag double-hull and double-bottom tankers, no major damage or pollution occurred. About 9 percent of all groundings and collisions result in pollution (NRC, 1991a). Accordingly, Congress required in the 1990 Oil Pollution Act that all tankers in U.S. waters have double hulls or else "provide protection . . . equal to or greater than that provided by double hulls." This requirement will be phased in over 25 years. The act also requires the transportation secretary to study new technologies that could improve or replace existing double-bottom and double-hull construction. According to the Tanker Advisory Center (1990), some 600 tankers—about 19 percent of all tankers in service worldwide—have double hulls. Twenty-five sail under the U.S. flag. Thus 81 percent of existing oil tankers have single hulls. Navy vessels, including half the oilers as well as chemical and liquid natural gas ships, have double bottoms. In late 1987, some 400 tankers, about 13 percent of the world tanker fleet, met this standard. As of early 1990, some 25 percent of the 225 U.S.-flag tankers had double bottoms.

Opponents of this measure argue that double hulls are too expensive for the pollution they prevent and will increase the risk of corrosion, fire, and overturning. An NRC report on tank vessel design, however, concluded that double hulls are cost-effective and that no alternative tanker design is better than the double hull in all accident scenarios (NRC, 1991a). The NRC estimates that fitting tankers with double hulls will add $700 million a year to oil transport costs, or about one cent per gallon. In return for this investment, double hulls will prevent an estimated 3000 to 5000 tons of oil per year from spilling in U.S. waters—and will reduce spillage wherever else those tankers travel. Double hulls would not prevent spills from high-energy collisions and groundings like that of the *Exxon Valdez*, but they would eliminate pollution from most low-energy accidents. Nor would double hulls reduce the risk of spillage from fires and explosions. But they would not increase this risk, nor would they be less stable than single-hull ships, the NRC found. On balance, the advantages outweigh the disadvantages. The benefits are to be found in a reduction of the tank-washing effort, lower cleaning costs, and improved pumping qualities.

This NRC committee emphasized that improved tank vessel design is only a "2½ percent solution" to the problem of marine oil pollution (pers. comm., Don Perkins, Marine Board). Accidental spills account for only about 5 percent of oil discharged to the sea, and safer tankers will probably cut these losses in half. Cutting oil discharge from routine maritime operations could reduce marine oil pollution by up to 20 percent. In particular, full compliance with MARPOL 73/78 could prevent oil discharges of several hundred thousand tons annually. Compliance today is incomplete, according to a U.S. Coast Guard study (1990a:26), because there are not enough adequate reception facilities for sludge; because not all nations that flag ships are party to MARPOL 73/78; because MARPOL 73/78 provisions are not sufficiently enforced; and because ships' crews are not always trained to follow MARPOL 73/78 requirements.

There is no doubt that adoption of LORAN C (long-range navigation) with its impressive accuracy and coverage is beneficial. Few would argue against the value of simulators in imparting a sense of realism to instruction of deck and engine crews. This technology, however, is no substitute for tighter structural and equipment requirements—that is, double bottoms and double hulls. Nor should traffic-management schemes be overlooked. Traffic Separation Schemes (TSS) and Vessel Traffic Services (VTS) are essential in heavy shipping lanes.

Human factors must be considered as well. Boredom, to some degree, has become a part of tanker operations because of tight scheduling. A tanker loads in the Persian Gulf in 12 to 18 hours before beginning its long voyage to its final destination. Tankers spend less than 10 percent of their time in port, and operational costs are based on a ship at sea 300 days per year. As a result, crew members frequently remain on board ship a year or more; even longer periods are on record. Nevertheless, the principal oil companies and responsible tanker operators try to give crews a fixed rotation of service and leave.

Because personnel account for 30 percent of tanker operational costs, it is in the shipowner's interest to reduce the size of the crew. Automation has been the usual answer. Paradoxically, tankers have increased in size while their complement has gone down. A T-2 tanker, for example, had a crew of about thirty-eight, while a 200,000-deadweight-ton vessel has nineteen or twenty. The reduction has had mixed effects on morale and efficiency. Curiously, automation bears with it further tedium. While easing some work procedures, automation also isolates the individual from the challenge of hands-on tasks. There is little quarrel with the general purpose of easing the work through integrated bridges and other systems—but only as aids, not simply to replace personnel. Yet the trend is toward more automation.

Oil pollution at sea is thus a consequence of economic growth based on continuing access to petroleum products—and the tankers provide the access. Segregated ballast tanks and protective locations to keep oil residues on board for discharge at terminals are in general use. Double hulls and double bottoms, especially the latter, have been slower to gain acceptance. The dismal record of oil outflows caused by tanker mishaps in 1989 has led to pending congressional legislation for adopting these features. The IMO's recommendations regarding improved tank cleaning and deballasting methods have met with widening approval. However hopeful these developments may be for protecting the marine environment, an aging world tanker fleet and imperfect adherence to responsible operating procedures undermine them. There can be no reliance on any single measure. Corrective actions must embrace technical, economic, and political means.

Evidence suggests that neither accidental oil spills nor chronic oil pollution now rank among the first tier of environmental problems—which would include ozone depletion and global warming, for example. This assessment, however, is no excuse for complacency. The short-term impacts of several large oil spills on coastal ecosystems have been catastrophic. The long-term effects of spills and chronic oil pollution are less severe, but not negligible. Further research may reveal harmful impacts we are not now aware of. A number of measures could substantially reduce the threat posed by oil to the marine environment, even as coastal populations rise and tanker traffic increases through the 1990s. These measures include safer tanker design, better spill response planning, stricter adherence to MARPOL 73/78 and other pollution regulations, more research on the sources, fates, and biological effects of oil in the sea, and increased monitoring and treatment of municipal and industrial sources of low levels of oil. Nature has always been efficient at cleaning up oil. Human beings could be far more adept at preventing it from spilling.

## NOTES

1. "Tons" refers to metric tons; 1 metric ton equals 2205 pounds, or 7.33 barrels, or 308 U.S. gallons, of Arabian light crude oil. Weight-to-volume conversions vary slightly for other kinds of crude oil, and somewhat more for refined products. For example, 1 ton equals 6.70 barrels of fuel oil and 7.80 barrels of kerosene. One barrel equals 42 U.S. gallons.
2. A ship's deadweight tonnage is defined as the displacement of the loaded ship minus the displacement of the unloaded ship. That is, deadweight tonnage includes the weight of cargo, fuel, and stores, but not the weight of the ship itself.

## REFERENCES

Abernathy, S. Allison (ed.). 1989. *Drilling and Production Discharges and Oil Spills in the Marine Environment.* Vienna, Va.: Mineral Management Service, Atlantic OCS Region.

American Petroleum Institute (API). 1991. *Proceedings of the 1991 International Oil Spill Conference (Prevention, Behavior, Control, Cleanup).* Publication No. 4529. Washington: American Petroleum Institute.

Drake, D. E., P. Fleisher, and R. L. Kolpack. 1971. Transport and deposition of flood sediment, Santa Barbara Channel, California. In *Biological and Oceanography of the Santa Barbara Channel Oil Spill 1969–70,* ed. R. L. Kolpack. Vol. 2. Los Angeles: Allan Hancock Foundation.

Efron, Sonni. 1991. Kuwait's oily nightmare slowly abates. *Washington Post,* 8 Jan., p. A15.

Environmental Protection Agency. 1991. *Summary of Effects of the Exxon Valdez Oil Spill on Natural Resources.* Federal Register (ISSN 0097-6326) V. 56 N. 70 14687-94 (4/11/91).

Gundlach, E. R., P. D. Boehm, M. Marchand, R. M. Atlas, D. M. Ward, and D. A. Wolfe. 1983. The fate of *Amoco Cadiz* oil. *Science* (8 July):122–129.

Horgan, John. 1991. Up in flames. *Scientific American* (May):17–24.

IMCO/FAO/UNESCO/WMO/IAEA/UN Joint Group of Experts on the Scientific Aspects of Marine Pollution (GESAMP). 1977. *Impact of Oil on the Marine Environment.* Reports and Studies No. 6. Rome: FAO.

International Maritime Organization. 1988. *Manual on Oil Pollution. Section IV: Fighting Oil Spills.* IMO Publication 569 88.11.E. London: IMO.

Lancaster, John. 1991. Weighing the gain in oil-spill cures. *Washington Post,* 22 April, p. A–3.

Leblanc, Leonard A. 1990. Advanced technology: Why oil decomposition rates are hard to improve. *Offshore* (April):17.

Lindblom, G. P., B. D. Emery, and M. A. Garcia Lara. 1981. Aerial application of dispersants at the Ixtoc I spill. In *Proceedings of the 1981 Oil Spill Conference.* Washington: American Petroleum Institute.

Lloyd's Register of Shipping. 1990. *Statistical Study of Oil Outflow from Oil and Chemical Tanker Casualties.* Report for the American Petroleum Institute. Technical Report STD R2-0590. Washington: American Petroleum Institute.

Mielke, James E. 1990a. *Double-Bottom/Double-Hull Tankers: Pro and Con.* CRS Report for Congress. Washington: Congressional Research Service.

―――――. 1990b. *Oil in the Ocean: The Short- and Long-Term Impacts of a Spill.* CRS Report for Congress. Washington: Congressional Research Service.

―――――. 1990c. *Oil Spill Response Technologies.* CRS Report for Congress. Washington: Congressional Research Service.

National Oceanic and Atmospheric Administration (NOAA). 1991. *Evaluation of the Condition of Intertidal and Shallow Subtidal Biota in Prince William Sound Following the Exxon Valdez Oil Spill and Subsequent Shoreline Treatment.* Washington: National Academy Press.

National Research Council (NRC). 1985. *Oil in the Sea: Inputs, Fates and Effects.* Washington: National Academy Press.

―――――. 1989. *Using Oil Dispersants on the Sea.* Washington: National Academy Press.

―――――. 1991a. *Tanker Spills: Prevention by Design.* Report of the Committee on Tank Vessel Design. Washington: National Academy Press.

―――――. 1991b. *Summary of the Planning Meeting on Oil Pollution R&D.* Washington: Marine Board, National Research Council.

Over 100 damaged wells in Kuwait brought under control. *Golob's Oil Pollution Bulletin,* 10 May 1991, pp. 1, 3–4.

Pritchard, P. Hap, and Charles F. Costa. 1991. EPA's Alaska oil spill bioremediation project. *Environmental Science and Technology* 25:372–379.

Roberts, Leslie. 1989. Long, slow recovery predicted for Alaska. *Science* (7 April):22–24.

Sanders, H. L., J. F. Grassle, G. R. Hannsson, L. S. Moore, S. Price-Gartner, and C. C. Jones. 1980. Anatomy of an oil spill: Long-term effects from the grounding of the barge *Florida* off West Falmouth, Massachusetts. *Journal of Marine Research* 38:265–280.

Senate hearing probes response to Gulf oil fires and spills. *Golob's Oil Pollution Bulletin,* 26 April 1991, pp. 9–10.

Tanker Advisory Center. 1990. *Guide for the Selection of Tankers*. New York: Tanker Advisory Center.

U.S. Coast Guard. 1990a. Update of inputs of petroleum hydrocarbons into the oceans due to marine transportation activities. Paper submitted to IMO Marine Environment Protection Committee 30, London.

————. 1990b. Assessment of success of tankships with double bottoms and PL/SBT in mitigating pollution due to casualties. USCG internal analysis. Washington: USCG.

U.S. Congress. Office of Technology Assessment (OTA). 1990. *Coping with an Oiled Sea: An Analysis of Oil Spill Response Technologies*. OTA-BP-O-63. Washington: Government Printing Office.

Year-end spill totals reveal downward trends. *Oil Spill Intelligence Report*, 21 Feb. 1991, p. 1.

# CHAPTER 6

# NUCLEAR POWER

## Christoph Hohenemser, Robert L. Goble, and Paul Slovic

NUCLEAR POWER currently accounts for 421 plants in twenty-six countries with a rated power output of 312 gigawatts (GWe), or about 16 percent of the world's generating capacity (*Nuclear News,* 1989; DOE, 1987a).[1] In 1989 there were 110 plants with full operating licenses in the United States, accounting for 97 GWe, or about 18 percent of generating capacity. After coal, nuclear is the largest contributor to U.S. electric supply (Figure 6.1), even though nuclear contributes only about 5 percent of all U.S. energy (DOE, 1987a).

According to nuclear power analyst and proponent J. J. Taylor (1989:318), "this tremendous block of power has been delivered, on the average, with greater safety, with less environmental impact, and

Chris Hohenemser wishes to thank the Energy and Resources Group at the University of California, Berkeley, for a semester of hospitality and stimulating discussion in the spring of 1991. The authors would also like to acknowledge the help of Jeanne X. Kasperson and Nick Rosov in preparing this chapter.

FIGURE 6.1. U.S. ELECTRIC POWER GENERATION BY FUEL TYPE.

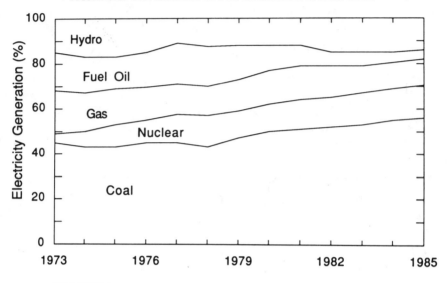

SOURCE: DOE (1987a).

FIGURE 6.2. PERCENTAGE OF ELECTRICITY GENERATED BY NUCLEAR POWER FOR VARIOUS COUNTRIES.

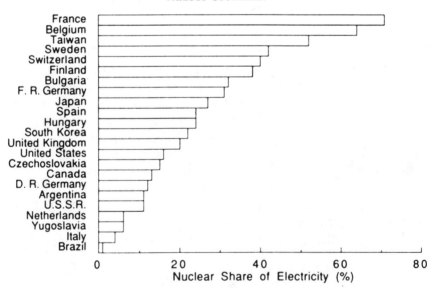

SOURCE: DOE (1987a).

at less cost than most prevailing methods of generating base-load electricity." Although this claim will be debated, any reasonable assessment of nuclear power would assign risks that are different (Hohenemser et al., 1983), but not more severe (Hohenemser et al., 1977), than competing technologies such as coal, Chernobyl notwithstanding. Even solar energy, which so far has not been developed to a significant extent, comes in for some unexpected and debated criticism (Inhaber, 1979; Holdren et al., 1979; Holdren, 1982).

Nevertheless, on a worldwide basis the growth of nuclear power has slowed dramatically, as annual growth has dropped from 5.3 percent projected as late as 1982 to less than 1 percent currently. Outside the Eastern Bloc, reactors are being ordered only in Japan, Taiwan, South Korea, and France (Forsberg, 1990). Along with this slowdown have come delays or abandonment of spent fuel reprocessing, the breeder reactor, and permanent disposal of high-level wastes. The abandonment of reprocessing and plutonium recycling is directly related to the perceived risk of a plutonium economy and the reduced demand for nuclear fuel (Albright and Feiveson, 1988). These developments follow a varied pattern among the world's major nuclear countries. France, with more than 70 percent nuclear electricity, continues to lead the world in proportion of nuclear electricity (Figure 6.2).

Although the United States has the largest total investment in nuclear power, only three plants are under construction, none is on order, and more than a hundred plants have been canceled (DOE, 1987a, 1991). Thus the United States currently finds itself stalled in a moratorium of nuclear power that is more extreme than that of any major nuclear country. Taylor (1989) and others describe this as a "period of introspection." During this period the U.S. nuclear industry is regrouping to assess what went wrong and why.

Several problems are frequently cited: inefficiency reflected by a low average capacity that remains at 60 percent (Taylor, 1989); regulatory delays, which have blocked opening of some reactors, like Seabrook on the New Hampshire coast and Shoreham on Long Island, for more than 10 years; lengthy plant construction times that currently average 14 years (GAO, 1989); lack of standardization, which lies at the root of many licensing delays (GAO, 1989); a continued failure to resolve the siting of a nuclear waste facility

(GAO, 1989); and a continuing crisis of public confidence (GAO, 1989; DOE, 1987a) in which opposition to nuclear power is expressed by well above 50 percent of the public.

The difficulties for nuclear power in the United States come at a time when the prospect of significant global climate change (Schneider, 1989; Houghton and Woodwell, 1989) should make nuclear power the darling of environmentalists (Hohenemser, 1989; Beyea, 1989). With the exception of hydropower and conservation (Lovins et al., 1981) nuclear power is the only currently available source of base-load electricity that does not produce significant greenhouse gases. (In the United States, nuclear power currently displaces about 7 percent of the total U.S. $CO_2$ output that would occur if equivalent electric energy were produced by coal.) Nevertheless, nuclear technology is rejected by most environmentalists because it is viewed at least in part in terms of presently existing plants and all their problems (Beyea, 1989), not in terms of the improved reactors currently envisioned by the industry in a hoped-for rebirth of the technology (Taylor, 1989). This means, in effect, that there is no escape from a troubled history.

Beginning with an inextricable and early link to the World War II nuclear weapons program, the history of U.S. nuclear power proceeded to early risk assessments in which catastrophic consequences were estimated while concomitant probabilities were described as unknowable (AEC, 1957). It continued with the issue of reactor safety, initially addressed via engineering arguments (Forbes, 1972) and later through probabilistic risk assessment that was designed to overcome the deficiencies of earlier risk assessments (NRC, 1975). It is a history in which the scope of emergency response plans is still debated 30 years after it should have been resolved and one in which the question of how nuclear reactors might be decommissioned has only recently entered our collective consciousness. Society continues to wrestle with the impact of nuclear power on nuclear weapons proliferation, an issue that is both old (Davidon et al., 1960; Hohenemser, 1962) and very current (Albright and Feiveson, 1988; Williams and Feiveson, 1990).

To understand these issues better, we review here the development of nuclear power and its institutional framework and attempt to shed light on two questions that appear to dominate the future: What can be done about managing the present lot of 110 U.S.

working reactors? And what is to be expected from a "rebirth" of nuclear power as presently envisioned by the industry?

## SETTING THE STAGE: THE PAST AS PROLOGUE

The stage was set in December 1938, when Otto Hahn and Fritz Strassmann culminated a 4-year search with the discovery in Berlin that uranium nuclei can be split by neutrons. When published in early 1939 and interpreted through Einstein's theory of relativity, the Hahn–Meitner result was immediately recognized as implying an energy release per atom about 100 million times larger than previously achieved in chemical reactions.

In the midst of rising tensions leading to World War II, the discovery of nuclear fission split the worldwide scientific community and resulted in two separate and secret efforts to develop nuclear weapons in the United States and Germany—a story that has been told and retold many times over (Rhodes, 1986). At the University of Chicago on 2 December 1942, a group working under the leadership of the Italian Enrico Fermi produced the first sustained nuclear reaction. The event was a critical step leading to production of plutonium and the nuclear weapon eventually released over Nagasaki on 9 August 1945. It also marked the forerunner of today's civilian nuclear power reactors. Thus nuclear power arose conceptually from the cradle of wartime weaponry and became inextricably linked in the perception of politicians and publics worldwide to the most fearful weapons ever produced.

### EARLY SMOOTH SAILING

Commercial nuclear power in the United States was born in December 1953 as a stepchild of the nuclear arms race when President Eisenhower proposed "Atoms for Peace" in a dramatic speech before the United Nations (Dennis, 1984). Beyond opening the discussion of disarmament, his goal was to counteract the martial image of nuclear energy by showing that it could be used for peaceful purposes. Though commercial reactors had been on the drawing board for several years, the direct result of Eisenhower's 1953 proposal was that a naval propulsion reactor was adapted. Thus the first commercial nuclear plant, with a rating of 72 megawatts (MWe), was brought on line in Shippingport, Pennsylvania, on 2 December 1957, exactly 15 years after the first chain reaction was achieved at the University of Chicago. Considered a demonstration unit, the plant was too small to be economically competitive or to be of much concern to the public, whose "nuclear" concerns focused on nuclear weapons testing. After a 7-year debate and extensive public demonstrations worldwide, the weapons issue was apparently settled via the atmospheric test ban treaty signed by the United States and the Soviet Union in 1963.

By 1965 the public's interest in nuclear issues had subsided to an all-time low, even though the 1963 nuclear test ban treaty had merely moved the arms race underground (Hohenemser and Leitenberg, 1967; Hohenemser et al., 1977). Commercial nuclear plants, with the startup in 1962 of the 170-MWe Indian Point I facility, 26 miles north of New York City, had managed not only to increase their size by a factor of 2.5 but were approaching economic parity with competing power sources. The future of nuclear power appeared assured because the electric power sector was expanding at a remarkable 7 percent per annum (doubling time of 10 years), while plans were made to build 1000 GWe of nuclear power by the year 2000.

With the public distracted by an escalating war in Vietnam, the issue of nuclear power's safety escaped the limelight during these years. Siting criteria had been considered by the Atomic Energy Commission (AEC) as early as 1950 (AEC, 1950) and led to the definition of an "exclusion zone" of radius $R = (0.01P)^{1/2}$, where $R$ is in miles and $P$ is the power measured in kWt. (Implied zone radii are 2 to 5 miles for the small reactors of the 1950s and 16 miles for the Three Mile Island reactor located 10 miles from Harrisburg, Pennsylvania.)

The consequences of hypothetical reactor accidents for a 150-MWe plant were discussed in a 1957 study known as WASH-740 (AEC, 1957) under the extreme assumptions of rainfall, a nearby large city, and 50 percent release of the entire core. This scenario produced an estimated zero to 3400 prompt deaths, zero to 43,000 injuries, and $0.5 to $7 billion in property damage. An "intermediate" case, considering the release of all volatiles and about 1 percent of the remaining fission products, led to an estimated zero to 900 prompt deaths, zero to 13,000 injuries, and zero to $0.5 billion in property damage. Like the 1950 siting study, WASH-740 received little public attention.

In rejecting the possibility of calculating probabilities of reactor accidents, the authors of WASH-740 made a prophetic statement that today, after extensive work on probabilistic assessment, is still believed by many:

> There is a common aversion to attachment of quantitative estimates to a phenomenon so vague and uncertain as the probability of occurrence of catastrophic accidents, particularly since such assignment of numerical estimations conveys the erroneous impression of the confidence or firmness of the knowledge constituting the basis of the estimate. Also, some hold the philosophic view that there is no such thing as a numerical value for the probability of occurrence of a catastrophic accident; that such a thing is unknowable. Thus, many decline to make even order of magnitude guesses of the probability of catastrophic reactor accidents.

A 1965 update of WASH-740 extended the methodology of the 1957 study to a 1000-MWe plant under the same assumptions. This study found a worst-case maximum of 45,000 prompt fatalities and property damage many times larger than WASH-740 found, but again without accompanying probability estimates. The AEC suppressed public access to this study "to avoid great difficulties in obtaining public acceptance of nuclear energy" (Mulvihill et al., 1965).

Public discussion of the results therefore did not enter the public consciousness until 1973, when a Freedom of Information intervention by the Union of Concerned Scientists unearthed the 1965 update (Ford, 1977). In retrospect, the remarkable fact about these "early quiet years" is not only the premonition of the AEC with respect to public fears, but also its control of both the development and safety management of nuclear power. The Nuclear Regulatory Commission (NRC), which was not formed until 1974, was unavailable as a counterweight, and public opinion, to the extent that it existed, was only mildly concerned. It thus appeared that AEC planners had little to worry about as they approached the nuclear safety problem through an overriding concern with engineered safeguards and quality control. The failure to address the probability of catastrophic accidents with any seriousness was, of course, a major omission.

## THE BUMPY EVOLUTION OF PROBABILISTIC RISK ASSESSMENT

The first significant attacks on commercial nuclear power came in the late 1960s when, as part of a growing environmental movement, prior concern about radioactive fallout from weapons testing was translated into fears about routine emissions of nuclear plants. Risk estimates for low-level radiation easily deflected this issue, since routine public exposure from nuclear reactors was shown to be less than 10 microsieverts per year, or about 1 percent of natural background (Eisenbud, 1987). But the discussion quickly moved on to a debate over emergency core cooling (Ford, 1977) and the prospect of a catastrophic core accident at a time when consequence estimates of WASH-740 and its suppressed update provided the only reference points.

To address the risk question, the newly formed NRC in 1973–1975 sponsored the first application of probabilistic risk assessment (PRA) to nuclear reactors in the Reactor Safety Study (NRC, 1975). Known as the Rasmussen Report, this study adapted fault-tree analysis from the aerospace industry and estimated the probability of a core melt, the likelihood of a significant release following a core melt, and subsequent health consequences. The report's "worst case" assumed that failure of containment would be restricted to materials ("source terms") consisting of largely volatile fission products, as in the "intermediate" case of WASH-740. Its principal findings were in the form of calculated risk spectra, $f(x)$, predicting the frequency (events per year) for which the number of prompt and delayed fatalities are expected to exceed $x$. The calculated form of $f(x)$ for 100 hypothetical reactors is shown in Figure 6.3. For a single reactor, $f(x)$ for $x =$ 45,000 prompt deaths (the consequences discussed in the WASH-740 update) is off the scale in Figure 6.3 but may be estimated by extrapolation to be in the neighborhood of $10^{-12}$ per reactor year. This extremely low value describes the estimated impact of a core melt (frequency of $5 \times 10^{-5}$) followed by rupture of containment and subsequent "worst-case" release of radioactivity.

FIGURE 6.3. PROBABILISTIC RISK SPECTRA FOR 100 LIGHT-WATER REACTORS AS
CALCULATED IN THE RASMUSSEN REPORT.

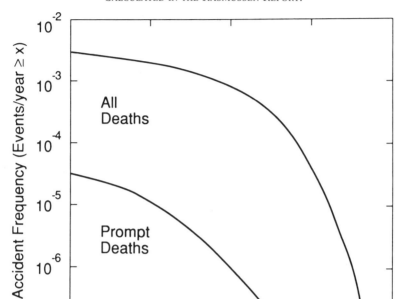

Note: *Not shown are factor-of-10 error bands around each spectrum.*

SOURCE: NRC (1975).

These results naturally surprised the critics and the public, which
had until then focused on the potential magnitude of "worst-case"
accidents without concomitant probabilities. How could the expected frequency of earlier assessments be one chance in 100 billion
per year? With 100 U.S. reactors it would take 1 billion years to
produce one such accident. What is the meaning of the earlier consequence studies in this context? Were the scientists deceiving us? Are
they deceiving us now with the new estimates?

The Union of Concerned Scientists and others immediately at-
tacked the Rasmussen Report on a variety of technical grounds, such
as inadequate treatment of uncertainty, common mode failure,

incompleteness in fault-tree analysis, unquantified human error, and unrecognized design inadequacy (Ford, 1977; Hohenemser et al., 1977). They also returned to the argument made in WASH-740 that calculation of extremely low probabilities is meaningless. The public reacted in kind: as they saw that scientists could not agree, many of them turned away from PRAs as a route to nuclear safety assessment. Thus grew one of the deep roots of the ensuing polarized debate.

More recently the technical community has found that insufficient attention has been paid to testing PRA assumptions (Lewis et al., 1979; Wilson et al., 1985; Sholly and Thompson, 1986), including the size of the appropriate source term (ANS, 1984). It has also been widely agreed that PRAs are best not interpreted in terms of absolute risks (Daniels and Canady, 1984), as suggested by the original Rasmussen Report. In effect, this conclusion concedes the WASH-740 argument that exceedingly small probabilities are unreliable if not meaningless. In the revised view, PRAs are most appropriate for judging the comparative advantages of risk-reducing strategies, using calculated risks as a guide. Since the publication of the Rasmussen Report there have been many PRAs for individual plants that together have led to a better understanding of the characteristics and limitations of the PRA method. (See Bernero, 1984; NRC, 1987a; Levine and Rasmussen, 1984; Daniels and Canady, 1984; Joksimovich, 1984; Garrick, 1984; Vesely and Rasmussen, 1984; Budnitz, 1984.)

The results of more than twenty PRAs may be summarized as follows (Sholly and Thompson, 1986; Joksimovich, 1984). First, most estimates of the probability $(p_c)$ of major core damage (level I PRA) lie in the range of $5 \times 10^{-5} < p_c < 5 \times 10^{-4}$ (Sholly and Thompson, 1986; NRC, 1987a). Second, the probability of a major release after core melt (level II PRA) is more variable, with a conditional probability $(p_r)$ of $0.003 < p_r < 0.3$ (NRC, 1975; Garrick, 1984; Joksimovich, 1984). Third, the health consequences of a major release (level III PRA) depend not only on containment but also on the details of the surrounding demography and social responses such as emergency evacuation. Among the three stages, assessments for levels I and III depend substantially on available data, whereas level II must rely on untestable computer codes.

After 15 years of development, PRAs have done little to resolve the deceptively simple question posed by the public: how safe is nuclear power? Instead they have provided technical tools for evaluating potential improvements in specific plants, while leaving the general public incredulous and confused. In this context it is important to note that the challenge of doing PRAs in the United States is particularly severe because there are fifty-four operators of 110 plants with minimal standardization. U.S. plants therefore provide extensive analytical challenges with little opportunity for generalization.

THREE MILE ISLAND AND CHERNOBYL

At 4 A.M. local time on 28 March 1979, Unit 2, an 850-MWe pressurized water reactor of the Three Mile Island (TMI) nuclear plant near Harrisburg, Pennsylvania, suffered an automatic shutdown of its steam generator because its feedwater pumps had stopped. Moments later, as pressure in the steam generator built up, the pressure relief valve on top of the steam generator opened automatically, as planned. Though the valve should have closed automatically soon afterward because cooling water was draining from the pressure vessel, it stuck in the open position. The result was a sequence of events during the following 7 hours that led to partial uncovering and melting of the core and involved substantial misapprehensions and misjudgments by operators who were, in fact, in a position to block the sequence but did not (Kemeny et al., 1979; Eisenbud, 1987). After 7 hours containment was still intact, the core was again covered, and some radioactive noble gases had been released to the environment.

In the days following this event, additional misjudgments by regulatory officials compounded the problem. For example, federal officials arranged for production and storage of potassium iodide and recommended its distribution to the public; the NRC also recommended an evacuation on the basis of helicopter radiation measurements taken above the plant stack that erroneously had been described as ground-level readings. Both recommendations were rejected by the state of Pennsylvania. This confusion stimulated self-evacuation of some 100,000 people (Kemeny et al., 1979) and induced psychological stress that is widely believed to exceed any possible radiation health effects (Kemeny et al., 1979; Moss and Sills, 1981; Davidson et al., 1987; Flynn, 1979; Bromet, 1980). Radiation exposure of the public was due to gaseous $^{133}$Xe and $^{85}$Kr that reached the outside environment in significant quantities in part due to intentional release. Through standard assessment methodology this exposure is estimated to produce at most one or two cancers in the next 50 years (Gerulsky, 1981; Eisenbud, 1987), although there is some residual uncertainty about the amounts released (Eisenbud, 1989; Goldberg, 1985).

The impacts of TMI, however, go much further than modest radiation exposure and possibly greater psychological effects. Since the time of the accident not a single new nuclear plant has been ordered in the United States. And whereas nuclear risk analysts believe the TMI accident sequence falls within the predictions of PRA methodology (Levine and Rasmussen, 1984), the accident is generally seen as a significant contributor to the continuing decline in the public's confidence in nuclear power. Even though no one was killed at TMI, the event indicated that serious accidents can happen. A partial core melt that almost got away imparted a sudden reality to the abstractions of probabilistic risk assessment. Additional evidence for the seriousness of this event is the fact that after 10 years, the $2 billion decontamination process of the TMI Unit 2 is still incomplete. Moreover, a large number of lawsuits have been filed by individuals who believe themselves affected by TMI. The settlement of the lawsuits has included a court-ordered financing of an alternative emergency plan.

At 1:23:44 A.M. local time on 26 April 1986, Unit 4 of the Chernobyl nuclear plant, a graphite-moderated pressure tube reactor located some 80 km north of Kiev in the Ukraine, suffered two explosions 3 seconds apart. As described by Grigori Medvedev (1990), the events leading to these explosions involved a dramatic sequence of faulty judgments and arrogance that has few parallels even in the events at TMI 7 years earlier.

The most prominent errors occurred when operators attempted to stabilize the unit at 6 percent of rated power, a region that is off limits according to stated operating rules because of the reactor's positive power coefficient at this level. To make matters worse, instead of shutting down and starting over, operators tried to increase the power from the forbidden 6 percent level by disabling substantial portions of the safety system. This act led directly to the fast critical excursion that produced the explosions (USSR, 1986; NRC, 1987b;

Medvedev, 1990). In essence, the excursion resulted from a runaway condition in which reactor power went from 6 percent to over thirty times rated capacity in 2 seconds.

The explosions sheared off the tops of all 1661 pressure tubes, lifted the 1000-ton cover off the core, ruptured existing containment, dislodged the refueling crane, discharged hot molten and pulverized fuel to an altitude of at least 7.5 km, and started as many as thirty fires (USSR, 1986; NRC, 1987b; DOE, 1987b; Hohenemser, 1988). The release of radionuclides, the largest ever recorded in a technological accident, continued for 10 days as Soviet officials attempted to smother the exposed core. Dispersed radioactivity reached all countries in the Northern Hemisphere and triggered unplanned emergency protection in many European countries (Hohenemser and Renn, 1988).

After substantial analysis by the Soviets (USSR, 1986) and others (NRC, 1987b; DOE, 1987b), the Chernobyl source term was shown to be similar to the "worst-case" hypothetical pressurized water reactor accident, PWR-2, described in the U.S. literature (Lewis et al., 1975). As shown in Figure 6.4, the Soviet release was equal to PWR-2 for the more volatile radionuclides and exceeded PWR-2 for the least volatile. Dry weather, favorable siting, evacuation of 116,000 people, and, above all, dispersal of radionuclides to high altitude contributed to holding prompt deaths to the small number of thirty-one.

The worldwide dispersal of core material has been estimated (DOE, 1987b; Hohenemser, 1988) to contribute a lifetime exposure of radioactivity equal to a total of 1.2 million person-grays. In 1987 the Department of Energy interpreted this to mean that there would be as many as 28,000 excess cancer fatalities resulting from this exposure. This interpretation was based on a 1980 consensus describing the dose/consequence relationship, as given in the report *BEIR III* by the Committee on the Biological Effects of Ionizing Radiation (NAS/NRC, 1980).

The most recent analysis, known as *BEIR V*, increases the estimated incidence by a factor of 3.5 (NAS/NRC, 1990). The increase comes primarily from better estimates of the doses received by atomic bomb survivors, continuing follow-up of these survivors, and changes in the cancer incidence models favored. *BEIR V* notes that "accumulation of the same dose over weeks or months is expected to

FIGURE 6.4. CHERNOBYL SOURCE TERM COMPARED TO THE WORST-CASE LIGHT-WATER-REACTOR SOURCE TERM (PWR-2) CONSIDERED IN THE WESTERN LITERATURE.

*The latter is shown as a solid line, the Soviet release fractions as open circles, and suggested DOE corrections in closed circles. The volatility classes are defined as follows: 1, noble gases; 2, iodines; 3, cesiums; 4, telluriums; 5, alkaline earths; 6, volatile oxides; and 7, transuranics.*

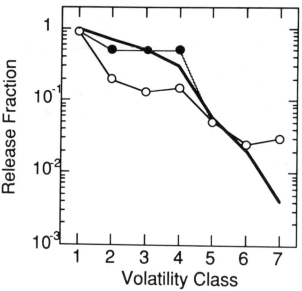

SOURCE: Hohenemser (1988).

reduce the lifetime risk appreciably, possibly by a factor of 2 or more" (NAS/NRC, 1990:6). Thus a plausible current estimate would be that as many as 50,000 to 100,000 cancer fatalities will result from Chernobyl exposures worldwide over the next 50 years. Moreover, studies of localized contamination continue in the Ukraine and elsewhere and can be expected to lead to further revisions of the potential delayed cancer toll of Chernobyl.

Because the Chernobyl reactor differs substantially from Western light-water designs, the rough equivalence of the Chernobyl and the worst-case U.S. source terms says nothing about the likelihood of a Chernobyl-scale accident in the United States. Yet from a wider perspective, Chernobyl has had a substantial impact. Chernobyl demonstrates that scenarios approaching worst-case assumptions are possible and that emergency planning is a problem of transnational dimensions that certainly extends beyond the 10-mile emergency evacuation zone and the 50-mile ingestion response radius used in the United States. Moreover, though projected cancer fatalities exceed 1 percent of background radiation only for the evacuees while amounting to about 0.1 percent of background for the rest of Europe (Hohenemser, 1988), the total burden of as many as 100,000 potential cancer deaths lies in a new region not heretofore touched by TMI or any other "peaceful" event. Finally, the accident at Chernobyl strongly reinforces public perceptions derived from TMI and has in fact entered the U.S. political process in the debate concerning licensing of the Seabrook and Shoreham plants.

Together, TMI and Chernobyl have added significantly to our nuclear perceptions. They have made real the abstractions of probabilistic risk assessment and have cast doubt upon the goals and achievements of the method. This doubt affects how the public and its political representatives view nuclear power worldwide (Hohenemser and Renn, 1988) and has contributed to direct and measurable impacts on the industry itself, such as high capital costs, regulatory delays, and the absence of new reactor orders in the United States.

## AN ECLECTIC MIX OF ISSUES

Beyond the bumpy evolution of probabilistic risk assessment and two major accidents in the United States and the Ukraine, nuclear power has been brought to its current hesitant state via an eclectic mix of issues. These involve deep-seated characteristics of economics, human institutions, and human fears. We turn to them next.

### ECONOMIC SHOCKS

The economic shocks originating in the 1973 oil embargo triggered a sea change in energy use and inflation that has reverberated throughout the world's economic system. The resulting economic climate has substantially weakened the development of nuclear

FIGURE 6.5. OIL SHOCKS SINCE 1973.

*The graph shows the price of oil expressed in constant 1985 dollars as a function of time.*

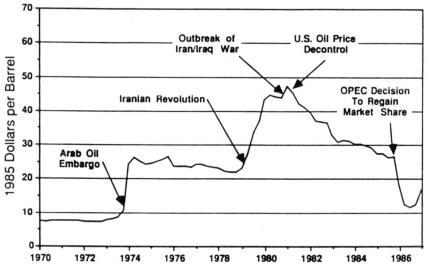

SOURCE: DOE (1987a).

technology quite apart from difficulties with risk assessment and reactor accidents. As seen in Figure 6.5, the price of oil has experienced three nearly discontinuous changes since 1973. The first two shocks marked initial victories by the OPEC cartel and increased oil prices by about a factor of 6. This price rise was followed by falling prices during 1980–1985 as OPEC slowly lost influence and then by a third shock in 1985 when OPEC dropped prices in an effort to recover market share. Since oil remains the single largest energy resource for the United States, its price has been associated with interest and inflation rates (GAO, 1989), amounting to 20 percent and 13 percent a year, respectively, at the peak of OPEC power in 1980.

In terms of energy consumption, the 1973 embargo and its sequels led to dramatic energy conservation. As shown in Figure 6.6, U.S. energy consumption in 1989 was less than 50 percent of that expected from extrapolating the historical trend prior to 1973. All U.S. energy has dropped from an annual growth of 2.8 percent to about zero, while electric energy has decreased from an annual growth rate of 7 percent to 2.8 percent (DOE, 1987a; Cook, 1971). As a result, the historically projected requirement for nuclear plants has been put into question, particularly the exuberant 1970 projection of 1000 plants by the year 2000.

The period of oil price shocks witnessed a dramatic rise in nuclear capital costs, but oil costs do not explain all of this increase. As shown in Table 6.1, capital costs increased by a factor of 10—from about $400/kWe to roughly $4000/kWe between 1973 and 1987 (GAO, 1989)—while general construction costs increased by less than a factor of 3 (Council of Economic Advisors, 1989). A major factor in the capital cost escalation has been the increase in the average con-

FIGURE 6.6. GROWTH OF ALL ENERGY AND ELECTRICAL ENERGY IN THE
UNITED STATES FROM 1880 TO THE PRESENT.

*Data show geometric growth for all energy for 100 years until the 1973 oil embargo, after which all energy has leveled off to near zero growth.*

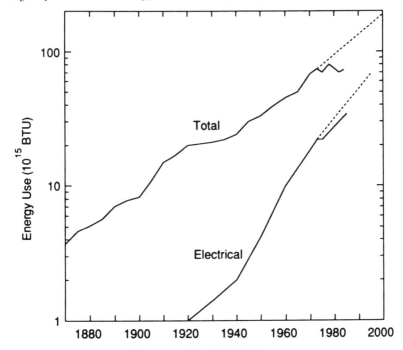

SOURCE: Data adapted from Cook (1971) and DOE (1987a).

TABLE 6.1. CAPITAL COSTS OF NUCLEAR REACTORS

| First year of commercial operation[a] | Cost per kWe (current $) | Cost per kWe (constant $) | Price deflator[b] for nonresidential structures | Years to complete construction |
|---|---|---|---|---|
| 1971–1974 | 388 | 388 | 1.00 | 6.8 |
| 1975–1976 | 564 | 424 | 1.33 | 8.7 |
| 1977–1980 | 670 | 392 | 1.71 | 9.8 |
| 1981–1984 | 1644 | 702 | 2.34 | 13.1 |
| 1985 | 2693 | 1108 | 2.43 | 14.4 |
| 1986 | 2933 | 1154 | 2.54 | |
| 1987 | 3776 | 1435 | 2.63 | |

[a] Data from DOE (1987a).
[b] Data from Council of Economic Advisors (1989: 312).

struction time from 7 years to 14 years, resulting in substantial increases in interest payment during construction.

The increase in construction time has in turn been attributed to the fact that during 1959–1970 the scale of the technology grew at a rate which was much more rapid than is normal for technological innovations (Hafele, 1985). This spurt of growth resulted in backfitting of safety features, poorly estimated capital costs, construction delays, and incorrectly projected operating costs quite independent of oil shocks and inflation. According to former NRC commissioner Peter Bradford (Adato et al., 1987:4):

> In 1968 the largest reactor in operation was only one-half the size of the smallest reactor under construction and was one-sixth the size of the largest plant under construction. As a result, an entire generation of large plants was designed and built with no relevant operating experience—almost as if the airline industry had gone from Piper cubs to jumbo jets in about 15 years.

To compound the economic problems of nuclear power, the breakup of OPEC created a 1989 oil glut with more than 10 million barrels per day of excess pumping capacity worldwide. This glut motivates utilities to consider fossil-fired capacity where possible, even when global warming and the maturing of nuclear power would argue against it.

The high capital costs of nuclear power come sharply to the public's attention when utilities pass on to the consumer their unexpec-

tedly high construction costs. Described as "rate shock," the resulting rate increases have led state utility commissions to disallow
major components of capital costs (GAO, 1989), thus destabilizing
utilities further and discouraging additional nuclear investment. But
prohibiting cost recovery during construction has also prolonged
efforts that might otherwise have been terminated. According to
Charles Bayless, chief financial officer of the bankrupt Public Service
Company of New Hampshire (*Boston Herald*, 9 November 1989):

> I really believe that if the law (prohibiting construction cost recovery
> before operation) had not been in effect, Seabrook would have been
> cancelled. . . . We would have written off part of it in the 1970s or early
> 1980s and received some return on our investment for stockholders
> and remained solvent.

Taken together, the economic impacts of the last 17 years are
probably sufficient to explain the present comatose condition of U.S.
nuclear power, quite independent of the role of the major accidents
we have discussed. At the same time, such an explanation aggregates
distinct factors such as too-rapid expansion, resultant retrofitting,
utility cost recovery laws, inflation during construction, and the
unpredictable price of oil.

## The Evolution of Reactor Safety Management

Safety management for nuclear reactors is built around the concept
of "defense in depth." Several barriers have been designed to block
public exposure and its consequences:

1. Remote siting—to reduce the population exposure in case of
   release
2. Engineered safety systems—to prevent initiation of accident
   sequences
3. Operator training—to moderate development of accident sequences
4. Containment—to prevent release of activity after a core accident
5. Emergency planning—to reduce exposure
6. Insurance—to compensate human and property losses

Together these stages may be envisioned in terms of a causal sequence or "hazard chain" running from "upstream" initiating events to "downstream" consequences, with regulation playing the role of event-blocking interventions at each stage (Hohenemser et al., 1982). A typical hazard chain for reactor safety, as shown in Figure 6.7, is not only useful for conceptualizing regulation but, in fact, summarizes the essential components of the Rasmussen Report.

The earliest regulatory interventions occurred in the years 1950–1957 when neither accident experience nor accident probabilities were part of the known world and no civilian nuclear reactors existed. During these years AEC regulations focused on upstream intervention—that is, remote siting—and downstream intervention—that is, evacuation and compensation—while leaving the middle of the hazard chain to future engineering development.

Siting and evacuation guidance were given in 1950 (AEC, 1950) and involved the exclusion zone criteria noted earlier. As large reactors were designed and built, these criteria were relaxed (Okrent, 1981) and then replaced by specific 1962 regulations (Eisenbud, 1987) that assigned the evacuation problem to reactor operators. Assuming that barriers 1 to 4 on the list would operate indepen-

FIGURE 6.7. STAGES OF REGULATION IN MANAGING THE HAZARDS OF NUCLEAR POWER.

*The diagram shows the causal sequence of hazards beginning with the choice of technology on the left and ending with consequences on the right.*

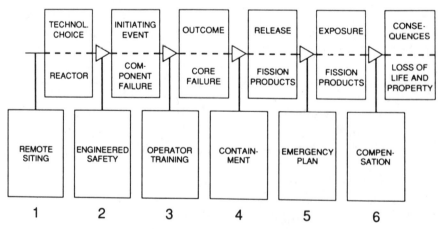

SOURCE: Adapted from Hohenemser et al. (1982) and Kates et al. (1985).

dently, the 1962 regulations defined "maximum credible" or "design-basis" accidents that survive today in certain regulatory settings. The associated radiation releases are substantially smaller than the intermediate case of WASH-740 or the worst-case PRA sequences developed later. At the same time, WASH-740 was instrumental in limiting accident liability to $560 million under the 1957 Price-Anderson law in exchange for government guarantee of compensation. The law, which has been subsequently updated, played a major role in encouraging utilities to invest in nuclear reactors (Dennis, 1984).

The full range of events and barriers in Figure 6.7 is, in principle, addressed in reactor licensing under the Atomic Energy Act of 1957 and subsequent revisions. It involves two stages: construction and operation. Because reactor construction and approaches to reactor safety have evolved concurrently, the operational stage has often raised issues not foreseen in the construction stage. This difficulty has been accentuated by inconsistent requirements in the two stages— construction uses "design basis" accidents in siting and plant design, for example, whereas operation demands consideration of larger accidents for retrofits and emergency planning.

With respect to retrofitting, the main controversies between the industry and the NRC have hinged on what is subject to regulation, what is generic and what is plant-specific, and to what extent cost considerations should enter. The NRC has defined a two-tier system of backfitting, the first deemed "essential to safety," the second subject to cost-benefit analysis (GAO, 1989). Plant-specific PRAs have been used in a "relative risk" mode to define priorities and determine which measures should fall in which tier. Although the process has been accepted by industry, the unpredictability of required changes has been a source of dissatisfaction (OTA, 1984).

Largely through concern arising from the Rasmussen Report (NRC, 1975), a 1978 NRC/EPA task force (NRC/EPA, 1978) was asked to define "the most severe accident basis for which radiological emergency response should be developed." The task force interpreted its charge narrowly by rejecting PRA in formulating these requirements. Instead it recommended the use of planning zones based on a "spectrum of accidents" including the severe accidents analyzed in the Rasmussen Report. Four months after the task force report, the TMI accident demonstrated that government agencies and the industry were unprepared for the potential scale of nuclear emergencies. When Congress demanded emergency planning regulation, the NRC responded (NRC/FEMA, 1980) by extending existing guidance to the larger accidents. The new requirements included participation by state and local officials (U.S. Congress, 1987), but this stipulation was dropped in 1987 as New York and Massachusetts (whose border with New Hampshire lies within the 10-mile emergency planning zone) claimed that effective emergency planning for the Shoreham and Seabrook facilities was impossible.

In rejecting risk performance standards, either absolute or relative, and providing no comparable basis, the 1980 NRC regulations failed to provide objective criteria for evaluating the effectiveness of emergency preparations. As a result, emergency plans read like checklists rather than a coherent activity (Golding and Kasperson, 1988) and risk experts have lost influence as the regulatory process has become encrusted with legal interpretation.

The present process of emergency planning is thus triply flawed: it combines the legacy of dual licensing, encrustment in litigation, and an approach that has no objective performance criteria. Particularly the third flaw may be attributed to the fact that objective criteria based on PRA run directly into the fundamental uncertainties of the method. In the cases of Seabrook, Shoreham, and TMI, the result has been long-lasting controversies that illuminate the failure of U.S. societal institutions in dealing with an unforgiving and distrusted technology.

Moreover, decommissioning and disposal of old reactors looms as an important future regulatory issue. The disposal of small military reactors, or DOE's scrapping of the 72-MWe Shippingport plant, cannot be scaled up to the much larger commercial plants (Pollock, 1987). Debate has already begun on financing, particularly the question of utility financing (Malko, 1987; Hendron, 1987; Ferguson, 1987; Sanderford, 1987). A related issue is the prospect of refurbishing instead of decommissioning plants. Given the high capital costs of new plants, this idea has intrinsic incentives (Spinrad, 1988) for which technical studies are just now beginning. If plants are refurbished, this may raise the issue of relicensing. Current nuclear reactor licenses expire after 30 years, and if nothing is done, this will lead to a continuous decline of licensed plants starting in the mid to late 1990s.

THE BACK END OF THE FUEL CYCLE

Beyond the safety and licensing of reactors per se, two of the most vexing regulatory problems concern spent fuel reprocessing and high-level waste disposal. Until 1977 these questions were considered part of the same issue in the United States because it was thought that plutonium would be separated from spent fuel and recycled as fresh reactor fuel while other radioactive material was sequestered as high-level radioactive waste. Since President Carter's abandonment of U.S. commercial reprocessing on economic and nuclear proliferation grounds (Dennis, 1984), U.S. nuclear power has been restricted to the "once-through" fuel cycle. The back end of this cycle involves only the sequestering of unprocessed spent fuel.

The 1977 Carter decision may be viewed as an important landmark that places the United States, along with Canada, Sweden, Russia, and China, among countries whose civilian nuclear fuel cycle poses no overt risk of nuclear weapons proliferation because it is currently not possible to divert or steal weapons-grade material. For France and Japan, in contrast, current plutonium recycling accounts for an annual production of about 8 metric tons a year. And for France, Britain, Germany, and Japan it may account for 28 metric tons, or 4000 nuclear weapons equivalents, per year by the end of the century (Albright and Feiveson, 1988).

Even without plutonium recycling, the back end of the fuel cycle is a nuclear issue that, as much as any other, divides U.S. society. From a technical point of view, dealing with nuclear wastes was from the beginning neglected by the Atomic Energy Commission because taking care of nuclear wastes was considered to be a trivial technical problem (Wilson, 1979). Throughout the 1950s, 1960s, and 1970s it was assumed that high-level nuclear wastes would be sequestered underground in stable geological formations—for example, salt domes in Kansas were discussed by the National Academy of Sciences in 1957. In early studies technical oversights such as the existence of oil-drill holes in the salt led to withdrawal of the Lyons, Kansas, site. Later, as significant state opposition grew, the federal government began to focus on federal lands. In 1976, the Carter administration considered two federal reservations associated with the military nuclear weapons effort: Hanford and the Nevada test site. By 1980 the search for a suitable geological site was further broadened, and the adoption of a particular technical solution was further delayed (Carter, 1987).

Seen retrospectively, the early history of radioactive waste management involved bureaucratic miscalculations in which a complacent federal government offered "solutions" that, upon examination, were found wanting (Colglazier and Langum, 1988; Carter, 1987). Since 1980 it has become increasingly clear that high-level nuclear waste, even without plutonium recycling, poses a substantial political problem in which technical experts confront an increasing public distrust and fear while the quest for a solution to the problem recedes. In this context, the Nuclear Waste Policy Act of 1982 established an elaborate comparative search process for an underground repository site; but a 1987 amendment abrogated the process by selecting Yucca Mountain in Nevada as a single site to be evaluated technically (Colglazier and Langum, 1988).

Throughout, the responsible institutions have been unable to achieve a balance between fairness and the lengthy process of evaluation—creating the impression that nuclear waste is an unmanageable problem and resulting in the neglect of other urgent issues beyond siting of a waste disposal facility. Thus there are no detailed plans for spent fuel storage at reactor sites, which of necessity must be the principal waste management mode for at least the next decade or two; similarly, cask design and transportation have languished. Exasperated by its lack of progress, the Department of Energy recently scuttled the work on Yucca Mountain with the intention to begin the process afresh (New York Times, 1989).

Currently 78 percent of Nevadans oppose the siting of a repository anywhere in Nevada and 80 percent want state leaders to oppose the siting with any means available (Las Vegas Review Journal, 1990). In view of insurmountable public opposition to a permanent underground repository, three current options appear untenable: a permanent repository (widely considered a costly and doomed effort); permanent on-site storage (a solution that is deemed unsafe); monitored retrievable storage (a solution that would be seen as a de facto permanent site). This leaves as the only viable option a delay in siting of a permanent repository for several decades, while wastes are left on-site in dry-cask storage (Slovic et al., 1991).

The regulatory landscape of the back end of the fuel cycle thus involves a litany of contradiction and delays, as well as some potentially positive developments such as the decision to forgo plutonium recycling. As in the case of reactor safety management, regulation continues to be beset by confrontation between public and expert perceptions for which there appears to be no ready solution. We turn to this issue next.

THE CRISIS OF PUBLIC ACCEPTANCE

Scientists and policymakers were slow to recognize the importance of public attitudes and perceptions in shaping the fate of nuclear power. In 1976, Alvin Weinberg (1976:19) observed:

> As I compare the issues we perceived during the infancy of nuclear energy with those that have emerged during its maturity, the public perception and acceptance of nuclear energy appears to be the question that we missed rather badly. . . . This issue has emerged as the most critical question concerning the future of nuclear energy.

During the 31 years of the nuclear age that preceded Weinberg's observation, the public developed a distrust of nuclear technology anchored in the fear of nuclear weapons (Hohenemser et al., 1977). Nuclear issues were raised in the context of nuclear deterrence, nuclear disarmament, and what came to be called the "nuclear holocaust." Major developments prior to 1976 included: the 1945–1949 diplomatic effort to place nuclear technology under international control (Noel-Baker, 1958); the 1956–1963 campaign for nuclear disarmament, which led to the Moscow atmospheric test-ban treaty; and the diplomatic and political efforts leading to the 1968 Nonproliferation Treaty (NPT) and the 1971 Strategic Arms Limitation Treaty (SALT I) with its antiballistic missile limitation (Craig and Jungerman, 1986). In all these efforts the word "nuclear" meant "nuclear weapons."

Why had perception and acceptance of nuclear power been "missed rather badly"? A plausible reason is simply this: by 1976 much of the public knew nuclear technology in terms of nuclear weaponry and, unlike the experts, had not seen nuclear power as fundamentally different. As nuclear power matured and the public

learned more about it, the public distrust did not subside. The problem of public acceptance is even more critical today. Public support
for nuclear power has declined steadily for a decade and a half (see
Figure 6.8), driven by a number of powerful forces and events. In
mid-March of 1979, the movie *The China Syndrome* had its premiere,
dramatizing the worst-case predictions of WASH-740. Two weeks
later, events at Three Mile Island made the movie appear prophetic.
Succeeding years have brought Chernobyl and other major technological disasters, most notably Bhopal and the *Challenger* accident.
The public has drawn a common message from these accidents:
nuclear (and other) technology is unsafe; expertise is inadequate;
government and industry cannot be trusted to manage nuclear
power safely. These dramatic accidents have been accompanied and
reinforced by numerous chronic problems involving radiation, such
as the discovery of significant radon concentrations in many homes,
the continuing battles over the siting of facilities to monitor and store
nuclear wastes, and the disclosures of serious environmental contamination emanating from nuclear weapons facilities (at Hanford in
Washington, Fernald in Ohio, Rocky Flats in Colorado, and Savannah River in South Carolina).

FIGURE 6.8. SHIFT IN PUBLIC ATTITUDE TOWARD NUCLEAR ENERGY IN THE
YEARS 1975–1986.

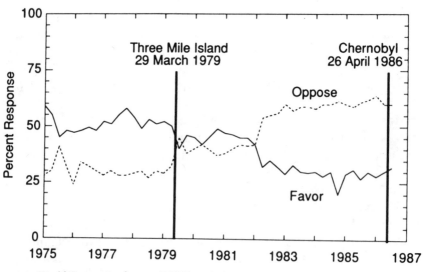

SOURCE: World Energy Conference (1989).

The structure of public acceptance in the United States can be gauged from public opinion polls and voting behavior, which was most recently summarized in the report of the World Energy Conference (1989:17–18):

1. In the early phase from 1970 to 1979 the proponents outweighed the opponents by 2:1, but many citizens were not yet decided. . . .
2. After TMI public opposition increased considerably, but did not form the majority of the population (with the exception of the immediate post-TMI period). The major change was the forming of an attitude . . . and a polarization between pro- and anti-nuclear groups.
3. After 1981 public opposition . . . increased slowly, but steadily. The reasons for this . . . change are not quite clear.

The report notes that these opinion trends are strongly dependent on the form and context of the question. Thus all fifteen referenda on nuclear energy in the United States proposing limitations on existing plants have been defeated (World Energy Conference, 1989), even while four referenda proposing limitations on new construction were approved (OTA, 1984). Moreover, polling questions referring to safety concerns invariably produce nuclear opposition greater than 50 percent.

Not surprisingly, the nature and determinants of public attitudes and perceptions regarding nuclear power have been the focus of considerable research. The psychometric approach to studying risk perception (Slovic, 1987) assumes that hazards can be characterized in terms of numerous characteristics analogous to the personality dimensions that characterize people. Nuclear power has a special distinction in the perception literature—it is, to date, the technological hazard with the most negative and most problematic constellation of traits. It stands apart in having qualities that make it fearsome and hard to manage socially and politically.

The mapping of nuclear power's "personality" began in the mid-1970s with a series of psychometric studies designed to understand why people were very concerned about certain hazards but not others and why these concerns often differed from experts' assessments of risk (Fischhoff et al., 1978; Slovic, Fischhoff, and Lichtenstein, 1980). An early study assessed perceived risk of death (for the United States as a whole) from thirty activities and technologies. Three groups of laypeople and a small group of risk assessment professionals took part in the study (Slovic, Lichtenstein, and Fischhoff, 1979). The results demonstrated the great concern regarding nuclear power (it had the highest perceived risk for two of the lay groups) and the discrepancy between the perceptions of laypeople and the perceptions of experts (whose ratings placed nuclear power twentieth on the list of thirty hazards). Some have argued that public concern about nuclear power reflects a concern about radiation risk in general, but the groups of laypeople in this study rated X-rays rather low in risk (seventeenth to twenty-fourth) and the experts rated them relatively high—seventh. Apparently it is not just radiation per se that was a concern to these people, but radiation as one may get exposed to it through the technology of nuclear power.

In an attempt to go beneath the surface of these global judgments, respondents were also asked to rate nuclear power, X-rays, and other hazards on a number of dimensions presumed relevant to perception and acceptance of risk. These ratings showed that nuclear power had the dubious distinction of scoring at or near the extreme negative end for most of the characteristics. Its risks were seen as involuntary, unknown to those exposed or to science, uncontrollable, unfamiliar, catastrophic, severe (fatal), and dreaded. Medical X-rays, in contrast, had a much more benign profile.

Nuclear power's perceived benefits were assessed, as well, and were found to be extremely low. These results have since been replicated with many different populations in numerous countries (Englander et al., 1986; Gould et al., 1988; Keown, 1989; Morgan et al., 1985; Rosa and Kleinhesselink, 1989; Teigen, Brun, and Slovic, 1988). Perceptions of risk and benefit associated with nuclear waste and its management are similarly negative (Kunreuther, Desvousges, and Slovic, 1988); Mountain West Research, 1989; Nealey and Hebert, 1983; Slovic et al., 1991). When asked to state whatever images or associations came to mind when they heard the words "underground nuclear waste storage facility," a representative sample of Phoenix, Arizona, residents could hardly think of anything that was not frightening or problematic (see Table 6.2). The disposal of nuclear waste is a technology that industry experts believe can be managed safely and effectively. The discrepancy between this view and the images shown in Table 6.2 is indeed startling.

TABLE 6.2. HIERARCHY OF IMAGES ASSOCIATED WITH AN
"UNDERGROUND NUCLEAR WASTE STORAGE FACILITY"

| Category | Frequency | Images included in category |
|---|---|---|
| Dangerous | 179 | dangerous, danger, hazardous, toxic, unsafe, disaster |
| Death | 107 | death, sickness, dying, destruction |
| Negative | 99 | negative, wrong, bad, unpleasant, terrible, gross, undesirable, awful, dislike, ugly, horrible |
| Pollution | 97 | pollution, contamination, leakage, spills, Love Canal |
| War | 62 | war, bombs, nuclear war, holocaust |
| Radiation | 59 | radiation, nuclear, radioactive, glowing |
| Scary | 55 | scary, frightening, concern, worried, fear, horror |
| Somewhere else | 49 | wouldn't want to live near one, not where I live, far away as possible |
| Unnecessary | 44 | unnecessary, bad idea, wasteland |
| Problems | 39 | problems, trouble |
| Desert | 37 | desert, barren, desolate |
| Non-Nevada locations | 35 | Utah, Arizona, Denver |
| Storage locations | 32 | caverns, underground salt mine |
| Government/industry | 23 | government, politics, big business |

SOURCE: Slovic et al. (1991).

The perception of nuclear power as a catastrophic technology was studied in depth by Slovic, Lichtenstein, and Fischhoff (1979). They found that, even before the TMI accident, people expected nuclear-power accidents to lead to disasters of immense proportions. When asked to describe the consequences of a "typical reactor accident," people's scenarios were found to resemble conditions in the aftermath of nuclear war. Replication of these studies after the TMI event found even more extreme "images of disaster."

People's fears about nuclear energy appear to be deeply rooted in our social and cultural consciousness. Weart (1988) argues that modern thinking about nuclear energy employs beliefs and symbols that have been associated for centuries with the concept of transmutation—the passage through destruction to rebirth. In the early decades of the twentieth century, transmutation images became centered on radioactivity, which was associated with "uncanny rays that brought hideous death or miraculous new life; with mad scientists and their ambiguous monsters; with cosmic secrets of life and death; . . . and with weapons great enough to destroy the world" (Weart, 1988:421).

The bombing of Hiroshima and Nagasaki linked this belief structure to reality. The sprouting of nuclear power in the aftermath has led Smith (1988:62) to observe: "Nuclear energy was conceived in secrecy, born in war, and first revealed to the world in horror. No matter how much proponents try to separate the peaceful from the weapons atom, the connection is firmly embedded in the minds of the public." During the past decade, research has shown that individual risk perceptions and cognitions, interacting with social and institutional forces, can trigger massive social, political, and economic impacts. Early theories equated the magnitude of impact to the number of people killed or injured, or to the amount of property damaged. The accident at TMI, however, provided a dramatic demonstration that factors other than injury, death, and property damage impose serious costs. Despite the fact that not a single person died at TMI, and few if any latent cancer fatalities are expected, no other accident in our history has produced such costly social impacts (Evans and Hope, 1984; Heising and George, 1986). Apart from its impact on the utility that owned and operated the plant, it also imposed enormous costs on the nuclear industry and on society. These costs came through stricter regulation, reduced operation of reactors worldwide, greater public opposition to nuclear power, reliance on more expensive energy sources, and increased costs of reactor construction and operation.

A theory aimed at describing how psychological, social, cultural, and political forces interact to "amplify risk" and produce ripple effects has been presented by Kasperson et al. (1988). An important element of this theory is the assumption that the perceived seriousness of an accident or other unfortunate event, the media coverage it gets, and the long-range costs and other higher-order impacts on the responsible company, industry, or agency are determined, in part, by what the event signals or portends. Signal value reflects the perception that the event provides new information about the likelihood of similar or more destructive future mishaps.

The informativeness or signal value of an event—and thus its potential social impact—appears to be systematically related to the characteristics of the hazard. An accident that takes many lives may produce relatively little social disturbance (beyond that caused the victims' families and friends) if it occurs as part of a familiar and well-understood system (say, a train wreck). But a small accident in an unfamiliar system (or one perceived as poorly understood), such as a nuclear reactor, may have immense social consequences if it is perceived as a harbinger of further and possibly catastrophic mishaps. The concept of accidents as signals helps explain our society's strong response to problems involving nuclear power and nuclear waste. Because these nuclear hazards are seen as poorly understood and catastrophic, accidents anywhere may be seen as omens of future disasters everywhere, thus producing resounding socioeconomic and political impacts.

## THE FUTURE OF NUCLEAR POWER

As already noted, one of the strongest arguments driving a revival of nuclear power is the fact that it can substitute for fossil fuels and reduce their most undesirable environmental impacts, such as acid precipitation and the potential for $CO_2$-induced climate change. While many environmentalists argue that conservation is preferable, the role of nuclear power in mitigating the greenhouse effect has become an undeniable issue worldwide, even though the greenhouse effect receives scant mention in President Bush's national energy strategy (DOE, 1991). It is important, therefore, to examine the necessary conditions for a significant nuclear revival.

If a substantial nuclear contribution is required to counter $CO_2$ buildup, the issue is arguable on grounds of probabilistic risk assessment alone. Assume, for example, that to slow climate change it is necessary to produce one-half of the world's future energy requirement via nuclear energy. Then, with $p_c = 5 \times 10^{-5}$ per reactor year (NRC, 1975), one projects one core-melt accident every 6 years. This is almost surely unacceptable. Thus one concludes that a necessary condition for deployment of nuclear reactors on a scale sufficient to contribute significantly to mitigation of the greenhouse effect is reduction of the core-melt probability considerably below Rasmussen's fiducial figure (Forsberg and Weinberg, 1990).

Without reference to global warming, it is eminently justifiable to improve nuclear safety based on the dismal state of public acceptance summarized here. In this case it is quite unclear what level of improvement will be needed. It seems important, however, that new safety efforts be clear and understandable to the general public, particularly the "skeptical elites" who influence and represent them (Weinberg, 1989; Forsberg and Weinberg, 1990).

## PASSIVELY STABLE REACTORS

To improve safety a number of proposals have been made, including designs with "inherent safety" that depend on the laws of physics rather than engineered interventions or human operators. In reviewing these options, Forsberg and Weinberg (1990) question the concept of inherent safety because with large fission-product inventories even the safest reactors can fail or be made to fail. Instead they prefer, with Taylor (1989), the concept of "passively stable" to describe designs that bypass interventions in order to assure that temperature, emission of radioactivity, and other parameters remain within acceptable levels. They consider several distinct light-water designs in their review of current options. Beginning with reactors directly derived from present designs, these options proceed in increasing novelty to fundamentally altered plants. In all cases the new plants are smaller than today's standard 1000-MWe reactors.

*Evolutionary designs* involve refinements and modernization, but no change in the basic light-water reactor (LWR) design. Their safety, like that of their predecessors, depends on a variety of active safety systems for which external generators must be operational throughout the entire accident sequence. The main advantage of evolutionary designs is that they are based on extensive experience.

*Evolutionary technology* involves redesign based on the principles of current LWR technology. This approach allows safety systems that require external power for initiation, but not for continued operation. Evolutionary technology is currently being pursued at General Electric and Westinghouse under U.S. government contract and has six things in common: water for heat removal in the primary system drains by gravity to the reactor core; large auxiliary power sources for emergency equipment are eliminated; emergency core flooding is assured by gravity flow of water that is stored above the core; cooling reactor containment during core-melt accidents is provided by passive systems; reduced power densities assure an increased "window" of safe operations; design simplification avoids operator/maintenance error. Overall, evolutionary technology permits plant simplification because designs are new and not just modifications of existing plants.

*PRIME designs*, in contrast to evolutionary technology, have no active safety systems. (PRIME is an acronym for "passive safety, resilient operation, inherent safety, malevolence resistance, and extended safety." Resilient operation refers to safety systems that do not interfere with normal maintenance; malevolence resistance describes the capability to withstand sabotage.) There are three types of PRIME reactors, each involving a distinctive means of decay heat removal during emergencies. In the Swedish PIUS design, typically limited to 700 MWe, the process occurs through boiloff of a large volume of water in the reactor vessel; in the German modular high-temperature gas-cooled reactor, limited to 135 MWe per module, the process occurs via conduction to the ground and surrounding air; for the molten-salt breeder and various aqueous-fueled models, the process depends on small quantities of fuel that are continuously processed.

So far, though the Department of Energy is supporting two evolutionary designs and two designs that include passive safety features (DOE, 1991), no advanced reactor has reached licensing stage (Forsberg and Weinberg, 1990) and none is expected to be operational until the end of the century. Although the economic viability of the second generation cannot be judged at this point, the technical improvements envisioned may avoid the economic disasters of the first generation by facilitating speedy construction with favorable capital, operating, maintenance, and fuel costs. Beyond improved safety and economics, issues for the new designs include high-reliability plants with a projected lifetime of perhaps 60 years and a design that is standardized and predictably licensable (Taylor, 1989).

A Second Coming

If improved safety, high reliability, and standardization are all achieved, it is conceivable that nuclear electricity will overcome its current economic, public acceptance, regulatory, and perceived safety problems. Among the potentially outstanding achievements of the new designs would be their immunity to catastrophic accidents (thus obviating the need for extensive emergency planning) and, through these innovations, an end to complex licensing processes. As noted in the new national energy strategy (DOE, 1991), achieving these benefits is currently the hope of the U.S. government. What, then, might prevent a rebirth of nuclear power?

First, second-generation nuclear plants are still in the research phase, and we do not yet know whether or when their goals can be achieved. Thus no new-generation plants can go on line at least until the end of the century (DOE, 1991). Second, new designs do not now receive the bulk of federal funding, which remains focused on improving the 110 first-generation plants in the field. Third, the new generation may differ in nuclear plant design but will have the same fuel cycle problems that were for a long time neglected in the first generation and are nowhere near solution today, including waste management and decommissioning. Fourth, it is not obvious that economic problems are solved by standardization and regulatory reform. One needs, in addition, to worry about the rate at which technology is installed and how the utilities are protected against technological surprises.

Perhaps most important, it is not likely that public trust, confidence, and acceptance can easily be regained. Although the consensus opinion of technical experts continues to assert that nuclear wastes can be sequestered with essentially no chance of any member of the public receiving a significant dose of radiation (Weinberg, 1989; Colglazier and Langum, 1988), public perceptions do not reflect this view. Indeed, they fall at the extreme opposite of the range of risk (Slovic et al., 1991).

How could the public come to believe that born-again nuclear power is inherently safe and acceptable? Weinberg (1989) argues that special-interest environmental groups (skeptical elites) could turn the tide of public opinion by siding with nuclear power as a solution to the greenhouse problem. It appears more likely, however, that environmentalists will embrace conservation and energy efficiency rather than nuclear power and that the public in general will continue to reject the proposed national repository for the disposal of high-level nuclear wastes. In their discussion of passively stable reactors, Forsberg and Weinberg (1990: 150–151) therefore conclude cautiously: "Proponents of passively stable reactors cannot prove that they are either necessary or sufficient for rebirth of nuclear energy. On the other hand, enough political support seems to have developed in favor of passively stable reactors that it seems likely that several such devices will be built and operated during this decade."

Beyond rebirth, an old problem dating to the genesis of nuclear weapons may rear its head. To assure an adequate fuel supply, as noted by Williams and Feiveson (1990), worldwide uranium shortages will eventually provide strong motives for reprocessing spent nuclear fuel into plutonium, making it widely available as a stimulant for worldwide nuclear weapons proliferation. So far we have been saved from this prospect because of the relatively slow development of civilian nuclear power, the establishment of substantial international monitoring, and the restriction of substantial reprocessing to a few countries (Albright and Feiveson, 1988). Nevertheless, born-again nuclear power may accelerate the proliferation threat and require unprecedented institutional changes worldwide.

## THE BUSINESS AT HAND: EXISTING PLANTS

Of more immediate concern are the existing lot of 110 U.S. plants, which, despite rapid nuclear developments in other nations, still constitute the single largest national investment in nuclear electricity worldwide. As our review has shown, the existing plants present society with a unique set of regulatory and assessment problems that demand to be addressed in a manner that at the very least takes into account past errors in siting, the limits of PRA, inconsistencies in the licensing process, the hazards of on-site waste storage, and the need to eliminate chronically troublesome plants. In this process it seems plausible that the federal government must take responsibility for nonviable utilities much as it has taken steps to bail out mismanaged savings and loan organizations. Some actions consistent with this philosophy might be the following:

1. An agreement to dismantle or put to nonnuclear use certain plants because their original construction licenses were errors that cannot adequately be corrected, particularly with respect

to emergency evacuation. Seabrook and Shoreham may be current examples, just as Indian Point I was a past example.

2. An agreement to dismantle chronically troublesome plants in which major structural or design problems are the cause of long-term operational disruption. Pilgrim may be an example.

3. An agreement to extend present NRC efforts to implement and maintain operating reforms where there has been chronic mismanagement of the plant.

4. An agreement on a systematic hazard management approach that limits the amount of on-site spent fuel storage in "swimming pools" (or alternatively dry-cask storage) so that present-day safety is not held hostage to delays in siting a permanent national spent-fuel storage facility.

5. An agreement in which emergency planning guidelines are based on an accepted "worst case" that depends neither on legal precedent nor on PRA quantification, but on consequence studies corresponding to the intermediate case of WASH-740 and its update.

In short, what is needed prior to the rebirth of nuclear power is radical surgery on an "out-of-control" regulatory system and major confidence-building actions including the physical removal or retirement of particularly troublesome or faulty systems. Even so, such reform of civilian nuclear power must be incomplete since it does not address the remarkable technical and process failures in the seventeen nuclear weapons plants in twelve states that were until recently shrouded in secrecy. The bill for this failure according to the Department of Energy is now expected to be in excess of $100 billion (Brumley, 1988) and should make some of the errors in the civilian nuclear reactor program appear benign.

Can the radical surgery we envision neutralize the issue of public acceptance, especially when the civilian and military pictures coalesce in the public's mind? No one can be sure. But we can be sure that without an attempt to correct the built-in errors of the present system, neither public acceptance nor a rebirth of civilian nuclear power in the United States will be achieved.

# NOTE

1. One kilowatt (kW) = 1000 watts; one megawatt (MW) = 1,000,000 watts; one gigawatt (GW) = 1,000,000,000 watts. GWe refers to *electrical* energy; GWt refers to *thermal* energy.

# REFERENCES

Adato, M., et al. 1987. *Safety Second.* A report by the Union of Concerned Scientists. Indianapolis: University of Indiana Press.

Albright, D., and H. A. Feiveson. 1988. Plutonium recycling and the problem of nuclear proliferation. *Annual Review of Energy* 13:239–265.

American Nuclear Society (ANS). 1984. *Report of the Special Committee on Source Terms.* Lagrange Park, Ill.: American Nuclear Society.

Bernero, R. M. 1984. Probabilistic risk analyses: NRC programs and perspectives. *Risk Analysis* 4(4):287–298.

Beyea, J. 1989. Is there a role for nuclear power in preventing climate disruption? Testimony before the House Subcommittee on Energy and Power, 15 March.

Bromet, E., et al. 1980. *Three Mile Island: Mental Health Findings.* Pittsburgh: Western Psychiatric Institute, University of Pittsburgh.

Brumley, B. 1988. Story on nuclear weapons plants. Associated Press, 13 Nov.

Budnitz, R. J. 1984. External initiators in probabilistic reactor accident analysis—earthquakes, fires, floods, winds. *Risk Analysis* 4(4):323–335.

Carter, L. J. 1987. *Nuclear Imperatives and Public Trust: Dealing with Radioactive Wastes.* Washington: Resources for the Future.

Colglazier, E. W., and R. B. Langum. 1988. Policy conflict in the process for siting nuclear waste repositories. *Annual Review of Energy* 13:317–377.

Cook, E. 1971. The flow of energy in an industrial society. *Scientific American* 224(3):134–144.

Council of Economic Advisors. 1989. *Economic Report of the President.* Washington: Government Printing Office.

Craig, P., and J. A. Jungerman. 1986. *Nuclear Arms Race.* New York: McGraw-Hill.

Daniels, T. A., and K. S. Canady. 1984. A nuclear utility's views on the use of probabilistic risk assessment. *Risk Analysis* 4(4):281–286.

Davidon, W. C., M. I. Kalkstein, and C. Hohenemser. 1960. *The Nth Country Problem and Arms Control*. Washington: National Planning Association.

Davidson, L. M., R. Fleming, and A. Baum. 1987. Chronic stress, catecholamines and sleep disturbance at Three Mile Island. *Journal of Human Stress* 13(2):75–83.

Dennis, J. (ed.). 1984. *Nuclear Almanac*. Reading, Mass.: Addison-Wesley.

Eisenbud, M. 1987. *Environmental Radioactivity*. Orlando: Academic Press.

———. 1989. Exposure of the general public near Three Mile Island. *Nuclear Technology* 87:514–519.

Englander, T., K. Farago, P. Slovic, and B. Fischhoff. 1986. A comparative analysis of risk perception in Hungary and the United States. *Social Behavior* 1:55–66.

Evans, N., and C. Hope. 1984. *Nuclear Power: Futures, Costs, and Benefits*. Cambridge: Cambridge University Press.

Ferguson, J. F. 1987. A case for internal funding. *Forum for Applied Research and Public Policy* 2(4):18–20.

Fischhoff, B., P. Slovic, S. Lichtenstein, S. Reid, and B. Combs. 1978. How safe is enough? A psychometric study of attitudes towards technological risks and benefits. *Policy Studies* 9:137–152.

Flynn, C. B. 1979. *Three Mile Island Telephone Survey*. NUREG/CR-1093. Washington: Nuclear Regulatory Commission.

Forbes, I. A., D. F. Ford, H. W. Kendall, and J. J. McKenzie. 1972. Cooling water. *Environment* 14(1):40–47.

Ford, D. F. 1977. *The History of Federal Nuclear Safety Assessment: From WASH 740 Through the Reactor Safety Study*. Cambridge, Mass.: Union of Concerned Scientists.

Forsberg, C. W., and A. M. Weinberg. 1990. Advance reactors, passive safety, and acceptance of nuclear energy. *Annual Review of Energy* 15:133–152.

Garrick, B. J. 1984. Recent case studies and advancements in probabilistic risk assessment. *Risk Analysis* 4(4):267–280.

Gerulsky, T. M. 1981. Three Mile Island: Assessment of radiation exposure and environmental contamination. *Annals of the N.Y. Academy of Science* 365:54–62.

Goldberg, J. F. (ed.). 1985. *Proceedings of the Workshop on Three Mile Island Dosimetry*. 2 vols. Philadelphia: Three Mile Island Public Health Fund.

Golding, D., and R. E. Kasperson. 1988. Emergency planning and nuclear power: Looking to the next accident. *Land Use Policy* 5:19–36.

Gould, L. C., G. T. Gardner, D. R. Deluca, A. R. Tiemann, L. W. Doob, and J. A. J. Stolwijk. 1988. *Perceptions of Technological Risks and Benefits*. New York: Russell Sage Foundation.

Hafele, W. 1985. A case study in the dynamics of technological evolution.

Presentation at the American Nuclear Society's Executive Conference on TMI-2, Hershey, Pa., Oct.

Heising, C. D., and V. P. George. 1986. Nuclear financial risk: Economy-wide costs of reactor accidents. *Energy Policy* 14:45–52.

Hendron, C. B. 1987. A case for external funding. *Forum for Applied Research and Public Policy* 2(4):14–17.

Hohenemser, C. 1962. The Nth country problem today. In *Disarmament, Its Politics and Economics*, ed. S. Melman. Boston: American Academy of Arts and Sciences.

———.1988. The accident at Chernobyl: Health and environmental consequences and implications for risk management. *Annual Review of Energy* 13:383–428.

———. 1989. Nuclear power in the greenhouse age. *Sanctuary* (Mass. Audubon Society), Sept., pp. 18–20.

Hohenemser, C., R. E. Kasperson, and R. W. Kates. 1977. The distrust of nuclear power. *Science* 196:25–34.

———. 1982. Causal structure: A framework for policy formulation. In *Risk in the Technological Society*, ed. C. Hohenemser and J. X. Kasperson. Boulder: Westview Press.

Hohenemser, C., R. W. Kates, and P. Slovic. 1983. The nature of technological hazard. *Science* 220:378–384.

Hohenemser, C., and M. Leitenberg. 1967. The nuclear test ban negotiations: 1957–1967. *Scientist and Citizen* 11/12:201–218.

Hohenemser, C., and O. Renn. 1988. Chernobyl's other legacy. *Environment* 30(3):5–11.

Holdren, J. P. 1982. Energy hazards: What to measure and what to compare. *Technology Review* 85(3):32–38.

Holdren, J. P., K. Anderson, P. H. Gleick, I. Mintzer, G. Morris, and K. W. Smith. 1979. Risk of renewable energy sources: A critique of the Inhaber report. ERG 79-3. Berkeley: Energy Resources Group, University of California.

Houghton, Richard A., and George M. Woodwell. 1989. Global climatic change. *Scientific American* 260(4):36.

Inhaber, H. 1979. *Risk of Energy Production*. Report AECB 1119, rev. 3, 4th ed. Ottawa: Atomic Energy Control Board.

Joksimovich, V. 1984. A review of plant specific PRAs. *Risk Analysis* 4(4):255–266.

Kasperson, R. E., O. Renn, P. Slovic, H. S. Brown, J. Emel, R. Goble, J. X. Kasperson, and S. Ratick. 1988. The social amplification of risk: A conceptual framework. *Risk Analysis* 9:177–204.

Kates, R. W., J. X. Kasperson, and C. Hohenemser (eds.). 1985. *Perilous Progress*. Boulder: Westview Press.

Kemeny, J. G., B. Babbit, P. E. Haggerty, C. Lewis, P. A. Marks, et al. 1979. *Report of the President's Commission on the Accident at Three Mile Island.* Washington: Government Printing Office.

Keown, C. F. 1989. Risk perceptions of Hong Kongese vs. Americans. *Risk Analysis* 9:401–405.

Kunreuther, H., W. H. Desvousges, and P. Slovic. 1988. Nevada's predicament: Public perceptions of risk from the proposed nuclear waste repository. *Environment* 30(8):16–20.

*Las Vegas Review Journal.* 1990. Public opinion poll in October.

Levine, S., and N. C. Rasmussen. 1984. Nuclear plant PRA: How far has it come? *Risk Analysis* 4(4):247–254.

Lewis, H. W., R. J. Budnitz, R. J. Kouts, et al. 1979. *Risk Assessment Review Group Report to the U.S. Nuclear Regulatory Commission.* Ad hoc risk assessment review report NUREG/CR-0400. Springfield, Va.: NTIS.

Lewis, H. W., R. J. Budnitz, A. W. Castleman, D. E. Dorfan, S. E. Finlayson, et al. 1975. Report to the American Physics Society by the Study Group on Light Water Reactor Safety. *Review of Modern Physics* 47(Suppl. 1):S1–S123.

Lovins, A. B., L. H. Lovins, F. Krause, and W. Bach. 1981. *Least-Cost Energy: Solving the $CO_2$ Problem.* Andover: Brickhouse.

Malko, J. R. 1987. Financing nuclear power plant decommissioning. *Forum for Applied Research and Public Policy* 2(4):5–13.

Medvedev, G. 1990. *The Truth About Chernobyl.* New York: Basic Books.

Morgan, M. G., P. Slovic, I. Nair, D. Geisler, D. MacGregor, B. Fischhoff, D. Lincoln, and K. Florig. 1985. Powerline frequency electric and magnetic fields: A pilot study of risk perception. *Risk Analysis* 5:139–149.

Moss, T. H., and D. L. Sills (eds.). 1981. *The Three Mile Island Nuclear Accident—Lessons and Implications: Report of a Conference Held on April 8–10 by the New York Academy of Sciences.* New York: New York Academy of Sciences.

Mountain West Research. 1989. *Interim Report on the State of Nevada Socioeconomic Studies.* Report NWPO-SE-022-89. Carson City: Nuclear Waste Project Office, State of Nevada.

Mulvihill, R. J., D. R. Arnold, C. E. Bloomquist, and B. Epstein. 1965. Analysis of United States power reactor accident probability. PRC R-695. Los Angeles: Planning Research Corporation. (Unpublished draft from the file "WASH 740 update," Public Documents Room, Nuclear Regulatory Commission, Washington.)

NAS/NRC. 1980. *Report of the Committee on the Biological Effects of Ionizing Radiation (BEIR III).* Washington: National Academy Press.

_____. 1990. *Report of the Committee on the Biological Effects of Ionizing Radiation (BEIR V).* Washington: National Academy Press.

Nealy, S. M., and J. A. Herbert. 1983. Public attitudes toward radioactive wastes. In *Too Hot to Handle: Social and Policy Issues in the Management of Radioactive Wastes*, ed. C. A. Walker, L. C. Gould, and E. J. Woodhouse. New Haven: Yale University Press.

*New York Times*. 1989. U.S. will start over on planning for Nevada nuclear waste dump. 29 Nov.

Noel-Baker, P. 1958. *The Arms Race*. London: Atlantic.

*Nuclear News*. 1989. 32(10):77–96.

Okrent, D. 1981. *Nuclear Reactor Safety*. Madison: University of Wisconsin Press.

Pollock, C. 1987. Decommissioning: An unresolved problem. *Forum for Applied Research and Public Policy* 2(4):23–30.

Rhodes, R. 1986. *The Making of the Atomic Bomb*. New York: Simon & Schuster.

Rosa, E., and R. Kleinhesselink. 1989. A comparative analysis of risk perceptions in Japan and the United States. Paper presented at the annual meeting of the Society for Risk Analysis, San Francisco.

Sanderford, P. R. 1987. A case in point: T.V.A. *Forum for Applied Research and Public Policy* 2(4):21–22.

Schneider, Stephen H. 1989. The greenhouse effect: Science and policy. *Science* 243:771–781.

Sholly, S., and G. Thompson. 1986. *The Source Term Debate*. Cambridge, Mass.: Union of Concerned Scientists.

Slovic, P. 1987. Perception of risk. *Science* 236:280–285.

Slovic, P., B. Fischhoff, and S. Lichtenstein. 1980. Facts and fears: Understanding perceived risk. In *Societal Risk Assessment: How Safe Is Safe Enough?*, ed. R. Schwing and W. A. Albers, Jr. New York: Plenum.

Slovic, P., J. H. Flynn, and M. Layman. 1991. Perceived risk, trust, and the politics of nuclear waste. *Science* 254:1603–1607.

Slovic, P., M. Layman, and J. Flynn. 1991. Risk perception, trust, and nuclear waste: Lessons from Yucca Mountain. *Environment* 33:6–11, 28–30.

Slovic, P., S. Lichtenstein, and B. Fischhoff. 1979. Images of disaster: Perception and acceptance of risks from nuclear power. In *Energy Risk Management*, ed. G. Goodman and W. Rowe. London: Academic Press.

Slovic, P., M. Layman, N. Kraus, J. Flynn, J. Chalmers, and G. Gesell. 1991. Perceived risk, stigma, and potential economic impacts of a high-level nuclear waste repository in Nevada. *Risk Analysis* 11:683–696.

Smith, K. R. 1988. Perception of risks associated with nuclear power. *Energy Environment Monitor* 4(1):61–70.

Spinrad, B. 1988. U.S. nuclear power in the next twenty years. *Science* 239:707.

Taylor, John J. 1989. Improved and safer nuclear power. *Science* 244:318–324.

Teigen, K. H., W. Brun, and P. Slovic. 1988. Societal risks as seen by a Norwegian public. *Journal of Behavioral Decision Making* 1:111–130.

U.S. Atomic Energy Commission (AEC). 1950. *Summary Report of Reactor Safeguards Committee*. WASH-3. Washington: Atomic Energy Commission.

————. 1957. *Theoretical Possibilities and Consequences of Major Accidents in a Large Nuclear Power Plant*. WASH-740. Washington: Atomic Energy Commission.

U.S. Congress. House Committee on Energy and Commerce. Subcommittee on Energy Conservation and Power. 1987. *Hearings on Emergency Planning at the Seabrook Nuclear Power Plant*. 99th Cong., 2nd sess.

————. Office of Technology Assessment (OTA). 1984. *Nuclear Power in an Age of Uncertainty*. Washington: U.S. Congress.

U.S. Department of Energy (DOE). 1987a. *Energy Security: A Report to the President of the United States*. DOE/S-0057. Washington: Department of Energy.

————. 1987b. *Health and Environmental Consequences of the Chernobyl Nuclear Power Plant Accident*. DOE/ER-0332. Springfield, Va.: NTIS.

————. 1991. *National Energy Strategy 1991/92*. Springfield, Va.: NTIS.

U.S. General Accounting Office (GAO). 1989. *Electricity Supply: What Can Be Done to Revive the Nuclear Option?* GAO/RCED-89-67. Washington: General Accounting Office.

U.S. Nuclear Regulatory Commission (NRC). 1975. *Reactor Safety Study*. WASH-1400, NUREG-75/014. Washington: Nuclear Regulatory Commission.

————. 1987a. *Reactor Risk Reference Document*. NUREG-1150. Washington: Nuclear Regulatory Commission.

————. 1987b. *Report on the Accident at the Chernobyl Nuclear Power Station*. NUREG-1250. Washington: Government Printing Office.

U.S. Nuclear Regulatory Commission and Environmental Protection Agency (NRC/EPA). 1978. *Planning Basis for the Development of State and Local Government Radiological Emergency Response Plans in Support of Light Water Nuclear Power Plants*. NUREG-0396. Washington: NTIS.

U.S. Nuclear Regulatory Commission and Federal Emergency Management Agency (NRC/FEMA). 1980. *Criteria for Preparation and Evaluation of Radiological Emergency Response Plans and Preparedness in Support of Nuclear Power Plants*. NUREG-0654, rev. 1. Washington: NTIS.

USSR. State Committee on the Utilization of Atomic Energy. 1986. *The Accident at the Chernobyl Nuclear Power Plant and Its Consequences*. Report compiled for the IAEA experts meeting, 25–29 August 1986. Vienna: International Atomic Energy Agency.

Vesely, W. E., and D. M. Rasmussen. 1984. Uncertainties in nuclear probabilistic risk analyses. *Risk Analysis* 4(4):313–322.

Weart, S. 1988. *Nuclear Fear: A History of Images.* Cambridge: Harvard University Press.

Weinberg, A. M. 1976. The maturity and future of nuclear energy. *American Scientist* 64:16–21.

————. 1989. Public perceptions of hazardous technologies and democratic political institutions. *Proceedings of Waste Management '89.* Tucson: University of Arizona.

Williams, R. H., and H. A. Feiveson. 1990. How to expand nuclear power without proliferation. *Bulletin of Atomic Scientists* 46(4):40–45.

Wilson, C. L. 1979. Nuclear energy: What went wrong? *Bulletin of Atomic Scientists* 35(6):13–18.

Wilson, R., K. Araj, A. Allen, P. Auer, D. Boulware, et al. 1985. Report to the American Physical Society of the Study Group on Radionuclide Release from Severe Accidents at Nuclear Power Plants. *Review of Modern Physics* 57:S1–S154.

World Energy Conference. 1989. *Report.* Vol. 1. London: World Energy Conference.

CHAPTER 7

# BIOMASS ENERGY

*Janos Pasztor and Lars Kristoferson*

WELL-MANAGED BIOMASS ENERGY SYSTEMS can form part of an energy
supply regime that is environmentally sound and sustainable. The
overall environmental impacts of bioenergy systems may be less than
those of conventional fossil fuels, since they are many in number but
localized and relatively small individually. The environmental impacts of bioenergy systems are thus more controllable, more revers-
ible, and consequently more benign. Furthermore, reliance on
bioenergy may often have beneficial side-effects, both locally and
globally.

Bioenergy systems, like all other energy systems, have an environmental impact. The impact often depends more on the way the
whole system is managed than merely on the fuel or the conversion
technology. Strictly speaking, the term "impact" implies either an
adverse or a beneficial result of a process outside the formal framework for analysis. Cost is not an impact, for example, since costs are
calculated as part of the classic analysis of energy systems. But the
effect of harmful pollutants dispersed into the atmosphere is

The authors would like to acknowledge the help of Duncan Brown in preparing this chapter,
which is based on material in J. Pasztor and L. A. Kristoferson, eds., *Bioenergy and the
Environment* (Boulder: Westview Press, 1990).

"external"—in that it is not generally calculated in the analysis—and is therefore considered an impact. Similarly, an energy forest may provide benefits in addition to energy production, such as erosion control, green space, and aesthetic beauty—benefits that, again, are external to the conventional economic analysis. Furthermore, the impact may be positive when compared with the effects of another technology. This fact is sometimes particularly relevant to bioenergy systems when compared with conventional fossil-based systems.

The distribution of impacts—in both space and time—is also important. In the case of cooking on an open fire in an unvented hut, the impact (the indoor air pollution) is acting on the same people who are enjoying the benefits. In many cases, however, this is not true. In the combustion of most fossil fuels, for example, the air pollution often harms people or ecosystems at considerable distances away. In the case of biomass use without replanting, the carbon dioxide produced during combustion is not absorbed by the growing biomass. The resulting climate change will affect people throughout the globe and not just present, but also future, generations.

One issue particularly relevant to bioenergy systems is the potential benefits. These could include the benefits of avoiding, by substitution, an adverse impact of a strongly polluting energy system. Replacing a well-managed coal-fired boiler (using the best abatement technologies and best environmental practice throughout the fuel cycle) with a well-managed wood-fired boiler will result in certain improvements in air quality. In some cases, however, in addition to the avoidance of adverse effects, positive effects can also be produced. A well-managed energy forest, in addition to supplying wood for that boiler, could provide green space, absorb carbon dioxide, improve soils, increase wildlife populations, and yield fruits and building materials. The challenge, therefore, is to incorporate all the adverse and positive impacts of energy systems into the overall cost-benefit analysis as a contribution to energy planning.

All of this planning takes place in a context in which the development needs of the developing countries will require considerable increases in energy inputs, even in cases where more energy-efficient scenarios are adopted for the future. Given present levels of energy consumption, it is not conceivable that bioenergy systems could substitute for all the fuels now used in the world. Much bioenergy use today has serious environmental consequences. Overexploitation of biofuels can be very harmful. Yet bioenergy systems have their niche in the energy supply matrices of most countries, and this niche could become much larger than is exploited today. For bioenergy's share to increase, however, it must be managed well. For the moment, this is far from being the case. Indeed, some of the most serious environmental problems today have their origins in the current patterns of bioenergy use, not least in developing countries. The transition to sustainable energy production and use is therefore just as important for biofuels as it is for other energy forms. In this respect, bioenergy and other forms of energy share the same problems.


## TRADITIONAL BIOMASS SYSTEMS


Most people in the developing countries, and especially rural households and the urban poor, rely on traditional biomass resources for their energy needs: fuelwood, charcoal, and various forms of agricultural and animal wastes. These fuels provide about half the energy used in developing countries. In many countries, such as Burkina Faso, Ethiopia, Malawi, Tanzania, and Uganda, the proportion is as high as 90 percent (Beijer Institute, 1982; Leach, 1987). In rural areas, most traditional biomass resources are collected by the users or come from informal, noncommercial markets—in contrast to traditional urban and industrial biomass resources, which are almost entirely bought and sold in commercial markets.

Because of the site-specificity of traditional biomass systems, it is impossible to characterize them universally. Nevertheless, some generally valid points can be made.


### TRADITIONAL BIOMASS COMBUSTION AND DEFORESTATION

Wood fuel has been blamed by the conventional wisdom of the past two decades as the primary cause of deforestation and consequent desertification in developing countries. The information emerging from more detailed, site-specific studies, however, indicates otherwise. Although a large proportion of the trees felled do end up as fuel, the most important driving force for the cutting of the trees, and consequently for the disappearance of forests, seems to be the need to open up land for agriculture and grazing, followed by other commercial uses of trees, including commercial wood fuel consumption by urban and industrial users. Rural fuelwood requirements rarely cause deforestation. It has been said that "if all fuelwood and charcoal use stopped tomorrow, deforestation rates would hardly be altered" (Leach and Mearns, 1988). Therefore the "second energy crisis" due to fuelwood shortages must be looked at, not merely as an energy crisis, but rather as an aspect of crises in land use and development patterns.

Rural collection of traditional wood fuels rarely causes significant environmental damage. Many traditional practices of tree management (such as pruning) and combining tree growing with food producing (agroforestry) are environmentally sound and may often result in increased wood and crop yields. Furthermore, rural fuel users generally learn to adapt in order to relieve scarcities of fuel. The efficiency of fuel use, for example, can be easily doubled—and consumption thus halved—by such measures as building fires in more sheltered places that are shielded from drafts, building smaller fires, and quenching fuel when cooking is done in order to save it for later. Alternative fuels, such as crop residues and dung, can be used. Cooking practices and staple foods can be changed; the villagers of Nyamwigura in the Mara region of Tanzania, for example, have largely replaced their former staple, beans, which require 3 hours of cooking, with rice and cabbage, which take only 30 minutes (Tobisson, 1980). While these practices may lower living standards and impart their own environmental effects (for example, when use of crop residues for fuel increases soil erosion or deprives livestock of fodder), they demonstrate that much of the conventional wisdom about rural fuel "crises" in developing countries has been based on faulty assumptions.

Exploitation of forests and woodlands by commercial loggers, as well as by expanding agricultural and grazing interests, often does have adverse impacts. Once the trees are removed, often forest soils are unable to sustain agriculture and become completely exhausted and barren after a few years. This, however, is not an energy issue but an agricultural one. The charcoal and fuelwood requirements of urban and industrial users usually result in major adverse environmental impacts. Forests and woodlands are indiscriminately cleared, without replanting, often in concentrated areas near major urban agglomerations and also frequently in more distant areas. Much current action, including tree planting and improved stove programs, has been based on wrong assumptions about the nature of the wood fuel problem and thus has failed completely or succeeded only partly. The programs may, however, have other environmental and social benefits that were initially considered secondary.

Problems related to the use of traditional biomass fuels are inherently local. Before any successful policy intervention can take place, therefore, the nature, size, and quality of biomass flows need to be known. The availability and price of fuels, time-series data, end-user profiles, and cross-elasticities are lacking in most places. Only from detailed, local knowledge of these facts can the problems and their causes be correctly identified and formulated. This is particularly true for the contributions of women, who are usually the most important providers and users of biomass but are more often than not outside the decision-making processes. In rural situations, it is impossible to separate the energy issues from the other development issues. Energy must be looked at as one element of an integrated whole, and policy must be developed accordingly. While these issues apply to all energy systems, in the traditional sector they are particularly relevant, given their specificity regarding site and culture.

One specific technical issue seems to be valid in most places. The industrial, institutional, and commercial demand for biomass in cities (for both fuel and food) is a major cause of deforestation in rural areas. Policies to improve the biomass supply-and-demand situation for these needs, including conservation and substitution measures, should be pursued by all countries. Generally, tree planting in rural and periurban areas to increase fuelwood supplies for urban areas tends to be very expensive. Such solutions, however, have worked well to supply specific industries, crop processing plants, and modern energy conversion systems with fuel.

## HEALTH IMPACTS OF TRADITIONAL BIOMASS COMBUSTION

The indoor combustion of biomass fuels in traditional settings (usually unvented cooking spaces) releases toxic gases, causing considerable health problems directly to the principal users, mainly women and children. Hundreds of millions of people in developing countries are thus exposed to high concentrations of carbon monoxide, nitrous oxide, and hydrocarbons and other organics, which are risk factors for several of the leading causes of illness and death in developing countries.

Dry wood and charcoal burn relatively cleanly. As the moisture content of wood increases, however, and as other biomass fuels of lower energy content (such as residues) are used, the emissions rise. Actual emissions and exposure levels have been subject to study, but standard methods for scientific comparisons have not yet been developed. Nevertheless, it is clear that concentrations resulting from open fires in badly ventilated rooms are generally higher than in the dirtiest outdoor urban environments. Table 7.1 lists some typical indoor concentrations of pollutants from traditional biomass combustion. For comparison, Table 7.2 lists applicable health-based pollutant concentration standards. The main observed health effects include acute respiratory infections, chronic obstructive lung disease, low birthweights, cancer, and eye infections, although there are few reliable statistics. Apart from the chronic health effects there are, of course, the risks of burns and of children falling into the open fires. Scarcity of biomass fuels in many places has also produced indirect health effects as a result—for example, cutting down on cooking times or failing to boil water for drinking.

The health impacts of traditional biomass combustion are not just an energy problem but a much broader problem of rural development. In designing strategies to reduce these problems, the positive side-effects of open indoor combustion must not be overlooked. These include lighting houses, repelling insects such as mosquitoes, and preserving thatched roofs. Nor should it be forgotten that approaches concentrating on single elements of the problem, such as stove efficiency or smoke emissions, have not in the past proved very useful. An integrated approach is required, one that evaluates the whole system of cooking or kitchen management. For example, the provision of alternatives for indoor lighting may be the key to acceptance of enclosed and vented stoves in many areas. Stove programs, while including many positive elements, must be liberated from being only energy programs. They must be converted into programs with much broader objectives, including health, hygiene, comfortable kitchens, and labor productivity as well as fuel efficiency and

TABLE 7.1. INDOOR AIR POLLUTION FROM BIOMASS COMBUSTION IN DEVELOPING COUNTRIES

| Location | Households | Duration | SPM[a] (mg/m³) | B(a)P[b] (ng/m³) | (CO[c]) (ppm) | Other |
|---|---|---|---|---|---|---|
| Nigeria, Lagos | 98 | ? | — | — | 940 | NO$_2$: 8.6 ppm<br>SO$_2$: 38 ppm<br>Benzene: 86 ppm |
| Papua New Guinea | | | | | | |
| Western highlands | 6 | all night | 0.36 | — | 11 | HCHO: 0.67 ppm |
| Eastern highlands | 3 | all night | 0.84 | — | 31 | HCHO: 1.2 ppm |
| Eastern highlands | 6 | all night | 1.3 | — | — | — |
| Kenya, highlands | 5 | ? | 4.0 | 145 | — | BaA: 224 ng/m³<br>Phenols: 1.0 µg/m³<br>Acetic acid: 4.6 µg/m³ |
| Kenya, sea level | 3 | ? | 0.8 | 12 | — | BaA: 20 ng/m³ |
| Guatemala, two villages | 180 | ? | — | — | — | — |
| Poorly ventilated | | | | | 26–50 | |
| Well ventilated | | | | | 15–31 | |

| | | | | | |
|---|---|---|---|---|---|
| India, Ahmedabad | | | | | |
| Wood | 5 | 15 min | 7.2 | 1,270 | — | NO$_2$: 318 µg/m$^3$ |
| | | | | | | SO$_2$: 169 µg/m$^3$ |
| Cattle dung | 4 | 15 min | 16.0 | 8,250 | — | NO$_2$: 144 µg/m$^3$ |
| | | | | | | SO$_2$: 242 µg/m$^3$ |
| Dung plus wood | 7 | 15 min | 21.2 | 9,320 | — | NO$_2$: 326 µg/m$^3$ |
| | | | | | | SO$_2$: 269 µg/m$^3$ |
| India, Gujarat | | | | | | |
| Boria A.M. | 10 | 45 min | 4.8 | 3,550 | — | — |
| Boria P.M. | 10 | | 8.2 | 3,550 | — | — |
| Denapura A.M. | 11 | 45 min | 2.7 | 2,220 | — | — |
| Denapura P.M. | 11 | | 4.3 | 3,210 | — | — |
| Meghva A.M. | 10 | 45 min | 4.9 | 6,070 | — | — |
| Meghva P.M. | 10 | | 10.0 | 2,620 | — | — |
| Monsoon conditions | 1 | | 56.6 | 19,300 | — | — |
| Rampura A.M. | 5 | 45 min | 6.2 | 5,410 | — | — |
| Rampura P.M. | 5 | | 5.6 | 3,040 | — | — |
| Nepal | 22 | 8–12 hr | 606.0[d] | — | — | — |

[a] Suspended particulate matter.
[b] Benzo(a)pyrene.
[c] Carbon monoxide.
[d] Respirable particulate matter.

SOURCE: Smith et al. (1983), WHO (1984).

183

TABLE 7.2. ENVIRONMENTAL HEALTH CRITERIA AND AMBIENT AIR STANDARDS
FOR SELECTED POLLUTANTS

| Pollutant | Concentration | Averaging time | Type of standard | Source |
|---|---|---|---|---|
| Particulate matter | 75 μg/m³ | 24 hr | Primary (annual geometric mean) | Stern et al. (1984) |
| $SO_2$ | 80 μg/m³ | 1 yr | Primary (arithmetic mean) | Stern et al. (1984) |
| B(a)P[a] | 10 ng/m³ | — | Derived permissible concentration | Jones et al. (1981) |
| CO | 55 μg/m³ | 30 min | WHO health criteria | WHO–UNEP (1988) |
| $NO_2$ | 190–320 μg/m³ | 60 min | WHO health criteria | WHO–UNEP (1988) |

[a] Benzo(a)pyrene.
SOURCE: Chadwick (1990).

low cost. Because the main users are women, it cannot be over-emphasized that their full participation in the development and implementation of the strategy is essential.

TRADITIONAL CHARCOAL CONVERSION

Charcoal is more popular than fuelwood in many areas in developing countries. Not only does it burn more cleanly than fuelwood, but it has a higher energy content per unit weight. It is also often more easily available. Most charcoal conversion is done in inefficient, traditional earthen kilns, although recent studies indicate that actual efficiencies tend to be considerably higher than previously thought. Whole trees, branches, and certain agricultural residues, such as coffee husks, can be used as feedstock.

Despite the higher conversion efficiency of charcoal at the point of use, typically one needs more than twice as much wood to deliver the same final energy as with fuelwood. This fact, by itself, has tremendous resource implications. Furthermore, charcoal use is almost entirely commercial, usually in urban and industrial areas. While much of the wood that is converted to charcoal comes from forests cleared for agriculture, often the producers cut trees specifically to produce charcoal. This practice can contribute directly to deforestation. In Zambia, for example, uncontrolled fuelwood cutting for charcoal production outside forest reserves, and encroachment on the reserves themselves, have deforested most of the areas surrounding urban centers. In the city of Lusaka, the effects of deforestation extend more than 200 km from the city center (Chidumayo, 1988). Kenya banned charcoal exports in 1975 because charcoal makers were causing severe deforestation, especially of the coastal mangrove forests (Kinyanjui, 1987).

The health impacts of charcoal use are considerably smaller than those of fuelwood or residues at the point of combustion, although invisible carbon monoxide production during incomplete combustion can be a problem. Given its clean burning and convenience, charcoal is a desirable fuel. Conversion efficiencies can be substantially increased by use of more sophisticated technologies. This is, for example, relatively easy in an energy forest (planted or natural) on a commercial scale. It is difficult to do on a small scale, however, since the livelihood of many small-scale charcoal makers may be threatened. Introduction of fixed kilns would also introduce a considerable transport problem. It is worth increasing the end-use efficiency of charcoal combustion through the development of more efficient charcoal cookers, as the widespread commercial consumption ensures a market for more efficient stoves even if they are more costly.

## Agricultural Residues

The term "agricultural residues" includes various wood crop residues, cereal remains, and green crop residues as well as crop processing residues. Those that are dry tend to be best suited for use as fuel. Green residues tend to be better as animal fodder or manure and can also be used as feedstock in biogas plants. Agricultural residues are increasingly important energy sources. On the one hand, they are seen as a major untapped resource with a variety of potential new energy applications (Strehler and Stützle, 1987). On the other hand, for families in much of the Third World they are second-rate energy

TABLE 7.3. PROPERTIES AND USES OF SELECTED AGRICULTURAL RESIDUES

| Type of agricultural residue | Gross calorific value[b] (MJ/kg) | SUITABILITY FOR DIFFERENT APPLICATIONS[a] | | | |
|---|---|---|---|---|---|
| | | Domestic fuel[c] | Animal fodder | Organic fertilizer | Building material[d] |
| Alfalfa straw | 18.4 | * | *** | *** | 0 |
| Coconut shell | 20.1 | *** | 0 | * | 0 |
| Coconut husk | 18.1 | ** | 0 | * | *** |
| Cotton stalks | 17.4 | *** | 0 | * | ** |
| Groundnut shells | 20.0 | * | 0 | * | 0 |
| Cornstalks | 16.7 | * | ** | * | ** |
| Corncobs | 17.4 | ** | * | * | 0 |
| Pigeon pea stalks | 8.6 | *** | 0 | * | *** |
| Rice straw | 15.0 | * | ** | * | ** |
| Rice husks | 15.5 | * | 0 | * | 0 |
| Soybean stalks | 9.4 | * | *** | *** | 0 |
| Wheat straw | 17.2 | * | ** | * | ** |
| Cow dung | 12.8 | * | 0 | *** | *** |

[a] Suitability ratings: 0 = not suitable; * = poor; ** = moderate; *** = good. These ratings are indicative only. In many cases the suitability for specific applications depends on the exact type of residue and how it is prepared and used.
[b] Calorific value figures from Barnard and Kristoferson (1985).
[c] Assumes traditional Third World cooking technologies and no pretreatment other than drying.
[d] Includes thatching and handicrafts.
SOURCE: Barnard (1990).

sources whose use is being forced by dwindling firewood supplies (Barnard and Kristoferson, 1985).

From the standpoint of energy density, many residues compare favorably with air-dried wood at 15 megajoules per kilogram (MJ/kg). Some are true waste products, not suitable for use in conditioning soil or preventing erosion. Others are more or less valuable as agricultural resources, and their use as fuel represents an environmental as well as economic loss. Table 7.3 summarizes the characteristics of some important residues.

The largest energy use of residues is as domestic fuel in developing countries. In some places, such as certain areas of China and Bangladesh, residues can provide as much as two-thirds of household energy. In Africa and Latin America, energy use of residues is smaller but increasing. As many as 800 million people may rely on agricultural residues as their main cooking fuel (Hughart, 1979). There has also been increased interest in the use of residues for energy purposes outside the traditional sector, as well as in industrialized countries. Considerable potential exists for increased residue use in a number of areas. One key constraint, apart from the serious environmental issue, is the competing uses for residues such as fertilizer and basket making.

The wide variety of agricultural residues implies a wide variety of potential environmental effects. There are, essentially, two areas in which environmental impacts may occur. First, the removal of residues may intensify soil erosion and rapid water runoff. Second, the combustion of the residues themselves may have specific hazards. There is generally a strong correlation between increased removal of residues from the soil and decreasing soil fertility and intensified erosion. The extent of this correlation, however, depends on the specific plant and residue in question, the type and condition of the soil, the climate, and the intensity of residue removal. In a few cases, plant residues can be removed from the land without serious environmental consequences. Often, however, the removal of crop residues will severely intensify soil erosion, resulting in the loss of soil nutrients, water-holding capacity, soil organic matter, and soil biota. Moreover, residue removal increases the rate of water runoff between tenfold and a hundredfold. Tropical or mountainous sites are generally more vulnerable than temperate or lowland sites. Detailed, site-specific analysis is necessary to predict the impacts of residue removal.

Tables 7.4 and 7.5 compare the soil nutrient contents of selected agricultural residues. Figures 7.1 to 7.3 show the estimated effects of

TABLE 7.4. NUTRIENT CONTENT OF SELECTED CROP RESIDUES

| Crop residue | NUTRIENT CONTENT (PERCENTAGE OF OVEN-DRY WEIGHT) | | |
| --- | --- | --- | --- |
| | Nitrogen | Phosphorus | Potassium |
| Rice straw | 0.5–0.7 | 0.06–0.1 | 1.1–2.5 |
| Cornstalks | 0.6–1.4 | 0.05–0.2 | 0.4–1.3 |
| Wheat straw | 0.4–0.7 | 0.04–0.08 | 0.8–1.0 |
| Sorghum stalks | 0.6–1.1 | 0.11–0.15 | 1.3–1.5 |
| Millet straw | 0.7 | 0.09 | 1.8 |
| Soybean straw | 2.3 | 0.22 | 1.1 |
| Groundnut leaves | 1.2–2.6 | 0.14–0.17 | 2.1–2.2 |

SOURCE: Pathak and Jain (1984), Kowalenko (1978), Posselius and Stout (1983), Jenkinson (1981), Balasubramanian and Nnadi (1980).

TABLE 7.5. NUTRIENT CONTENT OF ANIMAL MANURES

| | NUTRIENT CONTENT (PERCENTAGE OF FRESH WEIGHT) | | | *Typical moisture content* | NUTRIENT CONTENT (PERCENTAGE OF OVEN-DRY WEIGHT) | | |
|---|---|---|---|---|---|---|---|
| *Animal* | *Nitrogen* | *Phosphorus* | *Potassium* | *(%)* | *Nitrogen* | *Phosphorus* | *Potassium* |
| Cattle | 0.4–0.6 | 0.05–0.1 | 0.3–1.0 | 75–85 | 2.5–3.0 | 0.3–0.6 | 1.8–5.0 |
| Horse | 0.6–0.7 | 0.09–0.1 | 0.5–0.6 | 60–70 | 1.7–2.0 | 0.2–0.3 | 1.4–1.5 |
| Sheep | 0.6–1.4 | 0.10–0.2 | 0.4–1.0 | 65–70 | 1.8–4.0 | 0.3–0.6 | 1.1–2.9 |
| Pig | 0.2–0.6 | 0.10–0.14 | 0.2–0.4 | 75–90 | 1.8–3.9 | 0.4–0.9 | 1.6–2.7 |

SOURCE: Brady (1974), Cooke (1973), Jenkinson (1981), Winterhalder et al. (1974).

FIGURE 7.1. EFFECT OF CROP RESIDUE RECYCLING ON THE
EQUILIBRIUM CARBON CONTENT OF SOIL.

*Figures refer to top 23 cm of soil; 10 metric tons of carbon per hectare is equivalent to a soil with 86 percent organic matter.*

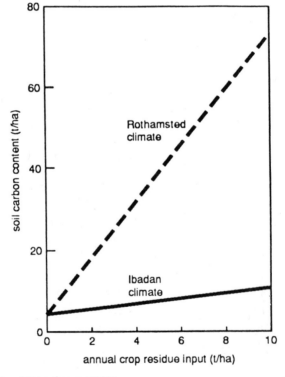

SOURCE: Barnard and Kristoferson (1985).

FIGURE 7.2. EQUILIBRIUM SOIL CARBON CONTENT ASSUMING DIFFERENT DEGREES OF CROP RESIDUE RECYCLING UNDER SIMULATED TROPICAL CONDITIONS.

*Yield assumptions for high-yielding crop, 7.5 metric tons (t) of straw and 2.5 t roots; for low-yielding crop, 2.5 t straw and 1.25 t roots (both per hectare per year). It is assumed that oven-dry biomass contains 40 percent carbon; soil carbon figures refer to top 23 cm of soil.*

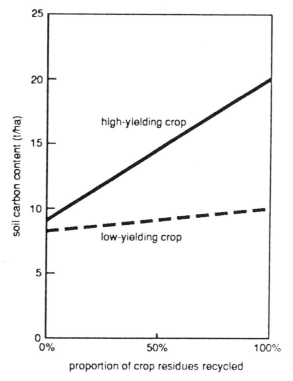

SOURCE: Barnard and Kristoferson (1985).

crop residue recycling on the organic matter content of soil and the effects of reducing the amount of residue recycled. Table 7.6 shows how much difference the recycling of crop residues can make in controlling erosion, using the U.S. Corn Belt as an example.

Combustion of residues has generally the same kind of health impacts as combustion of wood fuels. But since the energy content of residues is often lower, and the moisture content higher, than those of wood fuels, the impacts are much greater. The considerable health impacts of cooking in an unvented space are further aggravated by

FIGURE 7.3. EFFECT OF REDUCING CROP RESIDUE RECYCLING ON SOIL ORGANIC
CARBON CONTENT UNDER SIMULATED TROPICAL CONDITIONS.

*Assumptions are the same as in Figure 7.2.*

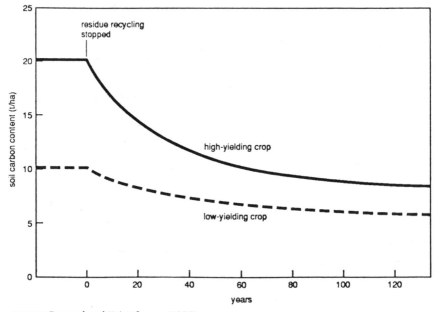

SOURCE: Barnard and Kristoferson (1985).

the fact that residues tend to be bulky and consequently need constant attention from the cook, thus increasing exposure. Smoke from
dung fires can be especially dirty. It has been linked, for example, to
the high rate of cor pulmonale (heart failure caused by chronic lung
disease) in northern and central India, where cow dung is an important fuel (Padmavati, 1974).

Combustion of agricultural residues in industry is highly dependent on the specific technology used. With appropriate emission
standards and mandatory emission control devices, the emission
levels can be minimized. The health impacts of indoor residue combustion are more difficult to control. From this standpoint, therefore,
residue use is not advocated. Nevertheless, many people in developing countries are obliged to depend on residues. For these people,
venting of smoke from the cooking space is the key technical problem to be solved.

TABLE 7.6. SOIL EROSION WITH DIFFERENT CROP RESIDUE MANAGEMENT PRACTICES IN THE U.S. CORN BELT

| Management practice[a] | Amount of crop residue recycled annually (kg/ha) | ANNUAL SOIL LOSS (METRIC TONS/HA) | |
|---|---|---|---|
| | | Range[b] | Average |
| Conventional tillage | 0 | 9.7–44.7 | 21.9 |
| Conservation tillage | 1680 | 7.8–33.8 | 15.6 |
| | 3920 | 4.5–19.2 | 9.1 |
| No-till management | 1680 | 6.2–25.5 | 12.1 |
| | 3960 | 4.0–16.0 | 7.0 |

[a] Conservation tillage refers to using a chisel plow rather than a conventional fall moldboard plow. No-till practices avoid all disturbance of the soil except in the seed row.
[b] Ranges indicate the extent of variation between different soil and site conditions.

SOURCE: Lindstrom et al. (1979).

## MODERN BIOMASS SYSTEMS

During recent decades, considerable research and development has focused on modern biomass systems, which have potentially wide applicability and lower environmental impact than the traditional systems just described. These include biogas, producer gas, alcohol fuels, and wood fuels.

### BIOGAS

Most wet biomass can be converted into a combustible gas through anaerobic digestion in a biogas digester. Typically, animal or human wastes or wet plant material are collected, mixed with water, and fed into an airtight container. The gas that is produced can be collected. The resulting sludge must be disposed of; for example, it may be used as fertilizer.

Biogas digesters come in various sizes—from small, single-family units up to large, industrial-scale facilities. Millions of small-scale biogas plants (producing 2 to 4 m³ of gas a day) have been built in developing countries, notably in China and India. Because of the need for water, as well as problems of cultural acceptability, biogas has been successful only in certain areas. For cooking purposes biogas can be burned directly. It can also be used to drive internal combustion engines, but traces of hydrogen sulfide must be cleaned out of the gas to avoid corrosion. Table 7.7 lists the typical constituents and properties of a cubic meter of biogas. Table 7.8 illustrates the energy value of the gas in some characteristic applications.

Figure 7.4 summarizes the environmental benefits and hazards involved. The most important impact is positive—namely that the digestion process destroys certain harmful bacteria and pathogens found in animal and human wastes. The technology of biogas, therefore, is generally an environmentally sound way of managing animal

TABLE 7.7. CONTENTS AND PROPERTIES OF 1 M³ OF BIOGAS AT 25°C AND 1 BAR

| Compound | Volume (m³) | Weight (kg) |
|---|---|---|
| CO₂ | 0.3–0.5 | 0.54–0.90 |
| CH₄ | 0.5–0.7 | 0.33–0.46 |
| H₂S | 0–0.01 | 0–0.01 |
| H₂, O₂, N₂ | traces | |
| H₂O | 0.03 | 0.02 |
| Total | 1 | 1–1.23 |

*Note:* 1 bar equals standard temperature and pressure. Calorific value: 16–23 MJ/m³; explosion risk limits: 9–23 percent biogas in air; octane rating: 120–140.

SOURCE: Ellegård (1990).

TABLE 7.8. USE OF BIOGAS

| Utility of 1 m³ of biogas | |
|---|---|
| Cooking | Three meals for a family of five (China) |
| Lighting | Equal to a 60-W bulb for 6 hr |
| Driving | A 3-ton lorry for 2.8 km |
| Running | A 1-hp engine for 2 hr |
| Generating | 1.2 kWh of electricity |

SOURCE: Ellegård (1990) after van Buren (1979).

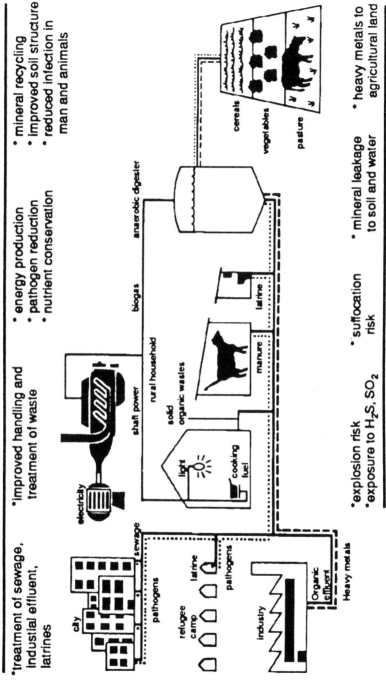

SOURCE: Ellgård (1990).

and human wastes that would otherwise present health hazards or entail considerable disposal costs.

The resulting waste sludge is an excellent fertilizer, sometimes better than the original wastes (certain seeds in the wastes are destroyed by the digestion process), especially if quickly returned to the soil. If wastewater is used in the digester, the value of the resulting sludge as a fertilizer will depend partly on the level and type of contamination in the original wastewater. If the feedstock is contaminated with pollutants such as heavy metals, the waste sludge too will contain the heavy metals and could represent a disposal problem, although the volumes may be lower.

Collection of animal waste to be used in biogas digesters, especially in decentralized rural settings, may result in a transfer of wealth from the land, where the collection takes place, to the owners of the digesters. Leaks from digesters, especially large ones, could contaminate drinking water supplies. Emissions from the combustion of biogas are essentially harmless, though the small amounts of hydrogen sulfide present may cause eye irritation. There is also a very small chance of explosion of the biogas in critical mixtures with air.

Biogas technology is an environmentally sound method for managing human and animal wastes. Moreover, a useful fertilizer and a combustible gas are produced. Overall the main impacts are positive. Given the water requirements of the process, however, biogas technology is not recommended in water-scarce areas unless wastewater is readily available. The relatively minor negative impacts cited here can be controlled with proper management. The social impacts of animal waste collection in rural areas must be considered on a case-by-case basis, particularly in view of the considerable labor requirements involved.

## PRODUCER GAS

Producer gas is generated by pyrolysis (incomplete combustion) of biomass, such as wood or crop residues. The gas can be used either for direct combustion in a boiler or to fuel internal combustion engines. Biomass feedstocks can be obtained directly from energy plantations or from commercial or noncommercial biomass sources.

Only a few thousand producer gas plants are in use worldwide.

They range from large industrial installations of 30 MW (thermal) to small biogas-fueled engines of a few kilowatts, used for pumping water and the like in remote areas of developing countries. The types of gasifiers used—updraft, downdraft, cross-draft, and fluidized bed (Figure 7.5)—are similar in their environmental impacts. They differ mainly in their feed specifications.

The supply of biomass feedstocks to the gasifier process has varying environmental impacts, depending on the source of the biomass. Following gasification, the gas consists mainly of carbon monoxide, hydrogen, carbon dioxide, water vapor, methane, and nitrogen. There are also small amounts of condensable organic vapors. The gas, if mixed with air, can cause explosions. The high content of carbon monoxide makes the gas highly toxic, and chronic exposures may lead to health problems or even death, especially of those exposed occupationally. Both problems can be reduced with adequate design and use of safety devices, as has been shown by the experience of industrialized countries during World War II.

The potentially harmful organic vapors are destroyed during direct combustion. If the gas is to be used in an engine, however, the organic vapors will condense as the gas cools. The condensate will then represent a certain hazard for the operator's health and the surrounding environment if not handled properly. This problem might become serious in large-scale operations. Disposal of the solid residue from the gasifier may lead to locally increased concentrations of heavy metals if residue dumps leak, creating a problem where there are many small installations or a few large facilities.

Combustion of producer gas results mainly in carbon dioxide and water vapor. There is also some nitrous oxide, which causes acidification and health effects similar to those of fossil fuels. Nevertheless, compared to fossil fuels, whose combustion usually results in considerable emissions of sulfur oxides, the effects are much smaller. Moreover, the emission of nitrous oxide during the combustion of producer gas is less than it would be with direct combustion of the original biomass feedstock.

The producer gas cycle may have a number of adverse impacts on land use, occupational and public health, and the natural environment. Table 7.9 summarizes the major environmental impacts of the use of producer gas. All of these impacts can be minimized with proper design, system management, and safety precautions. If these

FIGURE 7.5. FIVE TYPES OF BIOMASS GASIFIERS.

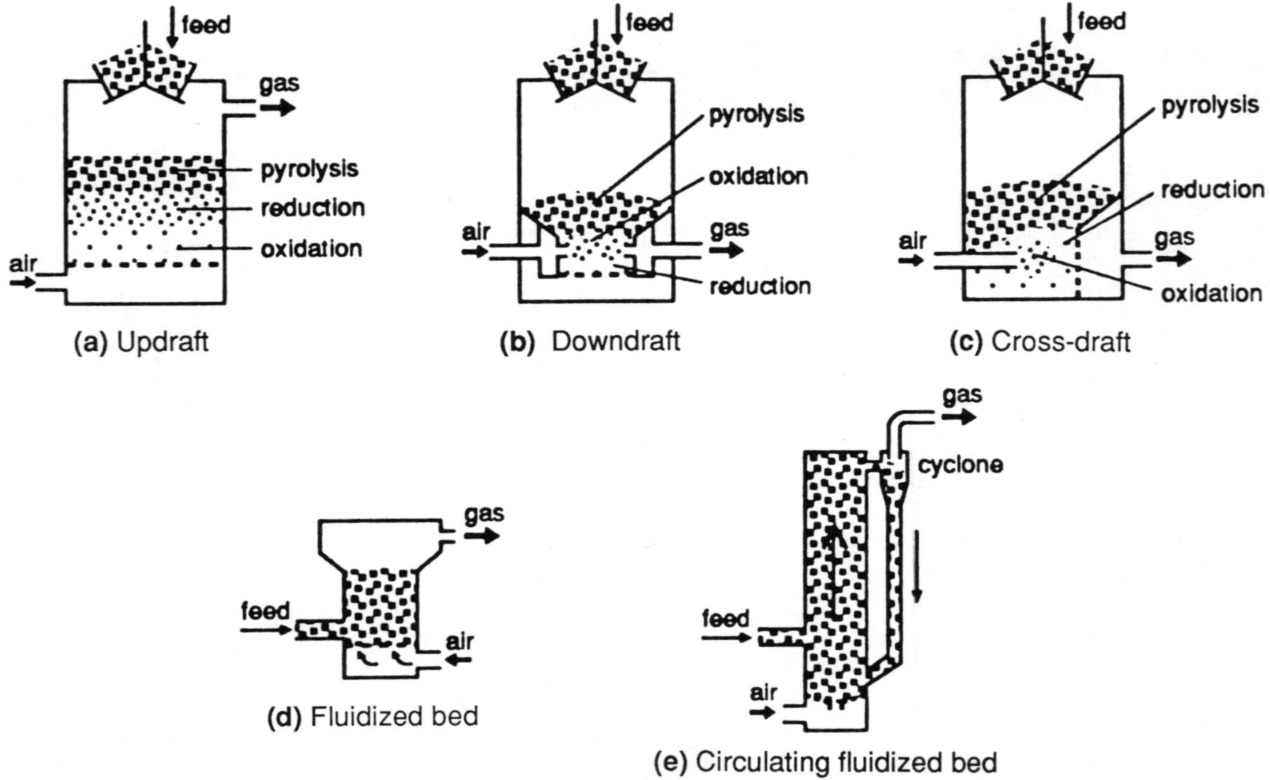

(a) Updraft

(b) Downdraft

(c) Cross-draft

(d) Fluidized bed

(e) Circulating fluidized bed

SOURCE: Kjellström (1990).

196

problems are taken care of, producer gas can sometimes be a viable and environmentally sound alternative to fossil fuels. Nevertheless, the maturity of this technology as well as the economics of producer gas generation remain problems.

ALCOHOL FUELS (ETHANOL)

A number of processes are available to produce ethanol from sugary and starchy plants, mainly sugarcane (as in Brazil and Zimbabwe) and corn (as in the United States). The basic processes involve fermentation and distillation, which require considerable energy inputs. This energy can come either from conventional sources, such as fossil fuels, or from the waste products of the sugar industry, known as bagasse. The alcohol produced can be used either neat (pure) or in various blends with gasoline (gasohol). The use of alcohols as a motor fuel has been increasing worldwide, although continuing low oil prices will make that trend difficult to sustain.

On the resource side, growing of the feedstock for alcohol production can have considerable environmental impacts, similar in quality and quantity to the environmental impacts of commercial agriculture or production of other energy crops. Depending on the particular crop, soil, climatic conditions, and chemical fertilizer use, sustained monoculture can reduce soil fertility if not well managed—as is generally the case with sugarcane plantations. Depending on the previous use of the land, the considerable land requirements of feedstock production can create serious problems—for example, subsistence farmers or multispecies tropical forests may be replaced.

The production of alcohol fuels entails two main environmental impacts: air pollution and water pollution. In the distillation process, considerable energy is required and thus air pollution problems must be controlled. Moreover, one of the by-products is the stillage waste, which has very high biological and chemical oxygen demand. Dumping this waste into local streams can cause serious environmental problems. It is practical to recycle this stillage as fertilizer back onto the sugarcane fields. Alternative uses of the stillage itself and the waste product (after anaerobic digestion) are also possible.

Combustion of pure ethanol yields water vapor, carbon dioxide, and small amounts of aldehydes and unburned alcohol vapor. The

TABLE 7.9. MAJOR IMPACTS OF PRODUCER GAS USE

| Impact | Agent or mechanism | Data for quantitative assessments/ possible control measures |
|---|---|---|
| *Potentially increased impact in comparison with use of petroleum fuels* | | |
| Acute health effects | Explosion in vessels containing combustible gas | The risk affects the operators. Accidents are only possible as a result of equipment malfunction or improper handling. |
| | Poisoning by inhalation of carbon monoxide | Risk can be controlled by use of adequate safety devices and correct operational procedures. |
| Delayed health effects | Poisoning by inhalation of carbon monoxide | The risk affects the operators. By adequate ventilation (plant design) and correct operating procedures, the concentration of CO on the premises can be kept below 20 to 50 ppm, which is considered harmless. |
| | Exposure to organic condensates (like phenol or polyaromatics) | Operators may be exposed to skin contact with tarry condensates with a potential carcinogenic effect. Risk can be avoided by use of protective clothing during maintenance and by correct procedures. |
| | | The general public may be exposed to contaminated drinking water if organic condensates from the plant are not collected and disposed of properly. |
| | | Amounts in condensates: phenols, 0.02−5 g/kg dry feedstock; polyaromatic matter, up to 0.2 g/kg dry feedstock. The amounts depend on gasifier design and operating conditions. |

TABLE 7.9. (*continued*)

| Impact | Agent or mechanism | Data for quantitative assessments/ possible control measures |
|---|---|---|
| Soil erosion, local climate change | Depletion of biomass resources | The conversion efficiency from biomass fuel to combustible gas is 65 to 90%, depending on system design. This amounts to the following biomass requirements for different applications: fuel gas generation, 60–85 kg DS/GJ(t); electric power generation, 1.0–1.7 kg DS/kWh(e). |

*Potentially reduced impact in comparison with use of petroleum fuels*

| | | |
|---|---|---|
| Acidification of soil and water | NO$_x$ in exhaust gases contribute to acidification | Approximate amounts of NO$_x$ in exhaust gases: combustion of producer gas, 0.04–0.3 g/MJ(fuel); producer gas engine, 0.08–2 g/MJ(fuel). |
| Global climate change | CO$_2$ in exhaust gas | No net addition of CO$_2$ to the atmosphere. |
| Ocean pollution by oil spills | Reduced use of oil will reduce risk | |

SOURCE: Kjellström (1990).

aldehydes can be controlled with catalytic converters. Emissions from the combustion of alcohol/gasoline blends depend on the concentrations. Generally there are no major differences between emissions of gasoline and gasohol. Table 7.10 summarizes the major air emissions of combustion of ethanol in automotive engines, both neat and in blends with gasoline. More studies should be made of these emissions, since precise effects are site-specific, depending on usage and climate. Certain mixtures of air and alcohol in fuel tanks may pose potential risks of explosion. Qualitatively, however, the problem is no different than that with gasoline.

Under proper management, growing feedstocks for alcohol production should not present serious environmental problems. The same care must be taken as during other commercial agricultural

TABLE 7.10. CHANGES IN EMISSIONS AS A RESULT OF CHANGING FROM GASOLINE TO ETHANOL FUEL AND BLENDS IN MOTOR VEHICLES

| Pollutant/fuel | Ethanol | Ethanol/gasoline (gasohol) |
|---|---|---|
| Hydrocarbon or unburned fuels | Few data; should be about the same or higher but less reactive. Expected reduction in polynuclear aromatic compounds (PNA) | May go up or down in unmodified vehicles; about the same when $\Phi$ remains constant;[a] composition may change; expected reduction in PNA. Evaporative emissions up. |
| Carbon monoxide | About the same; can be controlled; primarily a function of $\Phi$. | Decrease in unmodified vehicles (leaning occurs); about the same when $\Phi$ remains constant. |
| Nitrogen oxides | Lower, but not so low as with methanol; can be controlled. | Slight effect; small decrease when $\Phi$ is held constant, but may increase or decrease further from fuel leaning effect in unmodified vehicles. |
| Oxygenated compounds | Much higher aldehydes; particularly significant with precatalyst vehicles. | Aldehydes increase; most significant in precatalyst vehicles. |
| Particulates | Expected to be near zero. | Few data; no significant change expected. |
| Other | No sulfur compounds. | No data. |

[a] $\Phi$ = equivalence ratio.

SOURCE: OTA (1980).

activities. The land area required, however, may be considerable. Moreover, it is important to ensure that alcohol feedstock production does not replace essential food production. This is particularly important in the case of subsistence food production, since displaced growers will usually be unable to purchase their food requirements unless they are well employed in sugarcane production.

## MODERN WOOD FUEL RESOURCES

The large-scale harvesting of trees specifically for energy purposes is being practiced in a number of areas, and it is increasing worldwide. Some of the Brazilian steel and cement industry is fueled by charcoal from energy forests specifically planted for that purpose. Many natural forests are also being managed with the objective of supplying sustainable yields of wood for energy purposes. Other sources of biomass can also be grown specifically for energy purposes.

The two main options include management of natural forests—to supply wood for energy purposes at sustainable rates—and the planting of new forests. For the latter, fast-growing or slower-growing trees may be planted, usually monocultures, to be harvested on short or long rotations. The residues from conventional forestry or wood-based industries can also be used.

With regard to forest energy, the environmental impacts differ considerably among the different climatic zones, types of soils, and management regimes. In general, the use of residues and thinnings affects soil conditions, including possible erosion, nutrient loss, and acidification. If heavy equipment is used, the soil compaction may lead to increased runoff and erosion.

As a whole, the environmental impacts of forestry for energy are similar to what one may encounter in industrialized forest management. The complete removal and use of residues and thinnings, however, may significantly add to the adverse impacts unless carefully managed. Figure 7.6 summarizes the varied environmental impacts of modern silvicultural practices.

The effects of short-rotation forestry depend largely on the previous land use. If plantations are made on poor soils or abandoned arable land and appropriate species are selected, the forest may gradually lead to soil improvement and may also be positive in terms of wildlife habitat. When natural forests are cleared to provide space for fast-growing new species, the effect may be negative in the sense that diversity of the flora will be reduced and rare species may even be threatened. This is also true for many other types of land being transferred into intensive use, such as wetlands, meadows, and pasture.

Energy forests, depending on the species chosen, the ecoclimatic region, and the specific soil conditions, usually require the application of fertilizers, herbicides, and insecticides to achieve sufficient growth rates. These operations must follow sound agricultural and forestry management practices if they are to avoid unnecessary adverse impacts. Ecologically sound spacing of plantations is important.

FIGURE 7.6. MODERN SILVICULTURAL METHODS AND THEIR ENVIRONMENTAL IMPACTS.

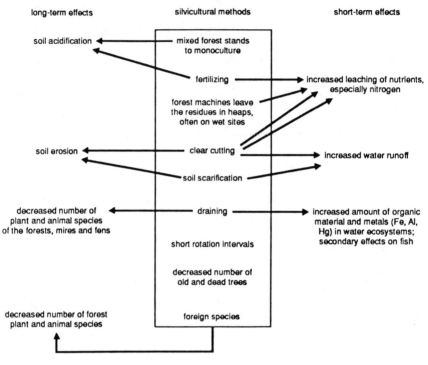

SOURCE: Stjernquist (1990).

Energy forests also come with an occupational hazard of their own. Since tree harvesting, in some countries, is one of the most hazardous of all jobs, the reduction of accident and injury rates in the forest industry is important. Appropriate protective clothing must be worn, safety procedures adhered to consistently, and continuous training of workers undertaken.

In areas where extensive wastelands are not available, the large land requirements of energy plantations are a critical issue. In developing countries, commercial plantations meant to produce fuelwood for rural consumption have not been found to be economic. The actual impacts on the soil and runoff into the watershed are all issues that, with careful management, can be controlled for benefit. Use of mixed species and planting blocks are examples of sound forestry practices.

## MODERN BIOMASS COMBUSTION

Biomass can serve as a feedstock for direct combustion in modern devices ranging from small domestic boilers, stoves, and ovens up to large-scale boilers and even multimegawatt power plants. The original biomass resource often must be processed before use in the combustion device. Since the energy content of biomass is usually lower than that of fossil fuels (coal, for example), considerable volumes may need to be transported to the point of use.

The combustion of biomass in domestic stoves and other appliances is highly variable. If fires are badly managed, emissions can be quite high. Particulates and carbon monoxide in the dwelling are often the main problems. Generally, the sulfur content of biomass is very small and emissions of sulfur oxides are insignificant. Combustion of biomass on a large scale has problems similar to those of domestic combustion, except that fires can be controlled much more effectively and pollution control equipment can be more easily installed and operated.

The combustion of biomass in modern stoves and boilers, where economic, is an environmentally sound substitute for conventional fossil fuels, particularly coal and oil. The carbon cycle is not affected if trees regrow or are replanted at the same rate as they are being burned and soil degradation does not occur. Sulfur emissions are much lower than from fossil fuels. The particulates can be controlled well at the source, particularly in large installations, as is the case with fossil fuels.

## BIOENERGY AND THE ENVIRONMENT

It is not enough to analyze individual bioenergy systems in terms of their environmental impacts. It is also essential to look at these systems collectively in the context of larger energy/environment problems—to discern the role played by the bioenergy systems both in an absolute sense and in comparison with other conventional energy systems.

LAND USE

There are few inhabited areas in the world where land is freely available. Biomass requires land to grow. Furthermore, biomass has relatively low energy densities, which means that relatively large areas of land are required to support bioenergy programs.

Land use, as a policy issue, is therefore crucial in the context of bioenergy programs. Do we have sufficient land, and of the right type, to grow a certain quantity of biomass? Will it conflict with existing food and fodder production? Will it conflict with the interests of farmers or industries?

A country may experience conflicting objectives concerning the use of land. On the one hand, a farmer may wish to attain security of food, fuel, and income. On the other, the state may wish to attain national self-sufficiency in food, balance exports and imports, or increase foreign exchange reserves. Sometimes these objectives reinforce each other, but often they are in conflict. When each side has a clear understanding of its objectives and there is possibility of dialogue, it is relatively easy to resolve such conflicts. In many areas this is not possible, and conflicts remain unresolved.

At the two ends of the spectrum, we have low-intensity and high-intensity farming systems. The former includes systems such as shifting cultivation, which are sustainable so long as sufficient fallow periods are possible. Increased populations, shifts in climatic conditions, and outside interference can lead to shorter rotation periods, gradual deforestation, devegetation, and soil degradation. In the high-intensity systems, farmers increase outputs by relying on external inputs to the system, including energy for the machines and chemical fertilizers and pesticides. Depending on the exact nature of these inputs, the environmental impacts can range from negligible to severe.

One of the principal areas of dispute is the "food-versus-fuel" conflict. While in smaller, on-farm systems it is generally easy to use agroforestry techniques to combine food and fuel production, in large-scale, industrial systems this is generally quite difficult and not much practiced. National programs to produce feedstocks for alcohol production, for example, can displace land on which subsistence food production is taking place.

Another important conflict for land resources arises between industries and communities living on the same land. Many communities living in or near forests, for example, depend on the multiple products provided by forests, such as food, fodder, medicines, and firewood. For certain industrial needs, however, monoculture of a generally fast-growing tree such as eucalyptus is more useful. An industrial concern clearing the forest and replanting with a single species will almost certainly create serious conflict. India's "social forestry" program, intended to involve local communities in regenerating forests, has produced such conflicts. In Karnataka, for example, the main result was that richer farmers planted eucalyptus as a cash crop on fields previously used for food crops, so that production of millet—the staple food of the rural poor—fell dramatically and millet prices tripled (Shiva and Bandyopadhyay, n.d.).

Sometimes the conflict can arise for ecological reasons. The planting of a certain species of forest on a certain soil type in a certain climatic zone may not be sustainable. As a general rule, it is essential to clarify the interests and resources of each relevant group. Before any activity can be implemented it is necessary not only to take into account these (often conflicting) interests, but also to seek and attain the full participation of the different groups. Only in this way can the conflicts be resolved.

When resources such as land are scarce, the first step should be to extend the use of existing resources. In our context, this implies concerted efforts to produce and use more rationally and efficiently biomass resources, including food and fuel. A number of options are available for increasing the efficiency of biomass conversion and use. The greatest efficiency improvements can be made in charcoal kilns, domestic stoves burning charcoal or fuelwood, and a number of formal and informal industrial activities. Table 7.11 lists several successful programs to improve the end-use efficiency of domestic stoves, with estimates of the resulting fuel savings. In areas where increasing efficiencies is difficult, substitution of conventional fuels such as kerosene and electricity should be considered. Fuel switching is particularly relevant for urban areas in developing countries but is difficult to implement cost-effectively unless socioeconomic conditions are favorable.

Improved management of biomass resources (whether food or fuel) is also crucial. Much of the food requirement in developing

TABLE 7.11. SOME SUCCESSFUL STOVE PROGRAMS

| Fuel | Country | Stove material | Total number | Period | Fuel saving (%) |
|------|---------|----------------|-------|--------|-----------------|
| Wood | Burkina Faso | Mud | 120,000 | 1985–1987 | 40 |
| Wood | Niger | Metal | 40,000 | 1984–1986 | 33 |
| Wood | Sri Lanka | Ceramic | 110,000 | 1985–1987 | 30–35 |
| Charcoal | Kenya | Metal/ceramic | 250,000 | 1982–1986 | 33 |

SOURCE: Newman and Hall (1990). Based on Jorez (1987), Damiba (1985), World Bank (1986), Amerasekera (1988), Young (1988), Walubengo (1988), Kapiyo (1986), and Khamati and Newman (1986).

countries is met from low-output, subsistence farming systems using considerable amounts of land. If productivity could be increased, the same land would feed many more mouths, so that less land would be required, thus reducing some of the pressures for deforestation and potentially increasing sustainable fuelwood yields from the forests.

One way of getting increased outputs from biomass systems is to aim increasingly to use multipurpose plants producing not just one but many products, including food, fuelwood, fodder, construction materials, and medicines. If this policy is carried out in agroforestry systems, where trees are grown together with food plants and grazing animals in some temporal or spatial allocation on the same land, even higher overall outputs can be achieved. The key word here is integration. (Figure 7.7 classifies some agroforestry systems.)

Overall, environmentally sound land-use planning should attempt to integrate all the concerns of all the relevant sectors. The importance of the full participation of all the relevant groups cannot be overemphasized.


IMPACTS OF AIR POLLUTION

The limited information available indicates that the emissions of pollutants during combustion of domestic fuelwood and alcohol fuels are considerably lower than for coal systems, except for polycyclic organic compounds, and that burning wood is generally less hazardous to health than burning dirty, low-quality coal, although more hazardous than burning clean, high-quality coal. Where important health impacts occur, they can be kept to acceptable levels with proper management and appropriate use of technology. Domestic fuelwood combustion in stoves and boilers is a relatively adverse activity from an environmental point of view. Indoor air pollution caused by unvented wood fires is the most critical impact, but this can be reduced by use of properly installed and operated stoves. There are increased risks of burns and other injuries, especially to children. Woodcutting and gathering, especially by non-professionals, carries very high injury risks (Table 7.12). Outdoor air pollution from residential fuelwood combustion is

FIGURE 7.7. AGROFORESTRY SYSTEMS AND THEIR COMPONENTS.

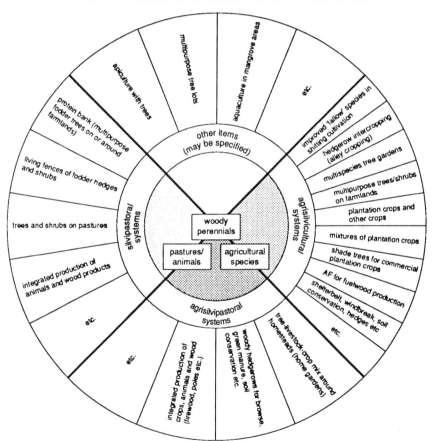

SOURCE: Nair (1985).

TABLE 7.12. ESTIMATES OF INJURY RATES IN
WOODCUTTING AND GATHERING

| Type | Injuries per $10^6$ GJ[a] |
|------|---------------------------|
| Commercial | 13 (10–17) |
| Noncommercial | 17 (16–18) |

[a] 95 percent Poisson confidence limits shown in parentheses.

SOURCE: Morris (1981).

relatively high compared to domestic oil furnaces or to large-scale wood combustion. The main pollutants of concern are organic particulate matter of inhalable size. Other pollutants include carbon monoxide and benzo(a)pyrene. The actual health impacts depend greatly on the geographical distribution of the emissions and the exposed populations.

Domestic fuelwood combustion in industrialized countries, however, has to be considered differently. In certain geographical and climatological locations, the local air pollution can be considerable and even cause for alarm. Missoula, Montana, in the 1950s and 1960s, because of emissions from paper mills, suffered daily airborne particulate levels of more than 300 $\mu g/m^3$, which were brought under control by the Clean Air Act. Then much of the population shifted to wood fuel for heating in the 1970s, and by 1980 daily levels of 500 $\mu g/m^3$ were not uncommon (Cannon, 1984). Yet in some areas, with properly installed stoves and furnaces, in spite of the relatively high injury risk of wood collection, wood fires can be—especially from an individual's point of view—a very desirable activity.

The health impacts of the use of alcohol/gasoline blends are generally positive compared to those resulting from gasoline combustion. There are usually increased levels of aldehydes, but these can be controlled with catalytic converters.

Another important local air pollution effect is the damage to structures caused by various constituents of photochemical smog—especially sulfur. Wood rarely contains more than 0.1 percent sulfur compared, typically, to 1 to 5 percent for coal. Other biomass, however, such as dung and certain residues, may contain more. Sulfur is also an important contributor to acidification, a major regional air pollution issue. Because of the smaller sulfur content of biofuels, replacement of fossil fuels by biofuels would considerably reduce the acidification problem.

Climate change induced by various greenhouse gases, especially carbon dioxide, may be the most important global effect of air pollution. Carbon dioxide is produced upon combustion of both fossil and biomass fuels. For a given production of energy, coal produces somewhat more and oil somewhat less carbon dioxide than wood. If biomass combustion is accompanied by an equal amount of regrowth or replanting, however, the overall effect of the wood cycle on the carbon dioxide balance of the atmosphere is negligible. Furthermore, large biomass plantations could serve as additional sinks of carbon dioxide produced by combustion of fossil fuels. Such plantations would also have auxiliary environmental benefits regarding local soil properties and water balances. The removal of biomass and vegetative cover causes soil erosion that will increase the release of greenhouse gases because of the resulting poor plant growth and rapid oxidation of exposed organic matter. A healthy, productive soil is essential for a favorable carbon dioxide balance.

As far as local impacts are concerned, the use of biomass is neither much better nor much worse than fossil fuels. The specific emissions depend on the fuel and the conversion systems being used. Although local use of biomass for domestic purposes may increase air pollution above the levels produced by centralized systems, increased use of biomass fuels would certainly improve the situation as far as regional air pollution is concerned—primarily because of the relatively low level of sulfur of most biomass, particularly wood. Overall the use of biomass as a fuel is expected to have a net benefit in terms of reducing air pollution.

## WATER

Water plays a crucial role in bioenergy development. Like land, it is one of the critical natural inputs to biomass growth, which is usually limited by geoclimatic as well as human constraints. Availability of water can be a limiting factor for biomass growth. Too little water can reduce and even stop biomass growth; too much can cause waterlogging and salinity, reduce growth, and eventually halt biomass growth. A crop like corn transpires more than 4.2 million liters of water per hectare in 3 months. The prime cause of water loss and erosion in agriculture is a shortage of vegetative cover of biomass. Figure 7.8 shows the chain of events by which increasing population and consequent growth in agriculture, grazing, and forest product consumption can combine to cause deforestation and waste of water resources.

The proper management of water supply, therefore, is vital in all areas. In arid and semiarid areas, where most water is already being used for agricultural, industrial, or domestic purposes, or where

FIGURE 7.8. RELATIONSHIPS AMONG POPULATION, FUELWOOD USE, WATER, AND THE ENVIRONMENT.

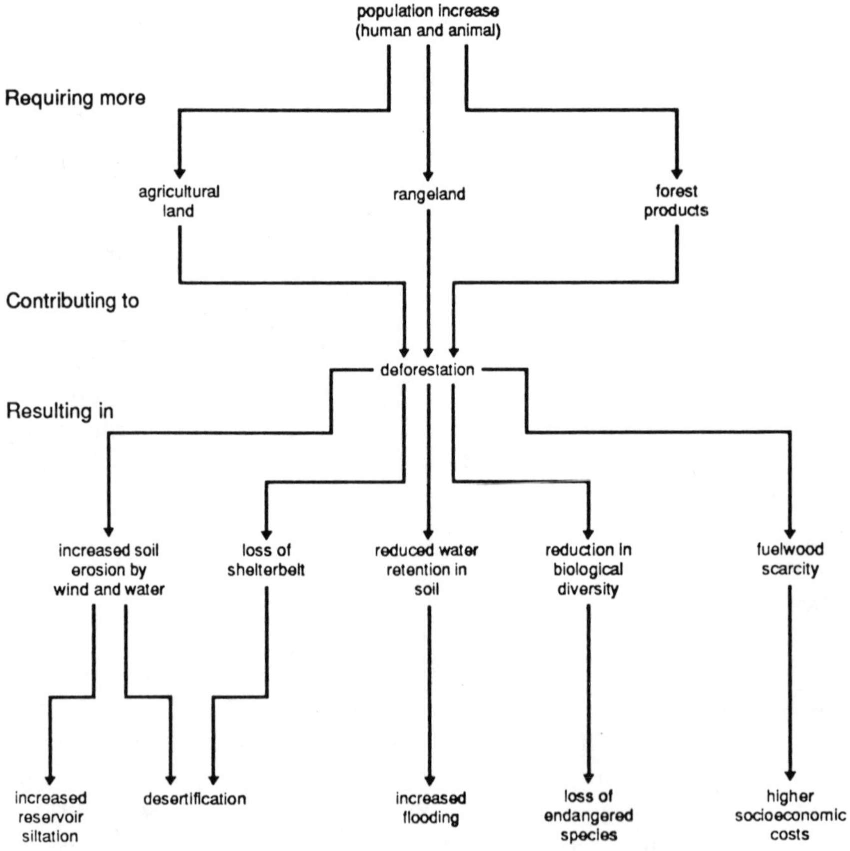

SOURCE: Biswas (1990).

FIGURE 7.9. IMPACT OF IRRIGATED AGRICULTURE ON BIOMASS USE.

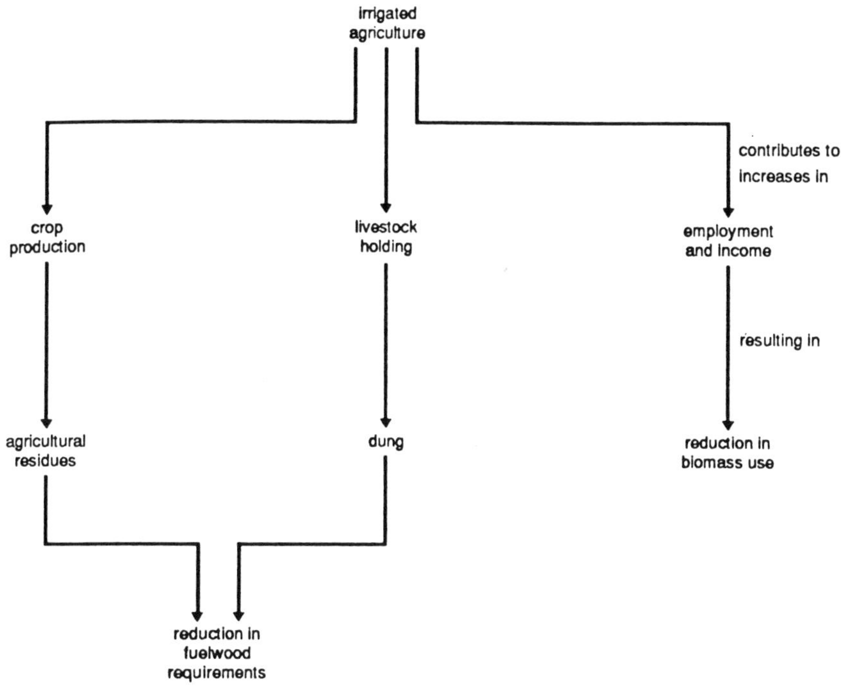

SOURCE: Biswas (1990).

expanded biomass production is envisaged, the parallel implementation of water conservation programs will also be central. In the case of properly irrigated agriculture, the amount of biomass produced will be more than in the case of unirrigated production. In addition to the increased yield of the economically valuable part of the crop, irrigation can result in considerably increased production of residues, many of which can be used as fuels. Irrigation has had this effect in many parts of the world, particularly with rice straw on the Asian subcontinent. Figure 7.9 illustrates this pattern. Table 7.13 gives an example from the Bhima irrigation project in Maharashtra, India.

A large energy plantation may itself have an impact on the local climate, through evapotranspiration, and consequently have impacts on the total amount of rainfall in the area. When planning such a plantation, therefore, all the possible positive and negative feedback effects must be considered. At the same time, both the production

TABLE 7.13. CHANGING PATTERNS OF BIOMASS FUEL USE AT BHIMA
AS A RESULT OF IRRIGATION

| | PERCENTAGE OF FUEL USED | |
| Fuel | Before irrigation (1980–1981) | After irrigation (1985–1986) |
| --- | --- | --- |
| Fuelwood | 66.3 | 53.2 |
| Cow dung | 19.2 | 23.8 |
| Agricultural residues | 7.4 | 13.1 |
| Other | 7.1 | 9.9 |

SOURCE: Biswas (1988).

and the conversion and use of biomass energy can have significant impacts, both physical and chemical, on local and even regional water bodies and ecosystems. A deforested mountain slope, if not replanted, quickly erodes, and the lost topsoil ends up in rivers, lakes, and hydroelectric reservoirs downstream. With no vegetation cover, the runoff on the slopes is much faster and causes flash floods in the valleys. Apart from loss of hydropower and irrigation potential, the silting up of water bodies is another cause of flooding. Dredging of waterways is expensive, usually more expensive than it would be to reforest the surrounding mountain slopes a few years before the problem occurs.

Biomass energy systems may also affect water bodies by releasing potentially harmful chemical compounds from various stages of the fuel cycle. Energy plantations—whether forests or other energy crops—may release pesticides, fertilizers, and other chemicals used to control growth. The chemicals and quantities concerned will depend on the production system being used, as well as the soil and climatic conditions. Moreover, different methods of harvesting the crop and collecting and using the residue will affect potential runoff to water bodies. Rapid water runoff and water loss, caused by a shortage of vegetative cover and biomass, is a major problem in world agriculture. As well, a lack of organic matter in the soil reduces its water-holding capacity. Thus crop residues should be removed only partially and only in specific situations.

Many biomass processing activities release pollutants into water bodies. Accidental spillage of manure or sludge from a large biogas plant can cause considerable pollution of local water. The effluents from an alcohol distillery contain high levels of nitrogen, phosphorus, and potassium with a high chemical and biological oxygen demand. If these are dumped directly into freshwater bodies, they will cause considerable pollution. With proper management, however, these problems can be minimized or eliminated. Properly constructed plants and careful adherence to safety procedures can reduce the risk of accidental spills to negligible levels. The effluents from alcohol distilleries can be used for irrigation or as fertilizer, or they can be broken down through anaerobic digestion to produce biogas. The careless disposal of solid residues from the gasification process, however, may also lead to the contamination of local water. As in the disposal of other chemicals, if procedures are followed, these contaminants can be kept below levels considered harmful.

Increased biomass production for energy, if not accompanied by environmentally sound water management efforts, can have negative impacts on water bodies, either by consuming too much water or by releasing physical and chemical pollutants to local and regional water. In areas of absolute water shortages, only well-planned conservation efforts will help make sufficient water available for expanded biomass production. Otherwise, all other potential environmental impacts to water bodies can be controlled with proper management techniques. In many places, the concept of integrated watershed management is increasingly being applied. All the impacts of bioenergy systems cited here must also be considered in the implementation of such management systems.

SOCIOECONOMIC IMPACTS

Despite the acknowledged importance of socioeconomic considerations in energy decision making, such issues have often been neglected in the case of bioenergy systems. Environmental, employment, and income distribution aspects of bioenergy systems must be integrated into a framework of comparative evaluation. The objective of economic evaluations, however, should not be to recommend or reject a particular system but rather to aid in the overall selection process. When certain costs and benefits are difficult to quantify, the evaluations should include different scenarios followed by sensitivity analysis to throw light on various assumptions and uncertainties.

Different methods may lead to different conclusions about comparative costs and benefits of different energy systems. Sometimes even minor differences in methods can significantly alter the results. Even when everything else is equal, the results of economic evaluations including bioenergy systems tend to be highly site-specific, reflecting particular conditions and assumptions. Table 7.14 shows that cost estimates for energy alternatives can be very sensitive to assumptions about technical characteristics and local conditions. The methodological details and the assumptions, therefore, must be made explicit.

The findings of the socioeconomic evaluation will help in formulating tax, subsidy, and pricing policies that may be required to promote investment in energy systems that are profitable from the viewpoint of society as a whole but are not attractive to private investors. Care must be taken, however, to ensure that such schemes do not lead to net transfers of resources from the society at large to higher-income households. Such transfers have benefited, for example, rich cattle owners in India in the case of biogas plants, rich families in the case of waterwheels in Nepal and solar photovoltaic installations in various countries, and the urban upper class in the case of the Brazilian ethanol program.

Many bioenergy systems are economically attractive for certain uses, regardless of fluctuations in world oil prices. The incorporation of environmental costs and benefits in the evaluations will improve the comparative ranking of small-scale, rural-oriented bioenergy systems, as well as improved stoves in urban and semiurban areas. When bioenergy systems are compared to conventional electricity supply systems, the analysis should try to include all adverse and beneficial impacts possible for both sides. When quantification is possible, it should be done. When only qualitative information is available, it should nevertheless form part of the evaluation. These impacts can be incorporated in various multidimensional comparisons using, perhaps, a weighting system for relative importance through sensitivity analysis.

Thus not only do bioenergy systems perform favorably in comparison with conventional systems in their environmental aspects, but their decentralized nature can be a significant socioeconomic advantage—especially in regard to their potential for promoting a more evenly distributed pattern of investment and resource use.

From an overall socioeconomic perspective, however, the environmental advantages of bioenergy systems are not a sufficient condition for their selection. When environmental advantages exist, whether quantifiable or not, they must be properly valued in the context of an integrated framework.

TABLE 7.14. COMPARISON OF PRESENT VALUES OF ECONOMIC COSTS OF ENERGY OPTIONS FOR WATER PUMPING IN GHAZIPUR, NORTHERN INDIA (IN 1986 U.S. DOLLARS)

| | FARM SIZE: 1 HA HEAD: 5 M | | | FARM SIZE: 0.4 HA HEAD: 5 M | | |
| | Crop rotation[a] | | | Crop rotation | | |
| Energy option | I | II | III | I | II | III |
|---|---|---|---|---|---|---|
| Electricity from grid | 2,130 | 1,570 | 2,250 | 1,550 | 1,500 | 1,600 |
| Diesel oil | 1,830 | 1,330 | 2,260 | 1,280 | 1,080 | 1,450 |
| Biogas (70%) and diesel | | | | | | |
| (30%) in dual-fuel engine | 1,810 | 1,500 | 1,920 | 1,480 | 1,420 | 1,530 |
| Producer gas (from wood | | | | | | |
| gasifier) and diesel | 2,310 | 2,070 | 2,530 | 2,040 | 1,940 | 2,120 |
| Solar photovoltaics | | | | | | |
| (A)[b] | 5,000 | 4,590 | 14,050 | 2,110 | 1,950 | 5,730 |
| (B)[c] | 3,740 | 3,440 | 10,430 | 1,610 | 1,490 | 4,280 |
| (C)[d] | 2,030 | 1,870 | 5,490 | 920 | 860 | 2,310 |
| (D)[e] | 1,550 | 1,430 | 4,100 | 730 | 680 | 1,750 |
| Solar thermal | | | | | | |
| (E)[f] | 5,820 | 5,380 | 16,230 | 2,550 | 2,360 | 6,690 |
| (F)[g] | 3,030 | 2,790 | 8,230 | 1,450 | 1,350 | 3,520 |
| Windmills | | | | | | |
| (G)[h] | 4,960 | 5,420 | 10,840 | 1,980 | 2,170 | 4,340 |
| (H)[i] | 6,060 | 6,620 | 13,260 | 2,420 | 2,650 | 5,300 |

*Notes:* Present value refers to the sum of capital and operating costs. Figures are rounded to the next 10 for easy comparison.
[a] Crop rotations: I (wheat and rice), II (wheat and bajra), and III (sugarcane).
[b] Current costs ($11.75/Wp); current efficiencies (3.4 percent overall instantaneous); Wp = peak watt.
[c] Current costs ($11.75/Wp); future efficiencies (4.6 percent overall instantaneous).
[d] Future costs ($4.5/Wp); current efficiencies (3.4 percent overall instantaneous).
[e] Future costs ($4.5/Wp); future efficiencies (4.6 percent overall instantaneous).
[f] $286/m² for the whole system and 0.5 percent daily average efficiency.
[g] $286/m² for the whole system and 1.0 percent daily average efficiency.
[h] Appropriate technology (Ghazipur-Allahabad multivane type).
[i] Hybrid technology (NAL sail type with centrifugal pump).
SOURCE: Bhatia and Pereira (1988).

## SUSTAINABLE BIOENERGY USE

Present exploitation patterns of food and energy biomass resources in developing countries cannot continue. The low productivity of the agricultural sector coupled with rising populations means that increasingly large areas of forest must be cleared for agricultural production and animal grazing. At the same time, the urban and industrial users are exerting an increasing demand for fuelwood and charcoal. The result is unabated deforestation. Often the easy availability of wood fuels disguises the serious shortages that follow. While there are considerable local variations on this theme, the problem is universal.

The environmental consequences of deforestation are well known: soil erosion, desertification, silting of reservoirs, flooding, increased atmospheric carbon dioxide concentrations, and many more. As the livelihoods of many people living in rural areas are being destroyed, sustainable development is put in jeopardy. To look at this narrowly as an energy problem is not only simplistic but wrong. Energy does have its share in it, however. Properly managed bioenergy strategies, while not resolving these environmental problems, will be able to contribute to their solution.

At the same time, the environmental impacts of continued use of fossil fuels, especially climatic change and the acidification of water bodies, soils, and forests, must signal a limit to the utilization of fossil fuels, at least with today's technologies. The global community may need to reduce fossil fuel combustion drastically in the next few decades, while considering all possible alternative strategies, including greatly increased energy efficiency, use of a variety of renewable energy sources, and possibly also nuclear energy if its current problems of acceptability can be overcome. Biomass could provide a considerable percentage of this combination of energy sources in an environmentally sound way.

The crucial element for successful and environmentally sound bioenergy programs is good management. Generally there are no inherent constraints to the use of biofuels. Most impacts can be eliminated or reduced to sufficiently low levels by proper agricultural and forestry practices, applying fertilizers or removing residues judiciously, keeping yields at or below sustainable levels, and following safety procedures. In most cases, if properly managed, the environmental impacts can be relatively easily internalized. Badly managed bioenergy systems, however, can have very serious impacts on the environment, potentially much worse than those of equivalent fossil-based systems.

Another key issue in achieving sustainable bioenergy programs is the integration of different components through systems analysis and management. It is unrealistic to look at bioenergy systems as technical systems isolated from their surroundings. Bioenergy programs work best when the people for whom they will be useful are fully integrated into their formulation and implementation. Local participation, therefore, is crucial.

But local participation alone is not enough. Bioenergy programs must be able to respond to the multiple needs of the users and the surrounding environment. Integration at a technical level, through the use of agroforestry techniques, as well as through multipurpose plants, is the key to many bioenergy programs. A biogas plant is viable in many situations—not simply because it produces a combustible gas, but because it also produces a good fertilizer and hygienically manages human and animal wastes. Similarly, an alcohol distillery can be made much more effective and, of course, economic, if the waste product is used to fuel the distillation—instead of brought-in fossil fuels—and the stillage is used as a fertilizer.

Such integration has tremendous implications for the economic analyses of the bioenergy systems discussed earlier. While there are various ways of analyzing the total costs, benefits, and profitability of bioenergy systems, most rely on calculating the perceived financial costs and benefits. But most bioenergy systems provide benefits that are difficult or even impossible to calculate financially. A newly planted energy forest on a previously unused wasteland, for example, will often have very positive impacts on the environment, including wildlife and fauna, local climate, soil erosion, and aesthetics. While these benefits are difficult to formulate in financial terms, they must somehow become part of the equation. If biomass energy systems are well managed, they can form part of a matrix of energy supply that is environmentally sound and therefore a significant contributor to sustainable development.

## REFERENCES

Amerasekera, R. M. 1988. National fuelwood conservation programme. *Boiling Point* 15:24–27.

Balasubramanian, V., and L. A. Nnadi. 1980. Crop residue management and soil productivity in savanna areas of Nigeria. In *Organic Recycling in Africa*, Soils Bulletin no. 43. Rome: FAO.

Barnard, G. W. 1990. Use of agricultural residues as fuel. In J. Pasztor and L. A. Kristoferson, eds., *Bioenergy and the Environment*. Boulder: Westview Press.

Barnard, G. W., and L. A. Kristoferson. 1985. Agricultural residues as fuel in the Third World. Technical Report no. 5. London: Earthscan.

Beijer Institute. 1982. *SADCC Energy Seminar: Country Background Reports*. Stockholm: Beijer Institute.

Bhatia, R., and A. Pereira (eds.). 1988. *Socioeconomic Aspects of Renewable Energy Technologies*. New York and London: Praeger.

Biswas, Asit K. 1987. Environmental concerns in Pakistan, with special reference to water and forests. *Environmental Conservation*, pp. 319–328.

————. 1988. Bhima revisited: Impact evaluation of a large irrigation project. *Water International* 13:17–24.

————. 1990. Water. In J. Pasztor and L. A. Kristoferson, eds., *Bioenergy and the Environment*. Boulder: Westview Press.

Brady, N. C. 1974. *The Nature and Properties of Soils*. New York: Macmillan.

Cannon, J. A. 1984. Air quality effects of residential wood combustion. *Journal of the Air Pollution Control Association* 34:895–897.

Chadwick, M. J. 1990. Air pollution. In J. Pasztor and L. A. Kristoferson, eds., *Bioenergy and the Environment*. Boulder: Westview Press.

Chidumayo, E. N. 1988. Woodfuel forestry issues: The Zambian experience. Lusaka: University of Zambia.

Cooke, G. W. 1973. *Fertilizing for Maximum Yield*. London: Crosby, Lockwood & Staples.

Damiba, T. E. 1985. *Étude comparative des performances du foyer 3 pierres-traditionel et du foyer 3 pierres-amelioré*. Ouagadougou, Burkina Faso: Institut Burkinabe d'Energie.

Ellegård, A. 1990. Biogas. In J. Pasztor and L. A. Kristoferson, eds., *Bioenergy and the Environment*. Boulder: Westview Press.

Hall, D. O., and R. P. Overend. 1987. Biomass forever. In D. O. Hall and R. P. Overend, eds., *Biomass: Regenerable Energy*. London: Wiley.

Hughart, D. 1979. Prospects for traditional and non-conventional energy sources in developing countries. Staff Working Paper no. 346. Washington: World Bank.

Jenkinson, D. S. 1981. The fate of plant and animal residues in soil. In D. J. Greenland and M. H. B. Hayes, eds., *The Chemistry of Soil Processes.* London: Wiley.

Jones, T. D., F. D. Griffin, C. S. Dudney, and P. J. Walsh. 1981. Empirical observations in support of "carcinogenic promotion" as a tool for screening and regulation of toxic agents. In M. Cooke and A. J. Dennis, eds., *Chemical Analysis and Biological Fate: Polynuclear Aromatic Hydrocarbons.* Fifth International Symposium on Polynuclear Aromatic Hydrocarbons. Columbus: Battelle Memorial Institute.

Jorez, Jean-Phillipe. 1987. Personal communication (CILSS, Ouagadougou, Burkina Faso).

Kapiyo, Raphael J. 1986. *The Jiko Field Test Study.* Nairobi: Kenya Energy Non-Governmental Organizations.

Khamati, B., and D. R. Newman. 1986. *Research and Development on Improved Stoves.* National Improved Stoves Workshop, 24–26 Sept. 1986. Nairobi: Ministry of Energy.

Kinyanjui, M. 1987. Fuelling Nairobi: The importance of small scale charcoaling enterprises. *Unasylva* 39:157–158.

Kjellström, B. 1990. Producer gas. In J. Pasztor and L. A. Kristoferson, eds., *Bioenergy and the Environment.* Boulder: Westview Press.

Kowalenko, C. G. 1978. Organic nitrogen, phosphorous and sulfur in soils. In S. U. Schnitzer and S. U. Kahn, eds., *Soil Organic Matter.* London and New York: Elsevier Applied Science.

Leach, G. 1987. *Household Energy in South Asia.* London and New York: Elsevier Applied Science.

Leach, G., and M. Gowan. 1987. *Household Energy Handbook.* Technical Paper no. 67. Washington: World Bank.

Leach, G., and R. Mearns. 1988. *Bioenergy Issues and Options for Africa.* London: IIED.

Lindstrom, M. J., et al. 1979. *Tillage and Crop Residue Effects on Soil Erosion in the Corn Belt.* Soil Conservation Society of America.

Morris, S. C. 1981. Residential wood fuel use: Quantifying health effects of cutting, transport, and combustion. BNL 30735R. Paper presented at American Public Health Association Meeting, Los Angeles.

Nair, P. K. R. 1985. Classification of agroforestry systems. *Agroforestry Systems* 5:97–128.

Newman, D. R., and D. O. Hall. 1990. Land-use impacts. In J. Pasztor and L. A. Kristoferson, eds., *Bioenergy and the Environment.* Boulder: Westview Press.

Padmavati, S. 1974. Cor pulmonale in India. In *Cardiovascular Diseases in the Tropics.* London: British Medical Association.

Pathak, B. S., and A. K. Jain. 1984. Characteristics of crop residues: Report on the physical, chemical and thermochemical properties of ten selected crop residues. Ludhiana, India: Punjab Agricultural University.

Pimental, D., A. Warneke, W. Teel, K. Schwab, N. Simcox, D. Ebert, K. Baenisch, and M. Aaron. 1988. Food versus biomass fuel: Socioeconomic and environmental impacts in the United States. *Advances in Food Research* 32:185–238.

Posselius, J. H., and B. A. Stout. 1983. Crop residue availability for fuel. In D. L. Wise, ed., *Liquid Fuel Systems*. Boca Raton, Fla.: CRC Press.

Scarlock, J. M. O., and D. O. Hall. 1990. The contribution of biomass to global energy use. *Biomass* 21:75–81.

Shiva, Vandana, and J. Bandyopadhyay (eds.). n.d. *Ecological Audit of Eucalyptus Cultivation.* Dehra Dun, India: English Book Depot.

Smith, K. L., A. L. Aggarwal, and R. M. Dave. 1983. Air pollution and rural biomass fuels in developing countries: A pilot village study in India and implications for research and policy. *Atmospheric Environment* 17:2343–2462.

Stern, A. C., R. W. Boubel, D. B. Turner, and D. L. Fox. 1984. *Fundamentals of Air Pollution.* 2nd ed. Orlando: Academic Press.

Stjernquist, I. 1990. Modern wood fuels. In J. Pasztor and L. A. Kristoferson, eds., *Bioenergy and the Environment.* Boulder: Westview Press.

Strehler, A., and W. Stützle. 1987. In D. O. Hall and R. Overend, eds., *Biomass: Regenerable Energy.* London: Wiley.

Tobisson, E. 1980. Women, work, food and nutrition in Nywamigura Village, Mara Region, Tanzania. Stockholm University, Sweden.

U.S. Congress. Office of Technology Assessment (OTA). 1980. *Energy from Biological Processes.* Vol. 2: *Technical and Environmental Analysis.* Washington: Government Printing Office.

van Buren, E. A. 1979. *A Chinese Biogas Manual.* IT Publications.

WHO. 1984. *Biomass Fuel Combustion and Health.* EFP/84.64. Geneva: World Health Organization.

WHO-UNEP. 1988. *Indoor Environment: A Guidebook on the Health Aspects of Air Quality, Thermal Environment, Light and Noise.* Geneva: World Health Organization and U.N. Environment Program.

Walubengo, Dominic. 1988. Improved stoves programmes in Kenya. *Boiling Point* 15:3–7.

Winterhalder, B., R. Larsen, and Thomas R. Brooke. 1974. Dung as an essential resource in a highland Peruvian community. *Human Ecology* 2(2):89.

World Bank. 1986. Niger improved stoves project: Mid-term progress report. Washington: UNDP/World Bank Energy Sector Management Assistance Program.

Young, Peter. 1988. Personal communication (ITDG, Rugby, UK).

# PART II

---

# THE
# BENEFITS OF
# ENERGY
# EFFICIENCY

CHAPTER 8

# ENERGY USE IN BUILDINGS

*Arthur H. Rosenfeld and Ellen Ward*

THROUGHOUT MOST OF THE WORLD buildings are energy-inefficient, using unnecessarily large amounts of electricity and fuel to provide basic lighting, space heating, and air conditioning. In 1989, U.S. buildings consumed about 33 exajoules (31.5 quadrillion Btu) worth $200 billion.[1] Buildings used over 35 percent of all primary energy consumed in the United States, more energy than was consumed by industry or transportation. Residences used $120 billion in energy while commercial buildings (offices, stores, schools, hospitals) used nearly $80 billion.

Many studies have quantified the vast "reserves" of energy being squandered by outmoded building technologies and processes in the United States. These studies show that improving the efficiency of U.S. energy end uses—refrigerators, lights, and HVAC systems (for heating, ventilating, and air conditioning)—could make as much energy available as would be supplied by a major new oil or natural gas field.[2] Energy-efficient retrofits (improvements in existing buildings) can affordably reduce demand in U.S. buildings by as much as

50 percent, saving the U.S. economy $100 billion per year (NAS, 1991). As replacing conventional sources of energy with alternatives (wind, biomass, solar, and others) could take as long as 20 years, energy-efficient improvements can bridge this time gap.

Energy use in buildings also takes a tremendous toll on the environment. Each year the energy that powers appliances and heats, cools, and lights U.S. buildings produces 500 million tons of carbon as $CO_2$, twice the 270 million tons produced by gasoline use. U.S. building energy use alone produces 2 tons of carbon for every person. Yet on average throughout the world, *total* energy use produces only 1 ton of carbon per person. Looking just at electricity, buildings consume 66 percent of all U.S. electricity.[3] Electric efficiency in buildings is thus a prime target for policies aimed at the mitigation of greenhouse gases.

Between 1973 and 1986, improvements in U.S. energy efficiency (spurred by the OPEC oil shocks) captured dramatic pollutant reduction and cost savings. With the collapse of OPEC in 1986, most leaders declared that their national energy problems were solved. (Japan is an exception.) With the persistence of low prices, progress has stalled and U.S. energy use once again nearly parallels growth in the GNP. Relying on energy prices to drive energy-efficiency improvements guarantees inactivity so long as prices are low, as they have been since the 1991 Gulf War.

Leadership and nonprice incentives can, however, reestablish the gains of 1973–1986. A strategy designed to capture available building energy savings will call for a variety of approaches:

- Reforming utility profit structures to reward those that sponsor demand-side management (DSM) programs and provide cleaner "least cost" and "least carbon" energy *services*, rather than raw energy.
- Leveling the tilted playing field that accepts much longer payback periods on investments in supply than on investments in efficiency.
- Increasing R&D for energy efficiency from 1/1000 of energy costs to at least 1/100.
- Strengthening building and appliance efficiency labels and standards.
- Implementing consumer-based financial incentives that strengthen the market for efficient technologies by rewarding

energy-efficient purchases and penalizing inefficient buying. These revenue-neutral incentives can also take into account the life-cycle costs of pollution "externalities" caused by energy use.

• Galvanizing a worldwide shift to rational energy pricing—that incorporates environmental "externalities"—and doing away with half the world's subsidies on energy.

Governmental leadership must do all it can to compensate for the shortsightedness of consumer decisions. Even though low energy prices and lack of political commitment have crippled the U.S. national energy strategy, certain proposals can still improve building-sector efficiency.

## U.S. ENERGY CONSUMPTION AND THE GNP

Prior to 1973 in the United States, oil use and the GNP were inexorably linked. Figure 8.1 shows the spectacular, and by now familiar, decoupling of GNP and energy consumption that occurred from 1973 to 1986. During this period, GNP grew 35 percent while total energy use remained constant and oil and gas use decreased 1.2 percent annually (Figure 8.2). Some 28 exajoules (EJ) of anticipated energy usage per year, worth $150 billion, never materialized. (Structural change is credited with one-third of the change.) When offset by the annualized cost of the required investments in efficiency, net annual energy savings are about $100 billion. In addition to the 35 percent dollar savings, roughly a 50 percent increase in carbon dioxide was avoided during this period. In fact, coal use during these years would have doubled without conservation. If the United States were still operating at 1973 efficiency levels, we would have dumped 50 percent more carbon dioxide ($CO_2$), sulfur dioxide ($SO_2$), and nitrogen dioxide ($NO_2$) into the atmosphere in 1986.

Improvements in U.S. electricity consumption were even more impressive than improvements in total energy consumption. Since buildings consume two-thirds of total U.S. electricity (three-quarters of all dollars spent on electricity), improvements in this sector contributed significantly to total electricity savings. Until 1973, total electricity use was growing at a rate of 7 percent a year (3 percent faster than the GNP). After OPEC, electricity use grew only as fast as

*Before the 1973 oil embargo, total primary energy use was growing at the same rate as GNP. Between 1973 and 1986, growth in energy consumption was halted by high oil prices and progressive energy policies. In 1986, projected primary energy use was 36 percent (28 EJ) higher than actual use. Electricity use followed a similar pattern. Prior to 1973, electricity use was growing 3 percent faster than GNP. Between 1973 and 1986, electricity use decreased, growing only 2.5 percent per year, or 3.2 percent per year less than projected by pre-1973 trends. (GNP-projected energy values are based on 1973 efficiency and GNP. The electricity projections include an additional 3 percent per year to account for increasing electrification.) In 1986, projected electricity use was 50 percent higher than actual electricity use, indicating a savings of 1160 BkWh.* Note: *Electricity use is given in terms of total equivalent primary energy input (exajoules—left-hand scale) and net consumption (1000 BkWh—right-hand scale).*

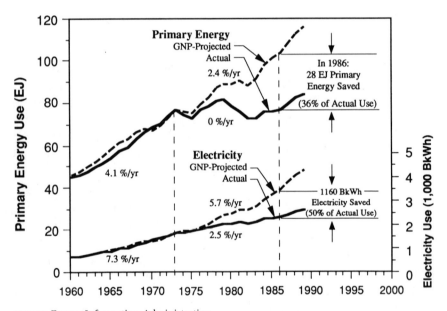

SOURCE: Energy Information Administration.

*The figure shows energy consumption by source. Oil and gas use dropped an average of 1.2 percent per year from 1973 to 1986, explaining why OPEC lost control of world oil prices. The upper dashed line is GNP, in 1982 dollars, scaled to go through 74.3 quads (1 quad = 10¹⁵ Btu) in 1973; it illustrates how GNP and energy use were in lockstep before 1973. Energy conservation and improved efficiency have made a significant contribution to the U.S. economy since 1973 and currently provide annual savings of more than 28 quads—worth $150 billion. The figure does not include about 4 quads of wood waste energy, but adding the contribution from conservation to the other nonfossil sources gives almost 37 quads—equal to one-third of the energy we would have used if we were still operating at 1973 efficiencies.* Note: *1 quad = 1.055 EJ. "Conservation since 1973" includes structural changes and improved efficiency.*

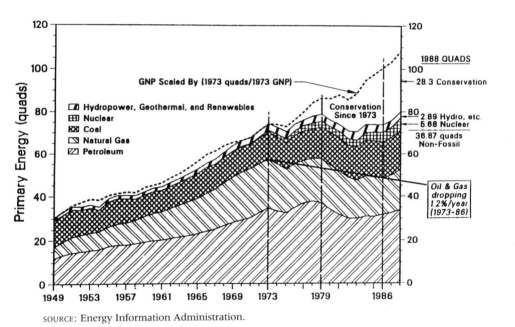

SOURCE: Energy Information Administration.

227

GNP, for an annual savings of 3.2 percent. Compared to projections of pre-OPEC trends, 50 percent of 1986 electrical usage was avoided between 1973 and 1986 (compared to 35 percent of all energy use). This savings—1160 BkWh per year—is equivalent to the annual output of 230 base-load (1000-MW) power plants, worth $85 billion per year.

## PERFORMANCE AND IMPROVEMENTS

Table 8.1 shows the performance of U.S. buildings in 1986, and Figure 8.3 shows improvements in the residential and commercial sectors starting in 1976. Energy use in these sectors demonstrates remarkable gaps between actual and GNP-projected energy use—gaps amounting to 2.1 percent a year in residences and 2.0 percent a year in commercial buildings. In 1986 some 8 EJ was saved, worth $50 billion and equivalent to 4 million barrels of oil per day (twice Alaskan oil production). This 8 EJ represents a savings of 28 percent of the primary energy (28.4 EJ) consumed in buildings in 1986. Moreover, a large amount of $CO_2$ emissions from power plants was eliminated because the United States avoided building dirty coal plants.

The energy efficiency of new buildings improved quickly in response to the energy shortage following the 1973 OPEC embargo. The space heat intensity (in megajoules per meter squared per year) of new buildings dropped by nearly 50 percent between 1973 and 1986—nearly 5 percent a year. The slower drop of 2 percent a year for the whole stock, shown in Figure 8.4, reflects the long turnover time of the stock and the slower retrofitting of existing buildings. Remarkable progress in reducing annual energy use and costs is shown in Table 8.2 per square foot of office space (Hafemeister et al., 1985).

Buildings built in 1991 are just as comfortable as 1975 structures (if not more so) and in fact command higher rents. More important, energy growth in the commercial sector is very small. The growth in energy consumption that one would expect to accompany a 2 or 3 percent annual growth in commercial floor space is mostly avoided, offset by a 1 percent annual rate of retirement of older, less efficient

TABLE 8.1. PRIMARY ENERGY CONSUMPTION BY FUEL TYPE: 1986 DATA FOR
COMMERCIAL AND RESIDENTIAL BUILDINGS (QUADRILLION BTU)

| Sector | Electricity[a] | Gas | Oil | Other[b] | Total | Percent |
|---|---|---|---|---|---|---|
| *Residential* | | | | | | |
| Space heating | 1.81 | 2.87 | 1.00 | 0.39 | 6.07 | 39.8 |
| Water heating | 1.61 | 0.82 | 0.10 | 0.06 | 2.58 | 16.9 |
| Refrigerators | 1.44 | | | | 1.44 | 9.4 |
| Lighting | 1.04 | | | | 1.04 | 6.8 |
| Air conditioners | 1.08 | | | | 1.08 | 7.1 |
| Ranges/ovens | 0.64 | 0.21 | 0.00 | 0.03 | 0.88 | 5.8 |
| Freezers | 0.45 | | | | 0.45 | 2.9 |
| Other | 1.19 | 0.54 | | | 1.72 | 11.3 |
| *Total* | 9.27 | 4.43 | 1.10 | 0.48 | 15.28 | 100.0 |
| *Commercial* | | | | | | |
| Space heating | 1.00 | 1.83 | 0.79 | 0.18 | 3.80 | 32.4 |
| Lighting | 2.96 | | | | 2.96 | 25.2 |
| Air conditioning | 0.98 | 0.12 | | | 1.11 | 9.4 |
| Ventilation | 1.49 | | | | 1.49 | 12.7 |
| Water heaters | 0.37 | 0.19 | 0.08 | | 0.64 | 5.5 |
| Other | 1.37 | 0.24 | 0.01 | 0.11 | 1.73 | 14.7 |
| *Total* | 8.17 | 2.38 | 0.89 | 0.29 | 11.73 | 100.0 |
| *Total residential and commercial consumption*[c] | 17.44 | 6.81 | 1.98 | 0.77 | 27.01 | 100.0 |

[a] Represents Btu value of primary energy inputs in production of electricity (11,500 Btu/kWh).
[b] For residential: coal and LPG. For commercial: coal, LPG, and motor gasoline (off-road use). Excludes estimated 0.8 quad of energy from wood fuel in residential sector.
[c] Totals for residential and commercial consumption from EIA State Energy Data Report (1960–1986). Distribution between end uses based on LBL Residential End Use Model and ORNL/PNL Commercial End Use Model. Latest period for which data are available is 1986.

SOURCE: *Analysis and Technology Transfer Annual Report 1988: Buildings and Community Systems,* May 1989 (DOE/CH/0016-H2).

TABLE 8.2. PROGRESS IN REDUCING ANNUAL ENERGY USE AND COSTS
(PER SQUARE FOOT OF OFFICE SPACE) IN 1985

| | Electricity | | Natural gas | |
|---|---|---|---|---|
| | kWh | Cost | kBtu | Cost |
| 1975 typical office tower | 30 | $2.25 | 170 | $3.40 |
| 1978 California Building Standard | 20 | $1.50 | 10 | $0.50 |
| 1991 California Building Standard | 10 | $0.75 | 5 | $0.25 |

buildings. Similar progress in the residential sector is discussed later in this chapter. In brief, homes built prior to 1975 use 8 to 10 Btu/ square foot/degree-day (°F). Homes built after 1975 use only 4 or 5 such units, and superinsulated homes use only 1.5. Growth in the number of residences since 1975 has exceeded the smaller growth in energy use resulting from more efficient buildings.

FIGURE 8.3. PRIMARY ENERGY USE IN U.S. BUILDINGS (1960–1989).

*Before 1973, total primary energy use in residential and commercial buildings was growing at 4.5 and 5.4 percent per year respectively. From 1973 to 1986, energy use in residential buildings leveled off at about 16 EJ, while commercial building energy use grew only 1.6 percent per year from 10 to 12.4 EJ. Since the collapse of oil prices in late 1985, total primary energy use for both residential and commercial buildings has grown about 10 percent, or greater than 3 percent a year. The residential GNP-projected curve is straightforward:*

$$\text{Projected energy (t)} = \frac{\text{energy } (1973)}{\text{GNP } (1973)} \times \text{GNP (t)}$$

*But for the commercial sector, the pre-1973 trend was for energy use to grow 1 percent faster than residential energy use (or GNP). Accordingly, the commercial GNP-projected curve has been tilted up by 1 percent per year to reflect this trend— that is, the formula has been multiplied by the factor 1.01 (t − 1973) ≈ 1 + 0.01 (t − 1973).*

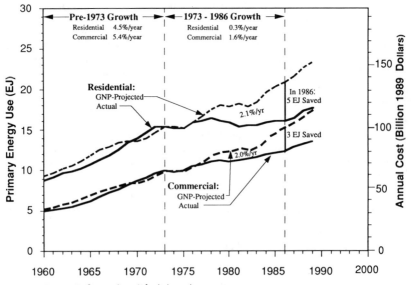

SOURCE: Energy Information Administration.

FIGURE 8.4. SPACE HEATING INTENSITY FOR NEW U.S. BUILDINGS (1960–1987).

*Since 1973, efficiency improvements in new homes and commercial buildings have reduced space heating intensity by almost 50 percent.* State-of-the-art new office buildings and superinsulated homes are now at 20 percent of their 1973 intensities. Note: *Data are for gas-heated buildings and the units are kilojoules (kj) of gas per square meter of floor area (m²), not external surface. HDD stands for Celsius heating degree-day. For commercial buildings, gas for hot water and other uses is also included.*

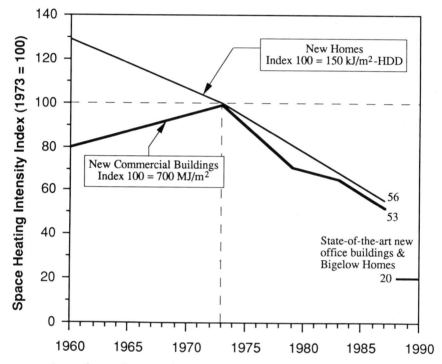

SOURCE: Source for new homes and new commercial buildings: U.S. DOE, Energy Conservation Trends, DOE/PE-0092, 9/89, figs. 19 and 21. Source for large new office buildings: ASHRAE SP 41, PNL Report 4870-18, 10/83. Source for Bigelow Homes: BECA 1991 Annual Report, Building Energy Data Group, LBL, 1991.

## BARRIERS TO IMPROVING ENERGY EFFICIENCY

Most building occupants fail to associate the services that energy provides in buildings—refrigeration, lighting, and heating—with the coal mines, oil rigs, and power plants that supply energy. Taking building energy use for granted creates many barriers to improving their efficiency. In most cases, market barriers prevent the implementation of at least half of all economically attractive energy efficiency improvements. Before 1991, utility-sponsored DSM programs in several states achieved only one-half to two-thirds of their calculated potential.[4] Now that utilities are being awarded significant extra profits for saving energy and money, bolder programs can achieve substantially more (Rosenfeld, 1991a).

### LIFE-CYCLE COSTING

*Barrier.* Unlike cars, which are replaced every 10 to 12 years, buildings last for 50 to 100 years. Thus energy efficiency improvements make tremendous economic and environmental sense, because of buildings' long lives and large energy use. Energy costs over a building's life span are comparable to initial construction costs, and the life-cycle energy costs of appliances and motors are usually many times greater than their initial cost. The current housing market is a remarkable anachronism, however, with almost no sense of life-cycle energy costing. It does not occur to most home buyers to demand the energy performance labels on buildings that are common on cars and appliances—indeed, many realtors resist labels as an intrusion on their industry. The typical consumer compares only the *first* costs of competing efficient and inefficient equipment, not their full life-cycle costs. Even those consumers who want to comparison-shop have little idea how to factor life-cycle costs into their purchasing decisions. The federal government is one of the major offenders in this regard with its contradictory policy (always ignored) of advocating buildings with the least *life-cycle* cost versus its practice (always followed) of awarding contracts based mainly on lowest *first* cost.

Corporations too seem unable to take advantage of obvious savings. Figure 8.5 shows that 50 percent of large businesses that spend at least $100,000 annually on energy bills only invest in energy efficiency when paybacks are 2 years or less. In contrast, investments in energy supply are normally tied up for 10 years or longer. Even for mandatory car, appliance, and building standards, our society tolerates only those requirements that can be met with investments that pay for themselves in 3 years or less.

FIGURE 8.5. MAXIMUM TOLERABLE PAYBACK TIME FOR
INVESTMENTS IN ENERGY EFFICIENCY.

*A survey of the 660 commercial customers of Potomac Electric Power Company was conducted to determine their typical payback thresholds for investments in energy efficiency. Some 33 percent of the customers responded that they did not know their preferred payback period, while 52 percent preferred payback periods of 3 years or less.*

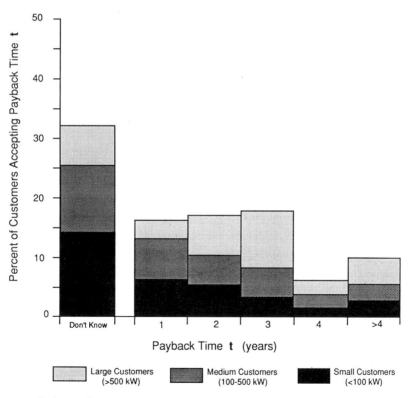

SOURCE: Barker et al. (1986).

*Solution.* Fee/rebate incentives for new residences and commercial buildings are one way to strengthen the market for energy-efficient technologies and processes by increasing consumer motivation to purchase energy-efficient structures. A "variable hookup fee" involves setting energy consumption and peak demand targets for each category of new building (office, school, and so forth). Buildings that exceed the average electricity intensity (watts per square foot) would pay a fee of $1000/kW,[5] which would be rebated to buildings that undershoot the average. The target would be adjusted annually to keep the account revenue-neutral, and a portion of the fees would be allocated to cover administrative costs for running the program. Bills proposing this policy have been introduced in both the Massachusetts and California legislatures by Representatives David Cohen and Charles M. Calderon respectively.[6]

Standards (augmented with incentives that reward beating the standards) are another way to ensure that energy-efficient products have a market. While mandatory in Western Europe and California, most of the United States has only voluntary building standards that are currently based on technologies with just 2-year or 3-year payback periods. The American Society of Heating, Refrigerating, and Air Conditioning Engineers (ASHRAE) regularly updates these standards, but their enforcement is spotty. Some states, however, are doing better. For example, California strictly enforces its "Title 24" standards for new buildings. The next step is enforcement in all states of new building standards and the implementation of standards for existing buildings whenever they change hands. In fact, some cities in California have already adopted such residential conservation ordinances (RECOs) and commercial building conservation ordinances (CECOs).

Rating systems bring the energy-efficiency characteristics of a building to the point of sale. Home Energy Rating Systems (HERS) certify a home's energy-efficiency characteristics, enabling buyers to estimate future energy bills and motivating builders to construct affordable, energy-efficient homes.

The Federal National Mortgage Association (Fannie Mae), the Federal Home Loan Mortgage Association (Freddie Mac), and the Veterans Administration (VA) recognize the reduced risk of loan default associated with the reduced life-cycle energy costs of energy-efficient buildings by offering bigger loans at lower interest rates for efficient properties. Realtors and banks, however, are not promoting these programs; only one in 10,000 home buyers takes advantage of them.

FRAGMENTED BUILDING INDUSTRY

*Barrier.* Efforts to improve energy efficiency are often obstructed by the dramatic fragmentation of the building industry. Even though energy costs in commercial buildings can easily reach 30 percent of their operating budget, lack of coordination between the architects/ designers, the building engineers and subcontractors, and the operations and maintenance staffs of large commercial buildings often results in a failure to capture potential energy efficiency savings.[7] Moreover, the ability of building designers to model the energy impacts of diverse architectural options is still limited. Without adequate design tools many efficacious combinations and design solutions may be missed.

*Solution.* HVAC and lighting typically constitute more than two-thirds of a building's energy consumption. A new breed of energy services companies (ESCOs) can design, install, and maintain sophisticated, automated, integrated control systems and HVAC and lighting systems or perform simple cleaning and calibration that dramatically improve a building's energy performance. Efficiency gains of 20 percent have been achieved in some commercial buildings simply by optimizing the performance of existing systems: replacing filters and cleaning coils in air conditioning systems, for example, and operating individual air conditioning units (chillers) in response to the time of day and outdoor air temperature (Bevington and Rosenfeld, 1990).

Large-scale retrofits, on the other hand, can easily involve a substantial capital outlay. Typically, ESCOs provide initial funding and a performance guarantee in exchange for 50 to 70 percent of the savings achieved by the retrofits, until their capital cost is repaid (Bevington and Rosenfeld, 1990). In nearly all major commercial building retrofits, a microprocessor-based energy management control system is installed to monitor and operate the HVAC and lighting systems. Strategically placed sensors feed data into the system, and skilled personnel can respond to any variations from preset parameters deemed appropriate for various locations: computer rooms can be kept cold, for example, and cafeterias can be supplied with greater airflow during meals. A multistage commercial retrofit may also involve windows, lights, and the HVAC system. Windows are either replaced or, for far less money, self-adhesive films that reflect solar heat can be applied to existing windows. Lighting fixtures can be retrofitted with reflectors that halve the number of lamps needed to deliver the same light levels. HVAC systems may be made more versatile with the addition of chillers of various sizes and cooling capacities.

As such integrated control systems evolve, they will come to incorporate self-diagnostic capabilities. Eventually, "smart systems" will reduce costs by replacing scarce, highly skilled personnel. In addition, simple design tools must be made available to ensure that designers maximize new building energy efficiency. Currently there are two types of computer programs geared for this function. DOE-2 is the national hour-by-hour simulation program that predicts a building's hourly, monthly, and annual energy use, using both weather and building design data. In contrast to DOE-2, which needs a trained operator, a user-friendly desktop design tool is also available. The ideal tool, however, would combine the technical sophistication of DOE-2 with the versatility of a desktop program. Research is currently under way to produce a program that, in less than an hour, is able to design a building that will beat energy efficiency code requirements (Selkowitz, 1991).

Building "commissioning" is another remedy for the problem of discontinuity within the building trades. To ensure that buildings perform to design specifications, third-party inspections should be

TABLE 8.3. R&D SPENDING COMPARISON

| R&D | Spending |
|---|---|
| Total U.S. | 3% of GNP |
| Nonmilitary | 2% of GNP |
| Mature industries | 1% of revenues |
| Public-domain energy efficiency | 0.04% of energy sales[a] |
| Building energy efficiency | 0.025% of utility revenues[b] |

[a] 2/5000 of bills of $500 billion a year.
[b] 1/4000 of building energy of $200 billion a year.

conducted for an initial period starting when the architect/builder hands the building over to the owner. Routine in many countries, this approach detects needed adjustments to the HVAC systems and monitors indoor air quality and lighting. In this way, systems and equipment that affect the energy demand of new buildings are de-bugged and made to perform as designed.

INADEQUATE R&D

*Barrier.* In 1989 total U.S. public-domain R&D for efficiency was approximately $200 million, less than 10 percent of public-domain R&D spending for all U.S. energy technologies. To set the scale, consider Table 8.3.

Figure 8.6 shows a time series of public-domain R&D spent on energy technologies in fields including efficiency, renewables, fossil fuel, nuclear power, and magnetic fusion. From 1978 to 1981 public energy efficiency R&D hovered around $200 million, dropping to one-twentieth of total energy R&D from 1982 to 1988 during the Reagan administration. Public energy efficiency R&D for buildings slid to 1/4000 of utility revenues, despite the dramatic economic growth achieved by efficiency improvements (demonstrated by Fig-ure 8.1), the escalating threat of global warming, and the huge successes of simple, affordable energy technologies such as those discussed later in Table 8.4. In 1991, under the Bush administration, $200 million in funding for energy efficiency did grow to $265 million, still extraordinarily out of step with the dramatic economic contribution of energy efficiency.

Figure 8.7 compares the energy performance (shaded) of various energy sources to their DOE funding (in black). Mainly because of its political muscle, fossil-fuel-related R&D is treated best, receiving $328 million per exajoule of new energy services. Nuclear R&D comes next at $78 million per exajoule supplied, largely because of devoted administrative and congressional support. Energy efficiency gets only $9 million per exajoule of services provided; it is probably neglected because there is no perception of political muscle (apart from transportation) either by the fragmented building sector or by the Congress. An added problem is that many policymakers honestly

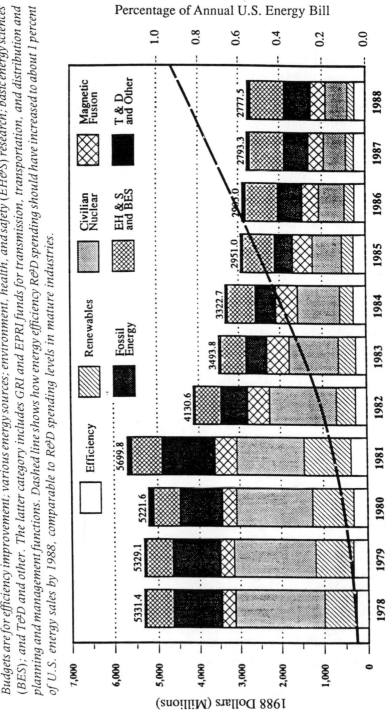

FIGURE 8.6. COMBINED ENERGY TECHNOLOGY R&D BUDGETS (DOE, EPRI, GRI, AND NRC) IN MILLIONS OF 1988 DOLLARS.

Budgets are for efficiency improvement; various energy sources; environment, health, and safety (EH&S) research; basic energy sciences (BES); and T&D and other. The latter category includes GRI and EPRI funds for transmission, transportation, and distribution and planning and management functions. Dashed line shows how energy efficiency R&D spending should have increased to about 1 percent of U.S. energy sales by 1988, comparable to R&D spending levels in mature industries.

SOURCE: Oak Ridge National Laboratory (1989).

238

Some 22 EJ of primary energy was saved as a result of energy efficiency improvements between 1973 and 1989. If efficiency is measured as the difference between
actual 1989 energy use and energy use projected by a constant 1973 E/GNP, then
full efficiency gain would be approximately 33 EJ. However, structural change is
credited with one-third of the change, resulting in an energy efficiency value of 22
EJ. Despite the fact that 22 of the 29.5 EJ of primary energy was provided through
energy efficiency, this source is allocated only $200 million of the $1.4 billion
budget.

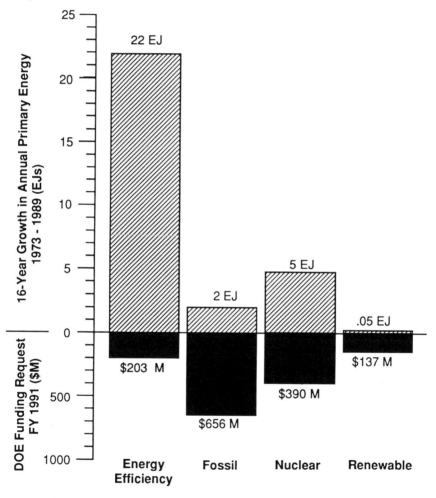

SOURCE: Funding data: DOE/MA-0400, *U.S. DOE Posture Statement and FY 1991 Budget Overview.* Energy data: DOE/EIA-0384, *Annual Energy Review 1989.* Structural change: Schipper et
al. (1990); OTA (1990).

feel that since efficiency is such a success compared to fossil fuel, nuclear, and renewables, it can take care of itself and needs little help from the government.

Of the total $200 million per year for efficiency shown in Figures 8.6 and 8.7, building-related R&D gets only $50 million per year. Compared with annual energy bills of $200 billion, that is only 1/4000 of revenues. Furthermore, the highly fragmented industry invests very little in private R&D—the majority of builders and component manufacturers are simply not large enough to do research or to lobby effectively for increased governmental R&D funding.

*Solution.* To optimize the potential benefits of energy efficiency in buildings, the federal government should follow the example of Sweden, which has a national council for buildings research whose funding is equivalent to $1 billion. The results of such a commitment are twofold: not only does Sweden lead the world in energy-efficient buildings, but its building sector runs an annual international trade surplus equivalent to $60 billion. The U.S. building sector's performance is dismal in comparison: a trade deficit of $6 billion per year (Bevington and Rosenfeld, 1990).

UTILITY SCOPE AND PROFITS

*Barrier.* Most utilities are still rewarded for providing raw energy rather than "energy services," even though it is cheaper today to invest in energy efficiency than to invest in new energy supply. As long as utilities' profits are linked to energy supply and energy sales, they will pursue supply over the efficient use of energy.

*Solution.* Recently the role of some West Coast, Northeastern, and Wisconsin utilities was dramatically redefined to allow them to *profit* from investments that improve the energy efficiency of their customers. Stockholders are now allowed to retain from 5 to 20 percent of the societal net savings resulting from efficiency investments implemented through utility-sponsored DSM programs. U.S. utilities already spend approximately $2 billion per year (1 percent of electric sales) on DSM and will probably raise spending to $5 billion per year within the next decade.

Least-cost (or "integrated resource") energy planning was mandated in 1990 in California. This approach involves ranking all energy sources (including improvements in energy efficiency) by cost (Cavanagh and Rosenfeld, 1991). An energy efficiency improvement must be implemented first—before the acquisition of new generating capacity—if the improvement costs less per kilowatt-hour *avoided* than per kilowatt-hour provided by new supply. A coalition of utilities, regulators, consumer groups, and environmental organizations met in California in the fall of 1990 to give teeth to the least-cost mandate. Based on their recommendations, the California Public Utilities Commission (CPUC) reformed its rate structure. It approved a 100 percent increase in the conservation budgets of investor-owned electric and gas utilities (Cavanagh and Rosenfeld, 1991). It also linked utilities' annual profits in part to the savings garnered by utility-sponsored DSM programs. By decoupling utility profits from volume of sales and allowing utilities to profit from reducing their demand, the CPUC provided a dramatic incentive for conservation. In response, Pacific Gas and Electric (PG&E), the state's largest utility with annual revenues of $8 billion, has established a conservation budget of $200 million (2.5 percent of revenues). PG&E plans to offset 2500 of 3300 projected megawatts of increased demand through efficiency improvements. As a result, 3 million tons of $CO_2$ emissions will be avoided (PG&E, 1990). It has also vowed to avoid adding any increases in fossil-fueled generating capacity for the next 10 years. Statewide, California utilities invested $350 million in energy conservation programs in 1991.

Across the country, at least six other states have instituted conservation profits for their utilities, and seventeen states have adopted laws that mandate least-cost energy planning. At least seven states have approved "all-sources bidding," which is another way to promote the delivery of least-cost energy services.[8] This utility-run competitive process pits providers of new conventional energy supply against providers of conserved energy (typically ESCOs). Awards are granted on the basis of least cost per kilowatt-hour supplied or avoided. Utilities in Maine, Massachusetts, and Virginia have already conducted all-sources bidding auctions and have awarded contracts to ESCOs to install lighting, HVAC, and other retrofits.

## ADVANCED TECHNOLOGIES AND
## SENSIBLE REDISCOVERIES

In this section we discuss some fruits of recent energy efficiency R&D: lighting, windows, and refrigerators. Table 8.4 shows that just these three examples can free up sixty-seven base-load power plants and avoid the need to burn fossil fuel equivalent to 2 million barrels of oil a day—about ten times the projected production from the Arctic National Wildlife Refuge (ANWR). Also considered here are advanced insulation and whole buildings. Not all energy conservation is high-tech. As we shall see, there is great merit in merely cleaning out inefficient deadwood such as the familiar black light-absorbing collars in recessed ceiling fixtures (which absorb about 5 GW of peak power) and high-flow shower heads (which waste about another ANWR).

### ADVANCED TECHNOLOGIES

During the past decade, significant strides have been made in the development of low-cost, environmentally sound, energy-efficient technologies.

### WINDOWS

Heat losses and gains through windows are responsible for 25 percent of all heating and cooling requirements in U.S. buildings (Rosenfeld, 1991a). The fossil fuel equivalent of this heat loss alone is the 1.8-million-barrel daily output of the Alaskan pipeline or that of prewar Kuwait. Leaky European windows west of the Urals similarly sop up the equivalent of all North Sea (or Iraq) oil.

Low-emissivity (low-E) windows are now available that resemble conventional double-glazed "thermopane" windows but block the radiation of indoor heat to the outdoors. These low-E windows have a thin film coating on at least one of their interior surfaces. The film is transparent but reflects up to 85 percent of room-temperature heat back into the building. These windows function at R-3 levels. (A single pane of glass offers only minimal resistance to heat conductance and is rated R-1; "thermopane" is R-2.) Recently manufacturers have further increased performance by filling the space between the two sheets of glass with gases such as argon and krypton. Argon raises the window's resistance 33 percent to R-4.

The economics of low-E windows are impressive, their payback time is only a few years, and they provide such comfort that they will nearly saturate the market. As shown in Table 8.4, at saturation 70 million low-E windows will be sold in the United States every year. The net annual savings from these windows will be $4 billion. In the period 1985–1990, some 50 million of these windows were sold in the United States; they have already saved $3 billion in cumulative avoided energy bills. A small low-E window (1 m²) costs $10 wholesale (or $20 retail) more than a typical thermopane window, but it saves 10 to 15 million Btu over its 20 to 30-year lifetime (equivalent to 80–120 gallons of gasoline), worth approximately $70 in avoided energy bills.

Although low-E windows were originally designed to reduce outward radiation of heat in cold winters, they equally reduce inward radiation during hot summers. As developing countries become urbanized and industrialized, their peak electrical demand moves to hot afternoons because of increased air conditioning and industrial requirements. If new buildings in these countries are constructed with low-E windows, air-conditioning demands can be significantly reduced. In the ASEAN countries of Indonesia, Malaysia, the Philippines, Singapore, and Thailand, urban areas are growing rapidly. In these countries, over 30 percent of the electricity generated is used by commercial buildings (about half of it for air conditioning) and this percentage is expected to increase (Levine, 1988). For a centrally air-conditioned office building in Bangkok, 1 m² of low-E window reduces the building's heat gain such that over 60 kWh can be avoided annually for a savings of approximately $5 (Gadgil et al., 1991). Even more striking, the net first cost is *negative*—that is, the savings from downsizing the air conditioner pays for the better windows.

For an investment of $10 million, a low-E window coating plant can be constructed in countries with high air-conditioning demands. Such windows should last about 30 years, so a plant can produce new windows for 30 years and then just keep up with replacements. In a year, one plant will produce about 2 million square meters of low-E windows. So, over 30 years, the same plant will

TABLE 8.4. ECONOMICS OF THREE NEW ENERGY EFFICIENCY TECHNOLOGIES AND APPLIANCE STANDARDS

| Costs and benefits | NEW TECHNOLOGIES | | | Total | Standards for refrigerators and freezers (1976 base case vs. 1985 CA stds.) |
| --- | --- | --- | --- | --- | --- |
| | High-frequency ballasts vs. core-coil ballasts | Compact fluorescent lamps[a] vs. incandescents | Low-E (R-4) windows vs. double-glazed windows per small window (10 ft²) | | |
| 1. Unit cost premium | | | | | |
| a. Wholesale | $8 | $5 | $10 | | |
| b. Retail | ($12) | ($10) | ($20) | | ($100) |
| 2. Characteristics | | | | | |
| a. % energy saved | 33% | 75% | 50% | | 66% |
| b. Useful life | 10 yr | 3 yr | 20 yr | | 20 yr |
| c. Simple payback time (SPT) | 2 yr | 1 yr | 2 yr | | 1 yr |
| 3. Unit lifetime savings | | | | | |
| a. Gross energy | 1330 kWh | 440 kWh | 10 MBtu | | 24,000 kWh |
| b. Gross $ | $100[b] | $33[b] | $70[b] | | $1800 |
| c. Net $ [3b − 1a] | $92 | $28 | $60[c] | | $1700 |
| d. Gross equivalent gallons[d] | 100 | 40 | 80 | | 1920 |
| e. Miles in 25-mpg car | 2500 | 1000 | 2000 | | 48,000 |
| 4. Savings (1985–1990) | | | | | |
| a. 1990 sales | 3 M | 20 M | 20 M | | stable sales |
| b. Sales 1985 through 1990 | 8 M | 50 M | 50 M | | (not a new |
| c. Cum. net savings [4b × 3c] | $750 M | $1.4 B | $3 B | $5 B/5 yr | product) |

| | | | | | |
|---|---|---|---|---|---|
| 5. Savings at saturation | | | | | |
| a. U.S. units | 600 M | 750 M | 1400 M | | 100 M |
| b. U.S. annual sales | 60 M | 250 M | 70 M | | 6 M |
| c. Annual energy savings [5b × 3a] | 80 BkWh | 110 BkWh | 0.3 Mbod | | 144 BkWh |
| d. Annual net $ savings [5b × 3c][e] | $5.5 B | $7 B | $4 B | $16.5 B/yr | $10 B |
| e. Equivalent power plants[f] | 16 "plants" | 22 "plants" | | | 29 "plants" |
| f. Equivalent offshore platforms[g] | 45 "platforms" | 60 "platforms" | 30 "platforms" | | 78 "platforms" |
| g. Annual $CO_2$ savings[h] | 55 Mt | 80 Mt | 18 Mt | 153 Mt | 100 Mt |
| 6. Project benefits | | | | | |
| a. Advance in commercialization | 5 yr | 5 yr | 5 yr | | 5 yr |
| b. Net project savings [6a × 5d] | $27.5 B | $35 B | $20 B | $82.5 B | $50 B |
| 7. Cost to DOE for R&D | $3 M | $0(i) | $3 M | $6 M | $2 M |
| 8. Benefits/R&D cost [6b/7] | 9000:1 | 6500:1 | 6500:1 | 14,000:1 | 25,000:1 |

[a] Calculations for CFLs based on one 16-W CFL replacing thirteen 60-W incandescents, burning about 3300 hours/year, assuming that a CFL costs $9 wholesale, or $5 more than the wholesale cost of thirteen incandescents. For retail we take $18–$8.

[b] Assuming a price of 7.5¢/kWh for commercial-sector electricity and a retail natural gas price of $7/MBtu (70¢/therm).

[c] For hot-weather applications where low-E windows substantially reduce cooling loads, air conditioners in new buildings can be downsized, saving more than the initial cost of the low-E window.

[d] Assuming that marginal electricity comes from oil or gas at 11,600 Btu/kWh, thermally equivalent to 0.08 gallon of gasoline.

[e] Net annual savings are in 1990 dollars, uncorrected for growth in building stock, changes in real energy costs, or discounted future values. See Geller et al. (1989:table 1).

[f] One 1000-MW base-load power plant supplying about 5 BkWh/year.

[g] One offshore oil platform = 10,000 barrels of oil a day (bod). To convert "plants" burning natural gas to "platforms": 1 "plant" × 27,000 bod × 2.7 "platforms." The Alaska National Wildlife Refuge, at 0.3 Mbod, is equivalent to about 30 "platforms."

[h] 1989 U.S. emissions of $CO_2$ were 5000 Mt.

[i] Descended from high-frequency ballasts (only DOE assistance was in testing).

SOURCE: Rosenfeld (1991b); adapted from Geller et al. (1989:tables 1 and 4).

have produced 60 million square meters of windows, each saving 60 kWh per year, for a total savings of 3600 million kilowatt-hours (3.6 terawatt-hours), which is equal to the output of a 700-MW base-load plant (Gadgil et al., 1991). For a smaller investment than is required for a coating plant, rolls of coated plastic could be imported and used in locally manufactured windows. (This discussion is based on Rosenfeld and Price, 1991.)

A new generation of "superwindows," rated from R-6 to R-10, pushes the performance of windows even farther. One way to step up a window's performance from R-4 to R-8 is to stretch a plastic film with a Heat-Mirror coat between the double glazing of an R-4 low-E window. The two low-E films face gaps that can be filled with argon or krypton. When the window frame is improved, as well, the advanced window's performance is increased to R-8. In field tests, superwindows outperform 6-inch-thick R-19 walls because they let sunlight into the building during the day and block heat loss to the outdoors at night. The superwindow is a net energy gainer, whereas the surrounding wall only prevents heat loss (Bevington and Rosenfeld, 1990). Moreover, superwindows minimize condensation and reduce the damage to interior furnishings by blocking ultraviolet rays. Although these windows can cost 20 to 50 percent more than conventional windows, their initial cost is repaid in about 4 years of avoided energy bills.

Selective coatings on advanced windows can replace the mirrored "solar control" glass used in office buildings. They are transparent and allow visible light to enter buildings, but reflect the near-infrared half of solar radiation that overheats interiors, raises air conditioning bills, and increases peak power demand. Because they are transparent, they permit daylighting of the interior spaces. Electronic and photosensitive windows are now under development. They switch from clear to white or dark under electronic or thermal control. Automobiles will probably be the first market, but as costs drop the new windows will penetrate the buildings market.

LIGHTING

Incandescent lighting consumes about 8 percent of all U.S. electricity, or the equivalent of forty base-load power plants' output. Compact fluorescent lamps (CFLs) are rapidly gaining acceptance in the

United States and are beginning to be used in the developing world, where they save both energy and scarce foreign exchange for capital-poor economies.

Fluorescent lamps are about four times more efficient than their 100-year-old incandescent ancestors. But before the advent of high-frequency ballasts in the late 1970s, inefficiency at each end of these lamps made it impractical to make the lamp shorter than about 2 feet. Development of the high-frequency ballast eliminated these losses and ushered in the CFL, which can be screwed into sockets normally occupied by incandescents. The economics of CFLs are shown in Table 8.4. An individual CFL rapidly pays for itself through reduced energy bills. For example, one 16-W CFL replaces a series of about one dozen 60-W incandescents since it burns twelve times longer than the incandescents. As shown in Table 8.4, this CFL would save 440 kWh and about $33 in electricity costs over its 40-month life in a commercial building.

A modern, automated CFL plant costs $7.5 million and can pro-duce 6 million lamps annually, each of which will save 440 kWh over its service life. The product of 6 million lamps per year multi-plied by 440 kWh saved per lamp is a total savings of approximately 2.5 BkWh per year, equivalent in the United States to the sales of a 500-MW intermediate or 500-MW base-load power plant that costs $1 billion to construct and $200 million per year to operate. It is expected that CFLs will soon penetrate enough of the U.S. market to save half of the 200 BkWh used annually by incandescents. As shown in Table 8.4, the net savings is $7 billion per year. The challenge now is to transfer this technology internationally—especially to developing countries where dramatic energy savings are possible in expensive peak power.

Many developing and Eastern European countries face explosive electricity demands and large end-use inefficiencies. For such coun-tries, construction of CFL production plants and installation of energy-saving CFLs are painless ways to conserve scarce electricity. In India, for example, demand for electricity is so great that 70 percent of the residences are not electrified and many industries suffer power shortages. Instead of meeting marginal electricity de-mand with capital-intensive, environmentally destructive generat-ing plants, India could construct CFL plants for a fraction of the cost. Due to the poor quality of power, incandescent lamps must run

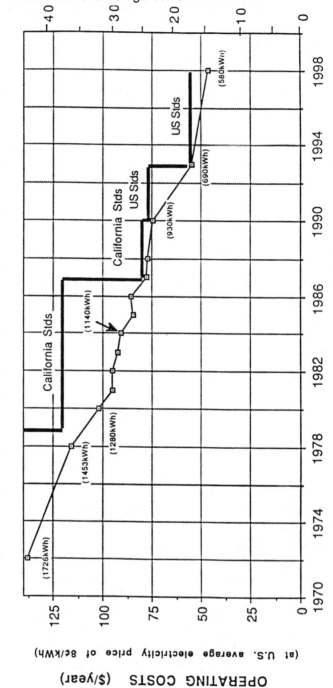

FIGURE 8.8. AVERAGE NEW REFRIGERATOR ANNUAL ELECTRICITY USE.

The 1998 estimate assumes the use of advanced insulation without CFC or HCFC and with an R value of 20. From 1977 to 1987, changes required an additional retail cost of $100: savings $40/year; 2.5 year payback time. From 1990 to 1993, the estimated additional retail cost is $66; savings $19/year; 3.5 year payback. In 1990, federal standards superseded California standards.

Typical 1000-MW baseload power plants needed to run 125 million U.S. refrigerators and freezers

SOURCE: Data from 1972 to 1988 are from AHAM. Data for later years are estimates by LBL.

inefficiently cool, so the alternative CFLs produced in 5 years by one CFL plant would save the equivalent of 3700 MW of installed peak capacity (as contrasted with savings equivalent to a 500-MW base-load power plant in the United States). Providing such capacity from traditional energy sources in India would cost between $2.8 billion for gas turbines and $5.6 billion for coal-fired thermal power stations—up to a thousand times more than the $7.5 million needed for the CFL production plant (Gadgil et al., 1991).

India is an example of the *twice*-tilted playing field in the competition between efficiency and supply. First, import taxes on components for power generation are only 30 percent, but CFLs are charged a customs duty of over 250 percent. Second, there is no residential market for CFLs because residential electricity is subsidized by commercial and industrial customers. To overcome these barriers, an experimental program promoting installation of CFLs in residences in Bombay began in late 1989. In the Bombay Efficient-Lighting Large-Scale Experiment (BELLE), the utility will purchase CFLs wholesale and lease them to customers for a modest 5¢ per month. Customers benefit because their electricity bills will be reduced by about 7¢ per month and they no longer have to purchase incandescent light bulbs. The community benefits because electricity saved by this project will be made available to other customers (those that either suffer from electrical shortage or are not electrified) and will reduce electrical demands that would otherwise have to be met by new power plants (Gadgil et al., 1991).

APPLIANCES

Figure 8.8 uses the refrigerator as an example of the spectacular efficiency gains in new appliances brought about by the 1978 California appliance standards (which raised refrigerator efficiency for the entire U.S. market) and the forthcoming 1993 federal standards. The improvements required by California between 1977 and 1987 saved about $60 of electricity per year for each refrigerator at a retail cost premium of $100. The simple payback time of this dramatic improvement is less than 2 years. For the current improvements required by the 1993 federal standard, the cost premium is $66 and the savings is about $20 per year (making the payback time about 3 years). The right-hand scale of Figure 8.8 is the number of base-load power plants it takes to run 125 million U.S. refrigerators and freezers. By the time the stock reaches the 1993 standard, we will have freed up nearly thirty base-load plants compared to 1971. Further economics are shown in the right-hand column of Table 8.4.

INSULATION

Optimum U.S. homes have walls with 6 inches of insulation (R-19) enclosed by an outer insulating sheath (for a total of R-24) in contrast to standard homes that use only a scant 4 inches of insulation (R-11). Some builders (notably the Chicago Bigelow Group) have so much confidence in the performance of optimum and superinsulated residences that they offer to pay the amount by which buyers' annual energy bills exceed $100 for townhouses ($200 for private homes). In 1989 the contest for the homeowner with the lowest heating bill was won by a customer who paid only $24 for heating for an entire year.

Concern over the ozone hole and global warming has caused U.S. manufacturers to discontinue the manufacture of chlorofluorocarbons (CFCs), used most significantly as the working fluid of refrigerators. Less well known is the fact that a typical U.S. refrigerator contains about one-half pound of CFCs in the compressor and twice that much in the CFC-blown insulating foam that fills the shell of the box. Hence there are now many parallel R&D efforts under way to replace this foam. At least four kinds of advanced superinsulation are currently being developed with the support of the Department of Energy. "Aerogel" is transparent or translucent and can therefore be used in windows and skylights as well as in appliances. In its opaque, evacuated form for refrigerators, it has reached R-30 per inch. Another form of superinsulation is an outgrowth of advances in low-E window technology. Gas-filled panels (GFPs) are an assembly of specialized components that simultaneously minimize radiative, conductive, and convective heat transfer. Multiple layers of highly reflective metallized polymer film compartmentalize the interiors of the GFPs, virtually eliminating radiative heat transfer. Much thinner, crumpled film is inserted between these parallel layers to minimize convection and further reduce radiative heat transfer. To reduce conduction, a low-thermal-conductivity gas (such as argon or krypton) or gas mixture is encapsulated at atmospheric pressure within sealed panels. These GFPs may someday be applied to pipe and duct insulation, hot water heaters, swimming pool and spa covers, refrigerated transport walls, airplane bodies, and even homes. Compared with fiberglass at R-3.5 per inch, GFPs have tested at R-7 to R-13 per inch and an R-15 performance is anticipated.

WHOLE BUILDING

Applications of energy-saving technologies and rediscoveries, such as those described in this chapter, can result in significant energy savings. California's PG&E has begun its ACT² (Advanced Customer Technology Test for Maximum Energy Efficiency) demonstration project aimed at determining the energy savings potential for building retrofits and new buildings. This project has just completed the design competition for its own 22,000-square-foot leased office building in San Ramon, California. Energy savings estimates from the competing design firms ranged from 67 to 87 percent of the building's current energy use (PG&E, 1991). These large savings will be accomplished in several ways: through exterior and interior "light-shelves" that make daylighting possible 24 feet into the office space; through conversion to "superwindows" that reduce heating and cooling loads; through improved lighting systems that automatically adjust to changing daylight levels; through replacement of office equipment with more energy-efficient models; and through downsizing the building's cooling system by at least two-thirds (Lovins, 1991).

SENSIBLE REDISCOVERIES

Table 8.4 shows that energy efficiency R&D for the building sector has a dramatic payback to the taxpayer—measured in *hours* rather than years. But to save several more ANWRs' worth of energy for free, we need only realistic energy prices, or standards, or enlightened utility programs to remind us to reinstate sensible practices abandoned since World War II when energy became contemptibly cheap. It is time to clean out the deadwood by implementing well-known technologies such as low-flow shower heads and reflective lighting fixtures that have been on the shelf for years. Beyond these obvious measures, whitening urban surfaces, planting urban trees, taking advantage of daylighting practices in buildings, and utilizing thermal storage techniques remind us of lessons learned long ago that have somehow been ignored.

### INSTALLING LOW-FLOW SHOWER HEADS

The United States would save natural gas equivalent to about 0.25 million barrels of oil a day (the expected yield of ANWR) if all the states adopted a $10 device: the 2.5 gallon per minute low-flow shower heads that are not only required in California but popular.

### ELIMINATING BLACK COLLARS IN RECESSED LAMP FIXTURES

In U.S. commercial buildings there are about 300 million recessed fixtures for incandescent lamps. The high benefit-cost ratio of replacing incandescent lamps with CFLs has been detailed already. But to add insult to injury, 100–200 million of the incandescent fixtures are still equipped with the all-too-familiar black "microgroove" collars that were designed (before World War II) to keep a bright filament out of sight. Viewing filaments has been a nonissue since World War II—either because bulbs are frosted or, even better, because unfocused lamps have been replaced by floodlights or spotlights.

Many lighting designers care much more about style than energy costs, however, so black collars persist. The catch is that an absorber which allows omnidirectional light to escape only downward in a 45-degree cone necessarily absorbs three-quarters of that light. Thus if the lamp is 60 W, the black collar probably absorbs 40 W. If the lamp burns 3000 hours per year, the collar absorbs 120 kWh, worth about $10 per year. Multiplied by 100 million, the annual cost of this wasteful technology is $1 billion per year. If designers were trained in energy efficiency, three or four base-load power plants would be liberated from powering the light and heat absorbed by such fixtures.

### WHITENING SURFACES AND PLANTING URBAN TREES

Most U.S. cities are 3 to 6°F hotter on summer afternoons than the forests or farms that surround them. These "urban heat islands" arise because heat-absorbing asphalt (on roads and the flat roofs of buildings) have replaced trees and fields. Downtown Los Angeles is 7°F hotter than in 1940 and is heating up by 1°F every 8 years.

*Direct Effect.* An unshaded U.S. home with a dark or terra-cotta tile roof typically needs 2 to 3 kW for air conditioning. This costs the average homeowner between $100 and $300 annually. In contrast, a shaded home with a light-colored exterior generally uses only half this amount of electricity.

*Indirect Effect.* Such simple, low-tech mitigation measures (planting urban shade trees and whitening surfaces of roofs, streets, and sidewalks) can garner tremendous savings when their impact is aggregated for an entire city, state, or region. In L.A. alone, $100,000 per hour in peak summer air-conditioning costs would be saved (Akbari et al., 1990). Lowering temperatures in urban heat islands by 5°F could save U.S. rate payers about $1 billion per year in avoided air conditioning. And since smog "cooks" slower at reduced temperatures, eliminating heat islands would significantly lower the levels of smog. Of course, such temperature reductions would also reduce (by 1 to 2 percent) the power plant $CO_2$ emissions associated with generating U.S. electricity.

DAYLIGHTING

Designing buildings that maximize sunlight to illuminate interiors is a low-tech approach to reducing a building's energy demand while meeting its lighting needs. This practice is widespread in Europe, where every office has an outside window, but it is still underutilized in the United States. Just 1 square foot of direct sunlight can actually illuminate 200 square feet of interior space if it is evenly distributed through a skylight or clerestory window (Bevington and Rosenfeld, 1990).

THERMAL ENERGY STORAGE

In a well-insulated building with low-E well-managed windows, thermal storage can economically save most of the energy now used for heating and cooling. There are several proven methods of storing heat.

*Passive Solar Heat.* In winter a superinsulated home with exposed masonry floors and perhaps a south-facing Trombe wall (Hafemeister et al., 1985) will hold heat for days. If its windows are concentrated toward the south and are well managed, it can get through the winter comfortably on the free heat from sun, appliances, and occupants, plus space heat corresponding to about 1 Btu/square foot/degree-day (°F)—compared with 8 units for a typical pre-1975 U.S. home or 4–5 units for newer homes.

*"Charging" with Nighttime Cool Air.* In the summer throughout most of the United States, the same superinsulated home—with a white roof, well-managed windows, and a whole-house fan to draw in cool night air—can remain comfortable all day with no air conditioning or just one or two small window units.

*Well-Insulated Office Buildings.* The same logic applies to offices, where in fact there is more free heat than in homes. In Sweden, a modern office building is designed to store free heat during a winter week, cool slightly over the weekend, and use the stored heat to warm itself on Monday morning. Accordingly, most offices purchase only minimal heat from the Stockholm district heating system. Similar strategies of night cooling are used during the summer and greatly reduce the demand for peak power. For further discussion of thermal storage in commercial buildings, see Hafemeister et al. (1985).

## CONCLUSION

It is crucial that we delay growth in annual carbon dioxide production during the 10 to 20 years it will take to better assess the seriousness of the threat of global climate change and introduce a nonfossil energy supply. As we have seen, however, new buildings are so efficient compared to U.S. stock that the building sector can grow 2 or 3 percent a year but use very little more energy as older, less efficient buildings retire at the rate of 1 percent a year. Indeed, with a combination of a vigorous energy policy, least-cost energy services marketed by utilities, and expanded R&D, we could actually reduce the energy needs of our growing building sector by 1 or 2 percent a year.

## NOTES

1. One exajoule ($10^{18}$ joules) equals $1/1.054$ quadrillion ($10^{15}$) Btu.
2. Windows in the United States leak as much heat every year as is produced from the annual output of the Alaskan pipeline (approximately 1.8 million barrels of oil per day).
3. In the United States, incandescent light bulbs and refrigerators each use 7 to 8 percent of the nation's electricity. In combination these two end uses alone require the output of seventy large (1000-MW) power plants.
4. A Michigan Electric options study characterized the state's conservation potential in residential lighting and appliances at 60 percent. This estimate shrank to 28 percent in actual savings from utility conservation programs (Rosenfeld, 1991a).
5. The value of $1000/kW is roughly equivalent to the cost of an avoided kilowatt of peak capacity.
6. Massachusetts House Bill 5277, "An Act Reducing the Greenhouse Effect by Promoting Clean and Efficient Energy Resources," was defeated in 1990 and 1991. The 1992 version of the bill is being introduced by Representative Cohen. In California, Senate Bill 1210, "Pilot-Project Variable Utility Hook-Up Fee for Energy-Efficient Buildings," sponsored by Senator Calderon, died in committee for lack of submission of pilot-project plans by a utility.
7. Clear-cut remedial efforts such as replacing worn-out HVAC and lighting systems with more energy-efficient technology can alone cut a building's energy bills by 30 percent.
8. California, Colorado, Connecticut, Maine, Massachusetts, New York, and Virginia have approved all-sources bidding. Michigan, New Jersey, and Vermont are in the process of doing so (Rosenfeld, Mowris, and Koomey, 1990).

## REFERENCES

Akbari, H., A. H. Rosenfeld, and H. Taha. 1990. Summer heat islands, urban trees, and white surfaces. *Proceedings of the American Society of Heating, Refrigeration, and Air-Conditioning Engineers Conference.* Atlanta: ASHRAE.

Barker, B. S., et al. 1986. *Summary Report: Commercial Energy Management and Decision Making in the District of Columbia.* Potomac Electric Company.

Bevington, R., and A. H. Rosenfeld. 1990. Energy for buildings and homes. *Scientific American* 263(3):77–86.

Cavanagh, R. C., and A. H. Rosenfeld. 1991. The threat of global warming: Statehouse effect combats greenhouse effect. *Journal of State Government* 64(4):94–99.

Gadgil, A., A. H. Rosenfeld, D. Arasteh, and E. Ward. 1991. Advanced lighting and window technologies for reducing electricity consumption and peak demand: Overseas manufacturing and marketing opportunities. Paper presented at the IEA/ENEL Conference on Advanced Technologies for Electric Demand-Side Management, 4–5 April 1991, Sorrento, Italy.

Geller, H. S., J. Harris, M. Ledbetter, M. D. Levine, E. Mills, R. Mowris, and A. Rosenfeld. 1989. The importance of government-supported research and development in the U.S. buildings sector. *Electricity: Efficient End-Use and New Generation Technologies and Their Planning Implications.* Lund, Sweden: Lund University Press.

Hafemeister, D., H. Kelly, and B. Levi. 1985. Energy sources: Conservation and renewables. *Proceedings of the American Institute of Physics Conference.* Vol. 135. Washington: American Institute of Physics.

Levine, M. 1988. ASEAN building energy conservation program. *Proceedings of the Conference on Energy Efficiency Strategies for Thailand.* LBL Report 26759. Berkeley: Lawrence Berkeley Laboratory.

Lovins, Amory B. 1991. Direct testimony and exhibits. Land and Water Fund of the Rockies, docket no. 91S-091EG, Phase I.

National Academy of Science (NAS). 1991. *Policy Implications of Greenhouse Warming: Report of the Mitigation Panel.* Washington: National Academy Press.

Oak Ridge National Laboratory (ORNL). 1989. *Energy Technology R&D: What Could Make a Difference?* Pt. I: *Synthesis Report.* ORNL-6541/V1. Oak Ridge, Tenn.: ORNL.

Office of Technology Assessment (OTA). 1990. *Energy Use and the U.S. Economy.* Washington: OTA.

Pacific Gas and Electric Company (PG&E). 1990. *Annual Report.* San Francisco: PG&E.

———. 1991. *Facts on ACT².* Issue 3. San Francisco: PG&E.

Rosenfeld, A. H. 1991a. Energy-efficient buildings in a warming world. *Proceedings of the Massachusetts Institute of Technology Conference: Energy and the Environment in the 21st Century.* Cambridge: MIT Press.

———. 1991b. *The Role of Federal Research and Development in Advancing Energy Efficiency: Hearing on DOE Conservation Budget Request.* Statement before James H. Scheuer, Chairman, Subcommittee on Environment, Committee on Science, Space, and Technology, U.S. House of Representatives.

Rosenfeld, A. H., and L. K. Price. 1991. Making the world's buildings more energy-efficient. Paper presented at the Technologies for a Greenhouse-Constrained Society Conference, 11–13 June 1991, Oak Ridge, Tennessee.

Rosenfeld, A. H., R. J. Mowris, and J. Koomey. 1990. Policies to improve energy efficiency, budget greenhouse gases, and answer the threat of global climate change. Paper presented at the annual meeting of the American Association for the Advancement of Science, 15–20 February 1990, New Orleans.

Rosenfeld, A. H., C. Atkinson, J. Koomey, A. Meier, R. J. Mowris, and L. K. Price. 1991. A compilation of supply curves of conserved energy for U.S. buildings. Paper contributed to National Academy of Science, *Policy Implications of Greenhouse Warming: Report of the Mitigation Panel*, app. B. Washington: National Academy Press.

Schipper, L., R. B. Howarth, and H. Geller. 1990. United States energy use from 1973 to 1987: The impacts of improved efficiency. *Annual Review of Energy*. Palo Alto, Calif.: Annual Reviews.

Selkowitz, S. 1991. Moving DOE-2 to the designer's desktop: A proposal to develop a new design-oriented, desktop version of DOE-2 in FY92. Lawrence Berkeley Laboratory, Berkeley, California.

CHAPTER 9

# ENERGY FOR
# TRANSPORTATION

*Marc Ross and Marc Ledbetter*

THE TRANSPORTATION SECTOR in the United States is an important focus
of both energy and environmental policy because it is a large consumer of energy, primarily oil, and a large source of air pollution
linked with both urban smog and global warming. About a quarter of
all energy and 70 percent of all oil consumed in the United States is
consumed in the transportation sector. Some 96 percent of the energy consumed in this sector is from oil, 61 percent of which is
attributable to automobiles and light trucks.[1]

Furthermore, transportation fuel use is responsible for about 35
percent of U.S. carbon dioxide emissions. Automobiles and light
trucks alone account for about 22 percent (280 million metric tons of
carbon) of these emissions, making them the single largest contributor of carbon dioxide emissions in the United States.[2] Cars and light
trucks also emit about 40 percent of U.S. carbon monoxide emissions
and large fractions of other important urban air pollutants (Gordon,
1991). Given their dominant role in both energy and environmental

This chapter is based on an ACEEE report (Ross, Ledbetter, and An, 1991).

problems in the transportation sector, this chapter focuses on light vehicles.

Since 1973, new car fuel economy doubled, from 14 to 28 mpg, and light truck fuel economy increased 60 percent from 13 to 21 mpg (Heavenrich et al., 1991). Had this improvement not occurred, cars and light trucks (which now consume 6.5 million barrels of oil per day) would be consuming an additional 4 million barrels per day—more than twice as much as is being produced in Alaska and over half our current level of imports. Avoiding 4 million barrels per day of oil consumption means the United States lowered its retail fuel bill by at least $60 billion per year, lowered its trade imbalance by at least $25 billion, and is emitting about 170 million fewer metric tons of carbon in the form of carbon dioxide. (About 1300 million metric tons of carbon are emitted annually as a result of fossil fuel combustion in the United States.)

But the 15-year trend of rising new vehicle fuel economy that has almost held fuel consumption by light vehicles in check has come to a halt. Low gasoline prices, as well as a halt in the upward trend of fuel economy standards and threshold for gas guzzler taxes, have taken the pressure off automobile manufacturers to improve fuel economy. In the last few years, average new car and light truck fuel economy fell about 3 percent (Heavenrich et al., 1991).

Making matters worse, the number of highway vehicle-miles traveled (VMT) in the United States continues to grow, seemingly inexorably. Since World War II, VMT has risen steadily and rapidly, with the two major oil crises of the 1970s represented by small, temporary shifts in the upward trend (Figure 9.1). Recent analysis of the factors driving this growth indicates that VMT should continue to grow about 2.5 percent a year through the year 2000 (Ross, 1989).

Yet another factor contributing to increased oil use is the growing number of light trucks in the country's light vehicle fleet. On average, new light trucks achieve 21 mpg, 24 percent below the new car average of 27.8 mpg (test values). In 1970 they represented only 15 percent of new light vehicle sales, but they now represent about one-third of these sales. Most light trucks are being used as passenger cars. A 1987 survey found that 81 percent of light trucks do not carry any freight—not even craftsman's tools (Bureau of the Census, 1990). Thus, for most people, a light truck serves the same purpose as a car but achieves much lower fuel economy.

FIGURE 9.1. VEHICLE-MILES TRAVELED, ALL VEHICLES: 1936–1989.

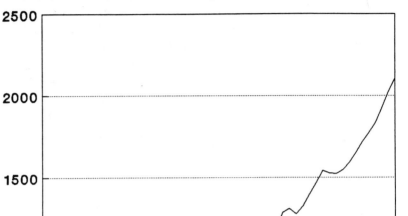

SOURCE: Federal Highway Administration.

Vehicles are also becoming much more powerful than they were in the early 1980s. Since 1982, average automobile 0 to 60 acceleration times have fallen from 14.4 to 12.1 seconds, a 16 percent drop (Heavenrich et al., 1991). We estimate that this move toward more powerful cars has reduced average new car fuel economy by almost 10 percent. Furthermore, the on-road fuel economy of new cars and light trucks is falling further behind their EPA-rated fuel economy. The on-road fuel economy is estimated to have been 15 percent lower than the EPA laboratory test value in 1982 (Hellman and Murrell, 1984). Analysts project that due to changes in driving patterns, primarily increasing traffic congestion, on-road fuel economy will fall 30 percent below test by the year 2010 (Westbrook and Patterson, 1989).

Taken together these developments—stalled fuel economy improvements, rapidly growing VMT, increasingly powerful vehicles, traffic congestion, and substitution of trucks for cars—are putting strong pressure on oil demand. Only the final stages of the replacement of inefficient cars of the 1970s with today's more efficient cars is temporarily keeping oil consumption in check. At stake are the nation's economic health, our energy security, and the earth's climate. Among the many actions that can be taken to improve the situation, increasing the fuel economy of light vehicles should have a high priority.

Major changes will be required to reduce light-vehicle energy use or even to check its growth. Substantially improving light vehicle fuel efficiency will have the single largest effect on fuel consumption. As we will see, it is within the range of technological and economic feasibility to improve the new car fuel economy to over 40 mpg within about 10 years. Similar improvements could be achieved in light trucks. Although highly important, these major improvements, if achieved, will be largely offset by fuel consumption increases caused by growing traffic congestion and vehicle-miles of travel. To achieve deep reductions in fuel use, the United States must not only improve fuel economies but slow the growth in vehicle-miles of travel and provide attractive alternatives to single-occupancy vehicles. Our focus here, however, is on the fuel economy of vehicles.

The effect of future fuel economy improvements on global warming will of course depend on the extent of these improvements. If U.S. new light-vehicle fuel economy is improved 40 percent by 2001, as now is being proposed in Congress, carbon emissions will be 120 million metric tons per year lower by the year 2005 than they would be if new light-vehicle fuel economy remains at today's levels. Although this is only 9 percent of the country's current carbon emissions, no other single improvement in end-use energy efficiency, or plausible switch to low-carbon, nuclear, or renewable fuel, will yield reductions as large by the year 2005.

If other countries were also to substantially increase their new vehicle fuel economy, much larger reductions in carbon emissions would be possible. A recent EPA report estimates that increasing the world's fleet fuel economy to 50 mpg by the year 2025 would reduce projected global warming by 5 percent in a future scenario that assumes rapid technological change and economic growth (Lashof and Tirpak, 1989). (This is not a 5 percent reduction in $CO_2$, but a 5 percent reduction in the projected average world temperature rise.) Although this figure may not seem large, one must consider the many greenhouse gases and their numerous sources. For perspective, EPA estimates that the 5 percent reduction is larger than the reduction in global warming that could be achieved through a nearly complete phaseout of CFCs by 2003 or through a rapid development of low-cost solar technology.

This chapter is organized into three sections. The first addresses the technical issues related to improving light vehicle fuel economy and determining how much it can be improved. The technologies considered here are especially effective at improving the mechanical efficiency of an automobile at part-load, an area that is particularly promising. The next section addresses the economics of improving fuel economy. In our analysis, which is limited to those technologies for which cost data are available, supply curves illustrate the cost and quantity of energy that can be saved by the year 2000 through improved automobile fuel economy. The last section addresses policy issues related to improving fuel economy and discusses alternative policies for reducing light vehicle fuel consumption.

## TECHNICAL ISSUES

Although this analysis addresses a time period of about 15 years, it is restricted to modifications of the present kind of vehicle: a vehicle with a gasoline-fueled, spark-ignition engine and today's size and power characteristics. This constraint is not due to lack of interest in alternative fuels, new light-vehicle propulsion systems, and alternative modes of transportation, all of which are important for addressing energy and environmental problems in the transport sector. Nevertheless, this country's petroleum-based personal transportation system involves an enormous investment in physical and human capital that will not be quickly replaced. Accordingly, moderate changes in light vehicles are a highly important means of addressing the dual issues of energy and environment.

Although many believe that reducing vehicle size and weight is the best way to substantially increase fuel economy, reducing the maxi-

mum power per unit of vehicle weight can also make a major contribution without degrading acceleration performance under most conditions. The subject of this analysis is yet a third kind of change—
technological improvement without reducing vehicle size or
performance—which was behind most of the progress from 1975 to
1988. We are in the midst of a remarkable period of increasing
technical capabilities. Electronic controls, new materials, and the
capability, through computers, to design a car in detail without
having to go through many stages of trial and error with real engines
and vehicles is making it practical to do the things only dreamed of
by earlier automotive engineers. Car designers can largely get rid of
vibration and noise, control engines and transmissions using optimized electromechanical systems rather than simple mechanical
linkages, and introduce much lighter materials to enable fast acceleration of vehicle parts. It is even becoming possible to control engine
performance from one revolution to the next. This technological
ferment can be sensed by reading papers of the Society of Automotive Engineers and attending their conferences.

The technologies available for improving vehicle fuel economy fall
into two broad categories. The first is load reduction—decreasing the
power required of the engine by reducing air drag, rolling resistance,
weight, drivetrain friction, and vehicle accessory loads. The second is
engine efficiency improvement—increasing the effectiveness with
which the energy in fuel is converted to useful work for powering the
car. The potential for improvement in both of these categories is
assessed by simulation analysis of a reference vehicle (AVPWR)
having a weight and power/weight ratio that are near average for the
1991 new vehicle fleet.

LOAD REDUCTION

In recent years the maximum engine power of new cars has increased. Not only has the average new-car power/weight ratio risen
from a low of 32 hp/1000 pounds for the period 1980–1982 to 40
hp/1000 pounds in 1990, but minimum acceleration times have
fallen. This increase in maximum power has apparently been a useful
marketing tool, for many customers are choosing the high-power
version of a model.

The power delivered to the drive wheels during typical patterns of
driving is relatively low, however. High power is required only in

unusual driving conditions, such as acceleration at high speeds and climbing mountains at speeds well over legal limits, situations that most drivers rarely encounter. Vehicles of average weight with modest engines, say 80 hp, can readily be used (with appropriate transmissions) to accelerate rapidly at moderate speeds, so acceleration at moderate speeds is not a rationale for high power.

Figure 9.2 is from our car simulation results: a typical distribution of time spent with respect to the engine load in the EPA urban driving cycle. For the car simulated in Figure 9.2 (AVPWR), the average engine power output is 5 hp. This is low compared to engine capabilities. At 5 hp output, more fuel is being used merely to overcome the internal frictions of a typical powerful engine than to provide the output power. Most of the time the engine operates with a part-load efficiency much lower than its best efficiency. This suggests the importance of engine downsizing coupled with load reduction and aggressive transmission management as key strategies for improving fuel economy.

For today's average car, leaving maximum engine power unchanged, we find that every 10 percent reduction in load results in a 3 to 4 percent reduction in fuel use in urban driving and a 5 to 6 percent reduction in highway driving. Load reduction technologies include reducing aerodynamic drag, reducing tire rolling resistance, weight reduction, and regenerative braking. Drivetrain efficiency can be improved through transmission technologies such as torque converter lockup, electronically controlled standard gearing, and reducing transmission friction. Accessory loads—the largest of which is air conditioning—can be cut by running them only when needed, improving component efficiencies, and reducing the need to run the accessories. Overall we find a near-term potential for 20 percent reduction in load, which would alone yield a 9 percent improvement in fuel economy.

THERMAL EFFICIENCY

Engine efficiency is the product of two factors: *thermal efficiency* (or optimal indicated efficiency), expressing how much of the fuel energy is converted into work to drive the engine and vehicle, and *mechanical efficiency*, the fraction of that work which is delivered by the engine to the vehicle. Even the best combustion-based engines

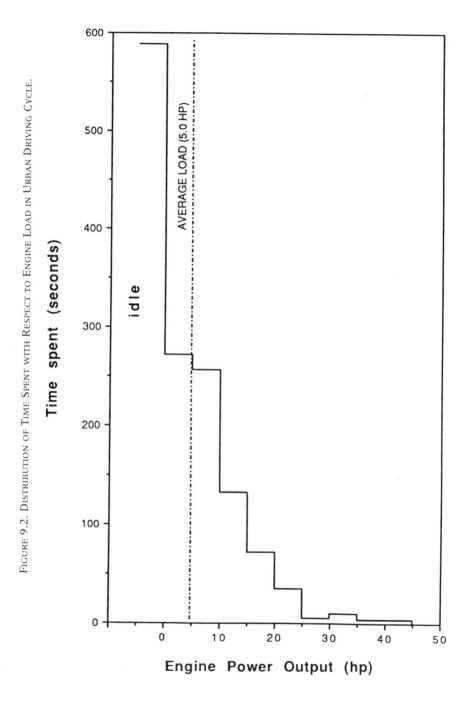

(standard electric power plants) have thermal efficiencies of only about 40 percent, and all of these engines are large, expensive, and stationary. About 50 percent will be achieved in electric power plants with the new combined cycle (and steam reinjection) technology for power plants, which involve energy recovery from the exhaust gases after the main energy conversion. (These efficiency estimates are based on the higher heating value of fuel.)

It is quite a challenge for a much less expensive mobile engine to achieve similar efficiencies. Nevertheless, certain changes in characteristics could be used to increase thermal efficiencies to approach these levels: increased compression ratio; lean burning (increased air/fuel ratio); recovery of work from the exhaust; faster combustion; effective control of working characteristics, such as the air/fuel ratio for each cylinder and each cycle instead of only for the average; and control of valve timing and enhancement of breathing so that intake and exhaust are optimized at each engine speed. The first three options are considered here (Amann, 1989).

In principle, compression ratios could be increased from today's typical value of 9 toward the best value of, perhaps, 15—resulting in a nominal efficiency improvement by a factor of about 1.15. In practice, not only does rubbing friction increase with compression ratio but wall effects (cooling and unburnt fuel associated with the surfaces) increase, so the improvement is not as good as simple analysis suggests (Muranaka et al., 1988). Moreover, high compression causes knock. Further knock inhibition than already achieved is possible—but not much more unless low-knock alternative fuels are adopted, such as alcohol, methane, or hydrogen.

Lean burning is advantageous in terms of efficiency because a gas of simple molecules when heated increases its pressure more than a gas of complex molecules (like vaporized gasoline); with complex molecules much of the thermal energy is diverted into internal motions. Radically increasing the air/fuel ratio by a factor of 2.5 (above the chemically correct stoichiometric value), for example, would increase efficiency again by a nominal factor of 1.15. Moreover, if the air/fuel ratio could be widely varied while still obtaining satisfactory combustion, this method could be substituted for a throttle to regulate engine power output. Significant additional mechanical efficiency benefits would result. But lean operation prevents the emission control system, based on a three-way catalyst, from reducing nitrogen oxides ($NO_x$), $NO_x$ emissions from the engine, at a given power level, may not be as low as one would hope; and lean mixtures can fail to ignite (misfire) or lead to incomplete combustion. Several engine manufacturers are working to overcome these drawbacks, and recent announcements by Honda and Mitsubishi indicate that they are having some success (Keebler and Maskery, 1991). Moreover, a more radical approach to lean burning is said to be achieving success: the two-stroke engine with modern fuel injection and controls.

The recovery of energy from the exhaust is analogous to the combined-cycle power plant just mentioned. Although the exhaust carries away about 40 percent of the fuel energy from the vehicle, the quality of this energy is low. The most work that could in principle be extracted from the exhaust is 15 to 20 percent of the initial fuel energy—and converting low-quality energy into work is costly and inefficient. The U.S. Department of Energy has, however, been conducting a development project on such "bottoming cycle" technology for large trucks. DOE is also supporting research on thermal insulation for the cylinder walls, which would increase the work recoverable, in principle, from the exhaust. It is too early to tell if there is any promise for automobiles.

In summary, improving thermal efficiency by a factor of about 1.25, from roughly 40 to 50 percent, is an important but challenging goal. One way to achieve it is to solve the environmental problems of the diesel and adopt modern direct-injection diesels such as those now in use in several European cars. Another way is to develop successful lean-burning engines. Still another way is to switch to a fuel with much simpler molecules and high octane—hydrogen, methane, or methanol, for example—designing a high-efficiency engine for that fuel. Achieving still greater improvements in thermal efficiency in internal combustion engines is likely to be impractical.

MECHANICAL EFFICIENCY

The mechanical efficiency of typical U.S. cars is roughly 35 percent· when averaged over the urban driving cycle and about 50 percent in the highway driving cycle. Over the composite cycle, the mechanical efficiency averages about 40 percent. It is lower for high-powered cars and higher for low-powered cars. The mechanical efficiency is zero when the engine provides no power output (an idling engine), and it is correspondingly low when the engine is operated well below its maximum power at wide-open throttle (at any engine speed). At wide-open throttle, mechanical efficiency is 85 to 90 percent. Unlike thermal efficiency, where it is not practical to achieve efficiencies more than about 50 percent, it is practical to achieve near-perfect mechanical efficiencies. We will show that it is a practical goal to increase the average mechanical efficiency close to 65 percent—a factor of 1.6 above today's average of about 40 percent in the composite driving cycle.

Improving mechanical efficiency at a given load requires reducing the power necessary to operate the engine—in particular, the energy used for pumping, for overcoming rubbing friction, and for driving engine accessories. (Pumping refers to moving the air and vaporized fuel into the cylinders and the combustion products out through the exhaust system.) There are many strategies for achieving this reduction: reducing engine size, reducing the sources of friction themselves, reducing engine speed, turning off the engine when it is not needed. These opportunities are the main subject of this section.

THE ANALYSIS

Our method of analysis is based on computer simulations of the fuel economy technologies in reference vehicles. The first is a vehicle simulation model (AUTOEFF) developed by Terry Newell for EPA in the late 1970s, modified by Frank Von Hippel and Barbara Levi, and further modified by the authors. The second is a simple algebraic approximation briefly discussed in the next paragraph. A combination of these two approaches is used for most of the analysis. A one-cylinder engine simulation model from General Motors Research is also used to study variable valve timing.

Part of this analysis rests on a simple approximation for fuel use as a sum of two terms: one to overcome engine friction, the other to provide output power. The validity of this approximation is exemplified by Figure 9.3, where fuel use per revolution is shown on the $y$-axis and energy output per revolution on the $x$-axis at different engine speeds. All the operating points lie essentially on one straight line. Thus the rate of fuel use has the form

$$aN + bP$$

where $N$ is the engine speed (rpm) and $P$ is the engine power output or load (in kW or hp). The constant $b$ is essentially the reciprocal of

FIGURE 9.3. VOLKSWAGEN JETTA 1.8-LITER (79 KW).

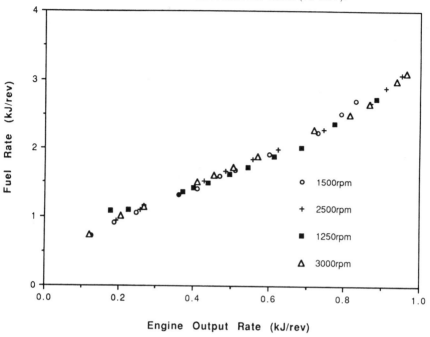

the thermal efficiency discussed above, and *a* is the fuel energy per revolution needed to overcome engine frictions. (The dependence of *a* on engine displacement and number of cylinders must be specified for the analysis but is not discussed here.)

Five different major technologies for improving mechanical efficiency are considered:

- Aggressive transmission management (ATM) to reduce average engine speeds
- Reduced displacement, or engine size, at constant maximum power
- Reduced rubbing friction and more efficient engine accessories
- Stop-start (idle-off)
- Reduced pumping (elimination of throttling)

AGGRESSIVE TRANSMISSION MANAGEMENT

When engine speed is reduced at a given power output, more energy is provided for each revolution but the frictional energy loss remains roughly the same, so that output relative to friction is increased. The mechanical efficiency is thus increased.

Modifying the transmission to reduce engine speed at a given power output is one of the best-established methods for improving fuel economy (Ludema, 1984). To implement it, more gears and lower gear ratios can be built into the transmission and, in driving, gears can be shifted up as soon as feasible. A feel for aggressive transmission management can be obtained by driving a Honda CRX HF with a shift indicator light on the dashboard. This car's relatively low first gear ratio requires that the clutch be slipped considerably in order to get the car moving smoothly. As the car accelerates, the upshift light comes on very soon. If one follows the shift light's suggestions, one shifts up at much lower engine speeds than is typical.

A critical consideration for fuel economy is the span—the ratio of the highest gear ratio to the lowest gear ratio. Consider a standard manual transmission. In the lowest gear (highest gear ratio), clutch slip is involved in getting the car moving, but the gear ratio must still be high enough to enable the engine to begin to move a stopped car up a grade (Stone, 1989). Good fuel economy performance in highway driving requires a low gear ratio in the highest gear. But the large span desired will not be feasible unless the ratios of adjacent gears are close enough to make shifting convenient. This requires many gears. For fuel economy, six gears are preferable to five in a manual transmission. For automatic transmissions with fluid coupling, fewer gears are needed because the lowest gear provides, roughly speaking, the function of the two lowest gears in manual transmissions. Four-speed automatics are now being widely used and five-speeds are being discussed.

Consider the following example of ATM. The base vehicle (AVPWR) is assumed to have a five-gear manual system. This is changed to an automated six-gear system in which the span is increased by a factor of 1.25 and the most energy-efficient gear is always selected in driving. Figures 9.4 and 9.5 show where the engine is operating at each of the approximately 800 seconds of the urban driving cycle when the engine of AVPWR is not idling. The official gear shift schedule of the Federal Test Procedure is used to generate Figure 9.4, and the ATM just defined is used to generate Figure 9.5. This example of ATM tends to move engine operations from the 1800–2800 rpm range to the 1250–2400 rpm range. The average engine speed falls from 2170 rpm to 1580 rpm. The combined effect of this ATM is found in our vehicle simulation to improve fuel economy 9 percent in the urban cycle compared to the base car and 15 percent in the highway cycle; the composite reduction is 11 percent.

FIGURE 9.4. TIME DISTRIBUTION IN URBAN CYCLE (JETTA).

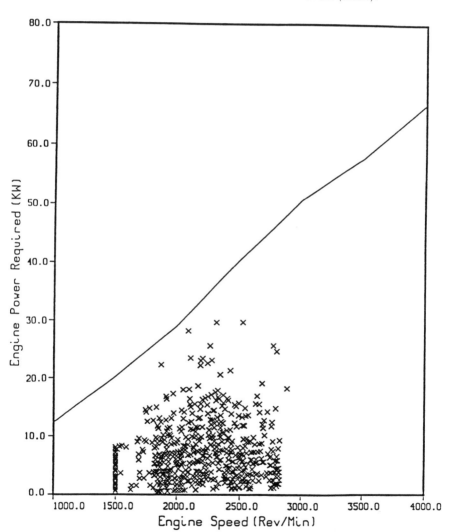

FIGURE 9.5. TIME DISTRIBUTION IN URBAN CYCLE (JETTA) (ATM).

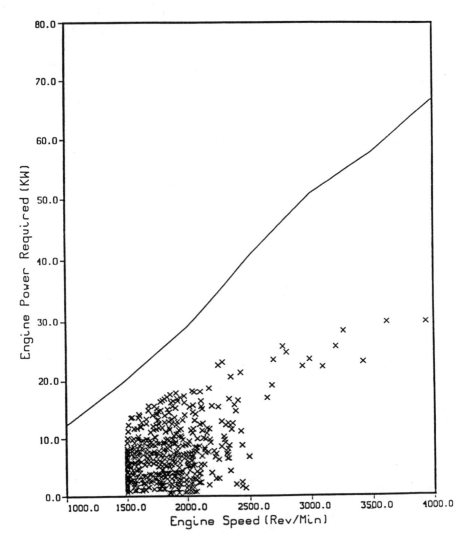

Aside from the issue of creating the transmission technology, manufacturers may be reluctant to reduce engine speeds for two reasons. First, the engine usually runs smoother, with less vibration at higher speeds. Second, relatively high power is almost immediately available at higher engine speeds. Consider the maximum power available to a driver without changing the engine speed, that

is, at wide-open throttle. Figure 9.6 is a sketch of an engine performance map; the $x$ and $y$-axes are engine speed ($N$) and power output
($P_{out}$). PWOT is the upper envelope, the output at wide-open throttle. If one is driving with a relatively high engine speed at point $A$,
then one has access to much higher power—for example, at point
$B$—merely by depressing the accelerator pedal and enabling more air
and fuel to enter the cylinders. If instead one is driving with a low
engine speed at point $A'$ (same $P_{out}$ as at $A$) under ATM, then the
fastest way to get substantially more power is to shift down to third
gear, speeding up the engine to point $A$, and opening the throttle to
achieve point $B$.

This process is familiar to drivers of many cars with a four-cylinder
engine and an automatic transmission, in which downshift and

FIGURE 9.6. ENGINE PERFORMANCE MAP.

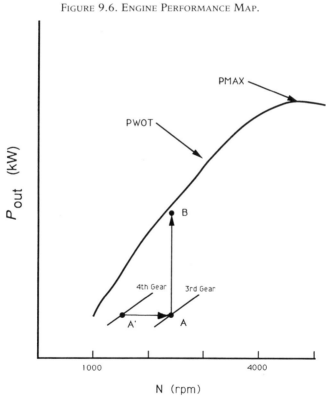

N (rpm)

Note: *The gear lines show engine use at constant velocity.*

engine speedup occur when the accelerator pedal is floored. The action of declutching, engine speedup, and reclutching takes time; if done well, however, it only takes half a second or less. Although this is a small loss of amenity, it is a loss of amenity. In this respect, the ATM technology is not analogous to many other technological changes that can be implemented without any loss of amenity, such as multipoint fuel injection or advanced engine friction reduction.

In a more sophisticated version of ATM called drive-by-wire, electronically controlled transmission management with a more complex algorithm than that just considered is used to reduce the amount of shifting (Ganoung, 1990; Huber et al., 1991). This technology rests on taking the driver's signal from the accelerator pedal electronically rather than mechanically. In addition, with drive-by-wire, the timing of power output and gear shifting is carefully controlled in order to make accelerations smoother. Almost the same energy savings is realized with drive-by-wire as with the example analyzed here.

REDUCED DISPLACEMENT

Here we address engine size reduction to reduce the power required for operating an engine while maintaining maximum power (and acceleration performance). Our analysis is based on replacing the conventional engine in AVPWR with an engine of 25 percent less displacement and substantially higher maximum power per unit of volume to compensate for the reduction in displacement (Ward, 1990). Other things being equal, reducing an engine's displacement by 25 percent reduces its power by roughly the same amount. A 33 percent increase in the maximum-power/displacement ratio would be needed to compensate ($1.33 \times 0.75$ is approximately equal to 1).

During the 1980s, maximum-power/displacement ratio increased about 3.5 percent per year on average. At this rate it would increase 33 percent in less than 9 years, enough to compensate for a 25 percent reduction in displacement. Moreover, simply replacing today's engines with average and below-average power/displacement ratio with today's better-performing engines would enable this 33 percent improvement. Among 1991 Big Three cars there are six four-cylinder and six-cylinder engines (without turbochargers) with a ratio greater than 61 hp/liter. (See Table 9.1.) The Quad 4 engine has a power/displacement ratio of 78 hp/liter, the highest among these

TABLE 9.1. HIGH-PERFORMANCE ENGINES IN 1991 BIG THREE CARS

| Car | Displacement (liters) | Cylinders | PMAX/V (hp/liter) | Engine manufacturer |
|---|---|---|---|---|
| Chev. Beretta (Quad 4) | 2.3 | 4 | 78 | GM |
| Chev. Lumina | 3.4 | 6 | 62 | GM |
| Saturn | 1.9 | 4 | 65 | GM |
| Dodge Stealth | 3.0 | 6 | 74 | Mitsubishi |
| Ford Escort GT | 1.8 | 4 | 71 | Mazda |
| Ford Taurus SHO | 3.0 | 6 | 73 | Yamaha |

*Note:* "High performance" is defined as PMAX/V > 61 hp/liter (1 hp/CID).

engines. The current average is about 50 hp/liter. Some of the techniques now being used to increase the power/displacement ratio are multipoint fuel injection, more valves per cylinder, double overhead cam valving, combustion chamber redesign, tuned intake manifold, variable intake manifolds for low-speed and high-speed operations, and higher engine speed capability (Amann, 1989). Variable valve timing, discussed later in the context of low power output, can be simultaneously managed to substantially increase maximum power (and torque at moderate engine speeds).

There are two broad approaches to reducing engine displacement: the engine can be scaled down, leaving its design roughly the same, or the number of cylinders can be reduced while keeping each cylinder's volume at least as large. A study by Muranaka et al. of Nissan suggests that scaling down the size of the cylinder may be less effective for fuel efficiency than hoped. As mentioned in the discussion on compression ratios, wall effects depress thermal efficiency— and wall effects increase with decreasing cylinder displacement. The loss of thermal efficiency roughly cancels the increase in mechanical efficiency achieved by reducing the size of the engine.

Our conclusion is that for six-cylinder engines and, perhaps, some eight-cylinder engines the route for downsizing is to adopt high-performance four-cylinder engines and, for most eight-cylinder engines, to adopt high-performance six-cylinder engines. For four-cylinder engines, high-performance three-cylinder and even two-cylinder engines should be designed and studied.

The main targets for downsizing to four-cylinder high-performance engines are six-cylinder engines with a relatively low power/displacement ratio. Examples of such engines are the GM 191, 204, and 231-cubic-inch engines used in many of GM's 1991 car lines. The ratio for these engines ranges from 43 to 48 hp/liter.

Both engine downsizing and ATM reduce the available torque at a given engine speed. Consider the combined effect of these two technologies applied to AVPWR. (See Table 9.2.) The 25 percent engine downsizing at a typical urban cycle engine speed of 2200 rpm leads to an estimated torque reduction of 19 percent for the engine considered. Application of ATM reduces the engine speed to about 1600 rpm, causing a combined torque reduction of 27 percent. (See the discussion in connection with Figure 9.6 for the way the transmission system is used to compensate for this reduced torque at low engine speed.) From an engineering standpoint one need not be satisfied with this reduction in torque at low engine speed. Depending on engine geometry, torque at low speeds can be enhanced by design of the intake manifold. Intake manifold tuning has been widely implemented, but not for the low engine speeds considered here. There is also a large potential for variable-geometry turbocharging to enhance low-speed torque (Bleviss, 1988; Inoue et al., 1989; McInerney, 1990).

The improvement in fuel economy associated with the 25 percent reduced displacement described here is 23 percent in the urban cycle, 19 percent in the highway cycle, and 22 percent in the composite. Because of the synergism between reduced displacement and ATM, it might be more meaningful to quote their combined effect on fuel economy: 33 percent in the city, 34 percent on the highway, and 34 percent in the composite driving cycle. This level of fuel savings will not be possible in all cars because some versions of a few models already incorporate high-performance engines that could not be downsized without significantly reducing power.

REDUCED RUBBING FRICTION

Most of the power required to operate the engine is to overcome rubbing friction and to run the fan, oil pump, fuel pump, fuel injectors, and coolant pump. If these accessories are operated electrically, the corresponding energy losses in the alternator and battery must be taken into account. Considerable improvement in accessory efficiency can be achieved by changing the drive for accessories to

TABLE 9.2. ESTIMATED TORQUES: AVPWR WITH ENGINE DOWNSIZING AND ATM

| Torque | ENGINE SPEED (RPM) | | |
|---|---|---|---|
| | 2200 | 2000 | 1600 |
| Empirical torque/disp. (N·m/liter) | | | |
| Current V6 engines[a] | 78 (est.) | 75 | |
| Quad 4[b] | 84 | 80 | 76 |
| Estimated torque (N·m) | | | |
| AVPWR[c] | 242 | | |
| Downsized AVPWR[d] | 195 | | 177 |
| Decrease in torque[e] (%) | 19% | | 27% |

[a] Torque for 204-CID Buick Century and 231-CID Buick LeSabre (*Consumer Guide*).
[b] Adapted from Thomson et al.
[c] 242 = 3.1 × 78.
[d] 195 = (0.75 × 3.1) × 84; 177 = (0.75 × 3.1) × 76.
[e] Relative to AVPWR at 2200 rpm.

electric motors, which run only when they are needed instead of being run continuously with a belt drive. Substantial savings could also be made by introducing efficient alternators.

Given the surprising paucity of published research and data on the topic of reducing rubbing friction, it is impossible to give a definitive estimate of the practical reduction potential. Examples in the literature suggest that a 10 percent reduction in energy requirements for engine friction and engine accessories is currently practical (Stone, 1989). The improvement in fuel economy associated with a 10 percent friction reduction is 6 percent in the urban cycle, 5 percent in the highway cycle, and 6 percent in the composite.

STOP-START OR IDLE-OFF

Volkswagen has fully developed a technology, with a diesel engine, that turns off the engine when no power is demanded and restarts it when the accelerator pedal is depressed or the manual-shift lever is moved. A test car fitted with this technology works very smoothly and without noticeable delay. The technology is now being tested and probably will be included in production vehicles.

VW is working on second-generation technology in which a second clutch is inserted between the engine and flywheel so that the flywheel continues to spin when the engine is stopped (Seiffert and

Walzer, 1991). Stored flywheel energy is used to restart the engine. The radical aspect of this technology is that VW has designed the normal flywheel to serve as the rotor of a motor/generator as well. This motor would be used as the starter (enabling elimination of a separate starter) and an alternator (enabling elimination of a separate alternator).

With stop-start, the use of vehicle accessories with heavy loads, especially air conditioning, is a serious issue. At times when the electrical load is high, the engine should not be stopped. Idle-off technology for gasoline engines should be managed so that the engine is not turned off in conditions where it may not start rapidly. The uses of this system should be optimized with electronic management. Here it is assumed that in 33 percent of all engine idling, stop-start would not be used. With this characterization, the fuel savings is found to be 15 percent in the city, 2 percent on the highway, and 10 percent in the composite cycle.

REDUCED PUMPING

A spark-ignition engine relies on a throttle to run at low power output. The throttle introduces fluid friction in the air intake to create a partial vacuum in the intake manifold and thus in the cylinders. In order to reduce power output, air intake must be reduced in proportion to fuel intake for two reasons. First, the flame propagates well if the mixture is near stoichiometric; second, the three-way catalyst works well only at stoichiometric.

A diesel, or compression-ignition, engine is quite different; the amount of fuel injected is reduced without reducing the air intake to achieve low power. Ignition occurs at fuel droplets, whatever their location, rather than by means of a moving flame front in a homogeneous air/fuel mixture. Ultra-lean-burn engines, like the new spark-ignition two-strokes, are also managed so that the air/fuel ratio can be varied, at least in part, to substitute for a throttle.

Variable valve timing (VVT) refers to controlling valve motion according to engine conditions, rather than having a fixed pattern of valve motion as in conventional designs. Variable early closing of the intake valve can be designed to substitute for throttling to achieve low power (Lenz and Grudent, 1988; Asmus, 1991). In this concept the intake valve is opened at the beginning of the intake stroke and then closed at a controlled time to admit the required amount of air. Since the main throttle valve is not used, the pressure in the intake manifold remains atmospheric. After closing the intake port valve, the piston continues to be drawn out, creating a partial vacuum in the cylinder. (The amount of air needed corresponds to a vacuum of about one-seventh atmospheric pressure at idle.) This work by the piston is recovered as the piston goes back in the first portion of the compression stroke. In this way pumping energy is largely eliminated. It cannot be wholly eliminated, however, since there remains some fluid friction at the valves.

One mechanism for actuating the valve is electrohydraulic, which uses oil under pressure with pressure release electrically controlled by a solenoid (*Popular Science*, 1990). This technology has been developed previously for high-pressure fuel injection for diesel engines. Other VVT mechanisms have also been developed (Demmelbauer-Ebner et al., 1991; McCosh, 1990).

Using a detailed single-cylinder simulation model, we estimate that 85 percent of the pumping energy can be eliminated with early intake valve timing in low-load conditions. Assuming a 200-W load to operate the technology, the fuel economy improvements in AVPWR are 8 percent in the urban cycle, 6 percent in the highway cycle, and 7 percent in the composite.

EFFICIENCY IMPROVEMENTS AND THEIR INTERACTIONS

The fuel economy improvements gained from applying each mechanical-efficiency technology independently are all in the 6 to 22 percent range. The improvement associated with application of all the technologies is 58 percent. Mechanical efficiency is improved from about 40 to 63 percent (composite cycle). This is more than half the way toward an ultimate goal of about 80 percent.

The efficiency improvement associated with a particular technology when other fuel economy technologies are already incorporated in a vehicle may be less than when it is applied alone (Table 9.3). One sees in the table that the savings of the added technologies are roughly the same in the two situations—that is, the interaction is not strong, except that the last in the sequence of five technologies creates less than half of the percentage benefit it creates when used alone. If one sums the percentage improvements of these interacting

TABLE 9.3. EFFECT OF INTERACTION AMONG MECHANICAL-EFFICIENCY TECHNOLOGIES APPLIED IN ORDER 1 TO 5 (AVPWR)

| Technology | FUEL ECONOMY IMPROVEMENT (%) | | |
| | Technology used alone (relative to base) | Technology used with others (relative to vehicle with previously applied techs) | Cumulative improvement (relative to 24.3 mpg) |
| --- | --- | --- | --- |
| 1. Displacement reduction | 22 | 22 | 22 |
| 2. Friction reduction | 6 | 5 | 28 |
| 3. ATM | 10 | 9 | 40 |
| 4. Idle-off | 11 | 9 | 53 |
| 5. Pumping reduction | 7 | 4 | 58 |

technologies, one obtains a total fuel economy improvement very close to the actual cumulative improvement. The sum of the five improvements is 56 percent, for example, compared with the actual cumulative improvement of 58 percent. Thus summing percentage improvements may be a good shortcut to estimate the cumulative fuel economy improvement of combining several technologies. (This method is used later when we consider the economics of fuel economy improvements.) On the other hand, noninteracting improvements affecting different aspects of the vehicle—such as three improvements, one in load, one in thermal efficiency, and one in mechanical efficiency—yield a fuel economy benefit that is the product of their independent improvement factors.

RESULTS

The fuel economy increase achieved by improving the mechanical efficiency of engines using the five technologies is calculated to be 58 percent (Table 9.4). This improvement is exclusive of any improvements in engine thermal efficiency or in load.

An interesting sidelight concerns the discrepancy between the EPA test values for fuel economy discussed in this report and the actual on-road fuel economy. The latter is now about 20 percent lower by our estimate. In urban driving, where congestion is leading to more stopping, the fuel economy measures evaluated in this report would be more effective than in the federal driving cycles. Thus the actual percentage improvements (as well as the absolute fuel savings) could be larger than estimated here.

The five technologies fall into two groups: those that represent current technology (reduced displacement, reduced friction, and aggressive transmission management) and those that require some development although the technologies are largely in hand (idle-off and fully variable valve timing). With the first group, the fuel economy improvement is estimated to be 36 percent. These technologies coupled with modest load reductions (such as reductions in drag coefficient, tire resistance, and accessory loads) would provide an increase of well over 40 percent in fuel economy without sacrifice of power or size. As the term is used here, current technology does not mean that there is no technological challenge in implementing such changes in millions of vehicles per year such that they are extremely reliable. It means that there need be no delay before beginning the design process.

## THE RELATION BETWEEN FUEL ECONOMY AND EMISSIONS

It is tempting for advocates of high fuel economy to believe that energy efficiency and emissions reduction are closely associated. This is true for carbon dioxide, of course. But in the first approximation,

TABLE 9.4. SUMMARY OF FUEL ECONOMY IMPROVEMENTS BY MODIFYING AVPWR WITH FIVE PART-LOAD-EFFICIENCY TECHNOLOGIES WITH AND WITHOUT LOAD REDUCTION

| | NO LOAD REDUCTION | LOAD REDUCTION[a] | |
| Technology | Cumulative increase (%) | Cumulative increase (%) | Cumulative fuel economy (mpg) |
| --- | --- | --- | --- |
| Base | 0 | 0 | 24.3 |
| Load reduction | — | 8.2 | 26.3 |
| Displacement reduction | 22 | 34.2 | 32.6 |
| Friction reduction | 28 | 41.2 | 34.3 |
| ATM | 40 | 56.0 | 37.9 |
| Idle-off | 53 | 72.2 | 41.9 |
| Pumping reduction (VVT) | 58 | 81.5 | 44.1 |

[a] 20 percent reduction in engine load assumed.

efficiency improvement in general is not important for reducing the other emissions of concern, because emission control systems play the dominant role in reducing emissions. A properly designed and operated three-way catalytic converter can convert 90 percent or more of the hydrocarbons, carbon monoxide, and $NO_x$ from the engine. Maintaining proper operation of the emission control system, which includes control of engine parameters like air/fuel ratio, has a major effect on emissions without having major impacts on fuel economy. Conversely, simply reducing fuel consumption as such has only a mild impact on emissions.

A critical aspect of the overall emissions picture is that for hydrocarbons and carbon monoxide, total emissions averaged per vehicle-mile for vehicles on the road are five to ten times higher than the tailpipe standards. Although these high emission rates are due in part to older vehicles still on the road, modern vehicles also contribute significantly. One major source is gradual degradation of vehicles' emission control systems; another is gross failure of these systems (although the incidence of failure is not well known because budgets for testing of in-use vehicles have been inadequate). It might be that, averaged over the vehicle's life, acceptable emission performance for most regions could be obtained with an efficient engine with low "engine-out" emissions such that its emission performance is robust over the engine's life. This might be achieved, for example, with a lean-burning engine. However, the approach being taken is emissions cleanup: catalytic conversion and increasingly tight monitoring, inspection, and maintenance systems.

Certain potentially important technologies can provide both energy and emissions benefits. The best known have already been mentioned: using a fuel with very simple molecular structure (and designing the engine for that fuel), as well as lean burning with conventional fuel. At least three other technologies are of interest: cycle-to-cycle control of engine characteristics; bringing cold engines rapidly up to warm operations; and using VVT to recycle some unburnt hydrocarbons into the cylinder during cold operations. Moreover, the elaboration of closed-loop control systems, based on measurement of both inputs and actual engine outputs, and the move toward recording and communicating data on engine performance relevant to emissions (on-board monitoring) should lead to both fuel economy and emissions improvements. The development of synergistic technologies such as these deserves support from public sources (R&D funding and other encouragement).

## ECONOMIC ISSUES

In this section we analyze the cost effectiveness of light-vehicle fuel economy technologies and the fuel savings that could result from their widespread use in the U.S. vehicle fleet. The technologies analyzed here do not exhaust the list that may be available for improving fuel economy. We have simply concentrated on those for which adequate cost information is available and which are relatively well understood. As the cost information necessary to analyze the five categories emphasized in the previous section is not available, in this section we analyze a more narrowly specific list of fuel economy technologies. Our analysis uses 1987 as the base year—that is, the technological characteristics of new cars in 1987 are the base to which technology changes are applied and from which fuel economy improvements are measured. Supply curves of conserved energy illustrate the results of the analysis.

### COSTS OF TECHNOLOGIES

Energy and Environmental Analysis, Inc. (EEA) has compiled a set of cost estimates for fuel economy technologies that the Department of Energy uses to analyze fuel economy policies. EEA derived its costs by estimating variable manufacturing costs for each technology and multiplying by an estimate of an industry average ratio between variable costs and retail vehicle prices. Costs used in this analysis are thus estimates of the change in consumer retail car prices that would result from use of these technologies.

The reader is cautioned not to consider these numbers as firm. For fuel economy technologies that are pieces of equipment added to a car, such as fuel injection, costs are relatively well determined. If the equipment serves more than one purpose, however, the portion of the equipment costs that should be allocated to fuel economy is subjective. For fuel economy technologies that are simply a new way of building an existing part of the car and require little or no extra materials, such as certain aerodynamic improvements, costs are more difficult to determine; often the costs for these technologies disappear as manufacturing practices evolve.

## TECHNOLOGIES ANALYZED

We have developed supply curves of conserved energy for two groups of technologies. Technology Group 1 is limited to technologies appearing in a paper summarizing recent DOE-sponsored research on automobile fuel economy (Difiglio et al., 1990). (See Table 9.5.) According to Difiglio et al., these are proven technologies that are already available in existing cars or prototypes; other technologies were omitted because "(1) they are not market-ready, or (2) they do not presently meet vehicle emission standards, or (3) they detract significantly from performance, ride, or capacity, or in some other way are not acceptable to consumers." Furthermore, the selected technologies "would not reduce performance, ride, or capacity over 1987 levels."

Technology Group 2 includes all the technologies in Group 1 plus idle-off and aggressive transmission management. Although these technologies were not included in the analysis by Difiglio et al., they are included here because they offer significant potential for improving fuel economy, they could be installed in production vehicles before the year 2000, and, like other technologies in this group, they do not significantly degrade ride, performance, or capacity compared to 1987 levels. These two additional technologies will change the feel of driving a car. With ATM, for example, more gear shifting will occur and a car will operate in higher gears more of the time, causing a slight delay for downshifts needed to accelerate quickly. Electronic transmission control can minimize the effect these changes will have on the feel of driving.[3]

## METHODOLOGY

All curves are calculated from a base year of 1987. (The average nominal, or EPA-rated, fuel economy of all domestic and import new cars sold in the United States in 1987 was 28.3 mpg.) Average

TABLE 9.5. FUEL ECONOMY SUPPLY CURVE FOR THE YEAR 2000: TECHNOLOGY GROUP 1

| Tech | Consum cost ($) | Annual cost ($) | Individ new car mpg incr (%) | Market share incr (%) | New car flt mpg incr (%) | Individ new car mpg incr (mpg) | New car flt mpg increase (mpg) | Actual new car flt mpg (mpg) | EPA-rated new car flt mpg (mpg) | Cost cnsrv fuel tech N ($/gal) | Avg cost cnsrv fuel tech 1 … N ($/gal) | Actual fleet mpg (mpg) | 2000 energy savings (quad Btu) |
|---|---|---|---|---|---|---|---|---|---|---|---|---|---|
| Base, 1987 mpg | | | | | | | | 21.7 | 28.3 | | | 21.6 | |
| 1 RCF | 15 | 2.06 | 1.5 | 37 | 0.56 | 0.33 | 0.12 | 21.8 | 28.4 | 0.25 | 0.25 | 21.7 | 0.03 |
| 2 OHC | 74 | 10.14 | 6.0 | 69 | 4.14 | 1.30 | 0.90 | 22.7 | 29.6 | 0.33 | 0.32 | 22.2 | 0.21 |
| 3 ADV F | 80 | 10.96 | 6.0 | 80 | 4.80 | 1.30 | 1.04 | 23.8 | 31.0 | 0.38 | 0.35 | 22.8 | 0.23 |
| 4 IVC | 80 | 10.96 | 6.0 | 75 | 4.50 | 1.30 | 0.98 | 24.7 | 32.3 | 0.42 | 0.37 | 23.3 | 0.20 |
| 5 FWD | 150 | 20.55 | 10.0 | 23 | 2.30 | 2.17 | 0.50 | 25.2 | 32.9 | 0.50 | 0.39 | 23.5 | 0.10 |
| 6 4V | 105 | 14.39 | 6.8 | 100 | 6.80 | 1.48 | 1.48 | 26.7 | 34.8 | 0.55 | 0.43 | 24.3 | 0.28 |
| 7 ACCES | 29 | 3.97 | 1.7 | 80 | 1.36 | 0.37 | 0.30 | 27.0 | 35.2 | 0.66 | 0.44 | 24.5 | 0.05 |
| 8 AERO | 80 | 10.96 | 4.6 | 85 | 3.91 | 1.00 | 0.85 | 27.9 | 36.3 | 0.70 | 0.47 | 24.9 | 0.15 |
| 9 TRANS | 39 | 5.34 | 2.2 | 80 | 1.76 | 0.48 | 0.38 | 28.2 | 36.8 | 0.74 | 0.48 | 25.1 | 0.06 |
| 10 MPFI | 67 | 9.18 | 3.5 | 56 | 1.96 | 0.76 | 0.43 | 28.7 | 37.4 | 0.83 | 0.50 | 25.3 | 0.07 |
| 11 CVT | 100 | 13.70 | 4.7 | 45 | 2.12 | 1.02 | 0.46 | 29.1 | 38.0 | 0.95 | 0.52 | 25.5 | 0.07 |
| 12 LUB/TIRE | 22 | 3.01 | 1.0 | 100 | 1.00 | 0.22 | 0.22 | 29.3 | 38.3 | 1.00 | 0.53 | 25.6 | 0.03 |
| 13 5AOD | 150 | 20.55 | 4.7 | 40 | 1.88 | 1.02 | 0.41 | 29.7 | 38.8 | 1.48 | 0.56 | 25.8 | 0.06 |
| 14 WT RED | 250 | 34.25 | 6.6 | 85 | 5.61 | 1.43 | 1.22 | 31.0 | 40.4 | 1.86 | 0.69 | 26.4 | 0.18 |
| 15 TIRES II | 20 | 2.74 | 0.5 | 100 | 0.50 | 0.11 | 0.11 | 31.1 | 40.5 | 2.05 | 0.70 | 26.4 | 0.02 |
| | | | | | | | | | | | | | 1.76 |

interior volume and acceleration capability are held at their 1987 levels.

The technologies and costs used in developing the fuel economy supply curves for the year 2000 are listed in Tables 9.5 and 9.6. (A key to the acronyms used to identify the technologies can be found in Table 9.7.) Some of the listed technologies are combinations (TRANS represents electronic transmission control and torque converter lockup), and some are not technologies in the sense of new devices or equipment (AERO represents an advance in design, not a new technology).

In Tables 9.5 and 9.6, the consumer retail costs estimated for each of these technologies are listed in the second column of the table (CONSUM COST) and are annualized in the third column (ANNUAL COST) using a 7 percent discount rate, a 10-year estimated useful life, and a distribution for miles driven per year, by car vintage, as estimated by the Department of Transportation.[4] The costs are approximates of those developed by EEA (with the exception of the costs for idle-off and ATM, which were independently estimated). All costs are stated in 1989 dollars.

Estimates of the fuel economy increase associated with each technology were also derived from Difiglio et al. (These estimates are listed under INDIVID NEW CAR MPG INCR.) The fuel economy increase associated with two technologies, ATM and idle-off, were independently estimated by the authors.

Values in the fifth column in Tables 9.5 and 9.6 (MARKET SHARE INCR) reflect the projected increase in market share—relative to the total new car market—for each technology. Values in the sixth column (NEW CAR FLT MPG INCR) reflect the new car fleet mpg increase expected from use of each technology to the extent projected in MARKET SHARE INCR. Estimates of market share increase were taken from Difiglio et al. except for idle-off and ATM, which were estimated by ACEEE. Market shares taken from Difiglio et al. were taken from their maximum technology scenario because we believe these rates of new technology penetration better reflect the future of the rapidly changing automotive industry, where competitive pressures are forcing manufacturers to redesign car lines much more rapidly than in the past.

Estimates of how much each technology can increase new car fuel economy are found in the ninth column (ACTUAL NEW CAR FLT

TABLE 9.6. FUEL ECONOMY SUPPLY CURVE FOR THE YEAR 2000: TECHNOLOGY GROUP 2

| Tech | Consum cost ($) | Annual cost ($) | Individ new car mpg incr (%) | Market share incr (%) | New car flt mpg incr (%) | Individ new car mpg incr (mpg) | New car flt mpg increase (mpg) | Actual new car flt mpg (mpg) | EPA-rated new car flt mpg (mpg) | Cost cnsrv fuel tech N ($/gal) | Avg cost cnsrv fuel tech 1 …. N ($/gal) | Actual fleet mpg (mpg) | 2000 energy savings (quad Btu) |
|---|---|---|---|---|---|---|---|---|---|---|---|---|---|
| Base, 1987 mpg | | | | | | | | 21.7 | 28.3 | | | 21.6 | |
| 1 TRANS | 60 | 8.22 | 9.0 | 75 | 6.75 | 1.95 | 1.46 | 23.2 | 30.2 | 0.18 | 0.18 | 22.4 | 0.35 |
| 2 RCF | 15 | 2.06 | 1.5 | 37 | 0.56 | 0.33 | 0.12 | 23.3 | 30.4 | 0.29 | 0.19 | 22.5 | 0.03 |
| 3 TCLU | 35 | 4.80 | 3.0 | 16 | 0.48 | 0.65 | 0.10 | 23.4 | 30.5 | 0.34 | 0.20 | 22.6 | 0.02 |
| 4 OHC | 74 | 10.14 | 6.0 | 69 | 4.14 | 1.30 | 0.90 | 24.3 | 31.7 | 0.37 | 0.25 | 23.0 | 0.19 |
| 5 IVC | 80 | 10.96 | 6.0 | 75 | 4.50 | 1.30 | 0.98 | 25.3 | 32.9 | 0.44 | 0.30 | 23.6 | 0.20 |
| 6 ADV F | 80 | 10.96 | 6.0 | 80 | 4.80 | 1.30 | 1.04 | 26.3 | 34.3 | 0.47 | 0.33 | 24.1 | 0.20 |
| 7 FWD | 150 | 20.55 | 10.0 | 23 | 2.30 | 2.17 | 0.50 | 26.8 | 34.9 | 0.56 | 0.35 | 24.3 | 0.09 |
| 8 4V | 105 | 14.39 | 6.8 | 100 | 6.80 | 1.48 | 1.48 | 28.3 | 36.9 | 0.62 | 0.40 | 25.1 | 0.25 |
| 9 IDLE OFF | 250 | 34.25 | 15.0 | 50 | 7.50 | 3.26 | 1.63 | 29.9 | 39.0 | 0.75 | 0.45 | 25.9 | 0.26 |
| 10 ACCES | 29 | 3.97 | 1.7 | 80 | 1.36 | 0.37 | 0.30 | 30.2 | 39.4 | 0.82 | 0.46 | 26.0 | 0.04 |
| 11 AERO | 80 | 10.96 | 4.6 | 85 | 3.91 | 1.00 | 0.85 | 31.1 | 40.5 | 0.87 | 0.49 | 26.4 | 0.12 |
| 12 MPFI | 67 | 9.18 | 3.5 | 56 | 1.96 | 0.76 | 0.43 | 31.5 | 41.0 | 1.00 | 0.50 | 26.6 | 0.06 |
| 13 CVT | 100 | 13.70 | 4.7 | 45 | 2.12 | 1.02 | 0.46 | 31.9 | 41.6 | 1.14 | 0.52 | 26.9 | 0.06 |
| 14 LUB/TIRE | 22 | 3.01 | 1.0 | 100 | 1.00 | 0.22 | 0.22 | 32.2 | 41.9 | 1.20 | 0.53 | 27.0 | 0.03 |
| 15 5AOD | 150 | 20.55 | 4.7 | 40 | 1.88 | 1.02 | 0.41 | 32.6 | 42.5 | 1.78 | 0.57 | 27.1 | 0.05 |
| 16 WT RED | 250 | 34.25 | 6.6 | 85 | 5.61 | 1.43 | 1.22 | 33.8 | 44.0 | 2.22 | 0.68 | 27.7 | 0.16 |
| 17 TIRES II | 20 | 2.74 | 0.5 | 100 | 0.50 | 0.11 | 0.11 | 33.9 | 44.2 | 2.44 | 0.69 | 27.8 | 0.01 |
| | | | | | | | | | | | | | 2.12 |

287

TABLE 9.7. KEY TO ABBREVIATIONS IN TABLES 9.5 AND 9.6

*Abbreviations*

| | |
|---|---|
| IVC | intake valve control |
| RCF | roller cam followers |
| MPFI | multipoint fuel injection |
| 4V | four valves per cylinder engines |
| AERO | aerodynamic improvements |
| TRANS | torque converter lockup and electronic transmission control (Group 1 only) |
| TCLU | torque converter lockup (Group 2 only) |
| OHC | overhead cam engine |
| FWD | front-wheel drive |
| CVT | continuously variable transmission |
| ACCES | improved accessories including electric power steering |
| ADV F | engine friction reduction |
| 5AOD | five-speed automatic overdrive transmission |
| LUB/TIRE | improved lubrication and tires |
| WT RED | weight reduction |
| TIRES II | advanced tires (improvements beyond LUB/TIRE) |
| ATM | aggressive transmission management (Group 2 only) |
| IDLE OFF | idle-off (Group 2 only) |

*Column Headings*

| | |
|---|---|
| TECH | fuel economy technology (or measure) |
| CONSUM COST | retail cost of each technology per car |
| ANNUAL COST | retail cost of each technology annualized over 10-year period at 7% discount rate |
| INDIVID NEW CAR MPG INCR (%) | % increase in fuel economy attributable to each technology |
| MARKET SHARE INCR | projected % increase (relative to total new car market) in use of each technology |
| NEW CAR FLT MPG INCR (%) | estimate of how much the technology can increase the new car fleet mpg, calculated by multiplying INDIVID NEW CAR MPG INCR by MARKET SHARE INCR |
| INDIVID NEW CAR MPG INCR (MPG) | same as above but expressed in mpg |
| NEW CAR FLT MPG INCR (MPG) | average new car fleet mpg increase caused by use of specified technology |
| ACTUAL NEW CAR FLT MPG | actual fuel economy of new car fleet after adoption of specified technology |

TABLE 9.7. (*continued*)

| | |
|---|---|
| EPA-RATED NEW CAR FLT MPG | Federal Test Procedure rating mpg of new car fleet after adoption of specified technology |
| COST CNSRV FUEL, TECH N | cost of conserved fuel of specified technology per gallon saved |
| AVG COST CNSRV FUEL, TECH 1 ... N | average cost of conserved fuel for specified and all preceding technologies per gallon saved |
| ACTUAL FLEET MPG | average actual fuel economy of all cars on road given new car fleet cumulative adoption of specified technologies |
| 2000 ENERGY SAVINGS | energy savings in the year 2000 |

MPG). These values are estimates of actual on-road fuel economy calculated by adjusting EPA-rated combined city/highway fuel economy to account for its growing overestimation of actual fuel economy. Year 2000 fuel economy levels in this analysis are taken to be 23 percent below the EPA-rated level (Westbrook and Patterson, 1989). All fuel savings estimates are based on the adjusted EPA fuel economy ratings.

The marginal cost of conserved fuel (COST CNSRV FUEL TECH N) was calculated by using a 7 percent real discount rate and miles driven per year, by vintage, as specified by the Department of Transportation (DOT, 1986). All technologies with costs lower than projected fuel prices are deemed cost-effective. The projected price of gasoline for the year 2000 was $1.32 a gallon (Energy Information Administration, 1989).

The values in the cost of conserved fuel column can roughly be interpreted as the social cost effectiveness of adopting the specified technologies in that the discount rate (7 percent) and the length of time over which fuel savings were estimated (10 years) more closely reflect a social perspective than a car buyer's.[5] A truer test of social cost effectiveness would value gasoline at a higher level by including such items as the environmental, security, and health costs of consuming gasoline. Levels of fuel economy deemed cost-effective here assume that automobile size and acceleration performance are held constant at their 1987 levels. Since both performance and size have increased slightly since then, this analysis assumes a small reduction of vehicle size and acceleration performance.

The fleet fuel economy in the next to the last column (ACTUAL FLEET MPG) was calculated by using a vintaging model based on survival probability data and annual-miles-traveled-by-vintage data (Davis et al., 1989). The model calculates fleet fuel economy on the basis of each vintage's new car fuel economy and assumes a fixed distribution of new and old cars for each year. The new car fuel economies specified in the ninth column (ACTUAL NEW CAR FLT MPG) are assumed to be achieved by the year 2000, after a steady increase in new car fuel economy over the period 1992 to 2000. For example, row 12 in Table 9.5 specifies a new car fuel economy of 29.3 mpg, so in 1996 the fuel economy would be 21.7 + (29.3 − 21.7)/2. Calculating the new car fuel economy similarly for all years from 1992 to 2000 and knowing or estimating the new car fuel economy for other vintages yields, after use of the vintaging model, the fleet fuel economy estimate of 25.6 mpg in the year 2000.

The energy savings associated with each technology (2000 EN-ERGY SAVINGS) are based on the assumption that light-vehicle-miles traveled in the United States will grow at the rate of 2.5 percent a year to the year 2000 (Ross, 1989). Two-thirds of vehicle-miles traveled in 2000 are assumed to be attributable to automobiles and the remaining one-third to light trucks. Cumulative energy savings are calculated relative to a vehicle fleet whose new car fuel economy is frozen at the 1987 level.


RESULTS

The marginal cost of conserved fuel estimates in Tables 9.5 and 9.6 is plotted in Figures 9.7 and 9.8. The supply curves in Figures 9.7 and 9.8 illustrate how much fuel could be saved in the year 2000 (horizontal axis) and the cost of achieving this level of savings (vertical axis). Each step on these curves represents a technology and reveals its cost and the potential savings associated with its adoption. As can be seen, the technologies are ranked in order of cost effectiveness. The technologies whose costs are less than $1.32 per gallon saved are cost-effective.

Care should be taken in interpreting these results. The order in which these curves suggest technologies be adopted is not neces-

FIGURE 9.7. CONSERVATION SUPPLY CURVE, AUTO FUEL EFFICIENCY, YEAR 2000 (TECHNOLOGY GROUP 1).

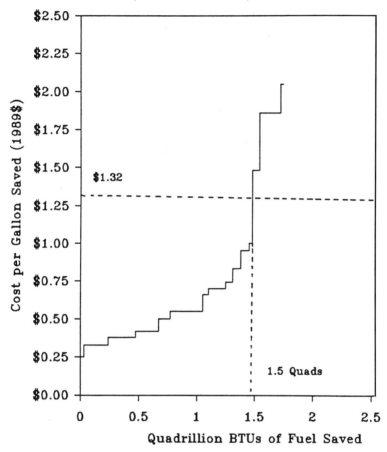

Note: *Assumes 7 percent discount rate.*

sarily reasonable. Schedules for vehicle redesign and introduction, amortization schedules for capital equipment, and other industry characteristics will probably dictate a different order of adoption. Furthermore, technologies not considered in the development of this curve are likely to become feasible and cost-effective by the year 2000, especially if the federal government mandates substantial fuel economy improvements in automobiles.

FIGURE 9.8. CONSERVATION SUPPLY CURVE, AUTO FUEL EFFICIENCY, YEAR 2000
(TECHNOLOGY GROUP 2).

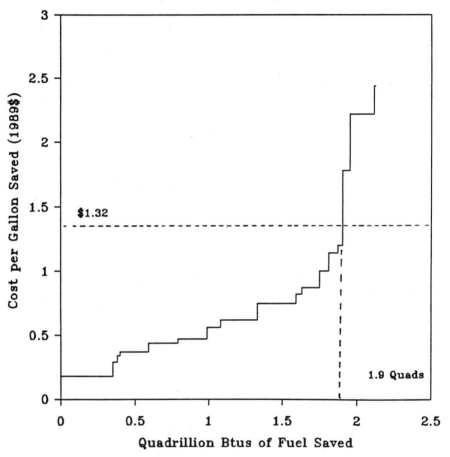

Note: *Assumes 7 percent discount rate.*

Table 9.5 shows that, using Technology Group 1, the maximum cost-effective level of new car fuel economy in 2000 is 29.3 mpg (38.3 mpg EPA-rated). Only three technologies on the list—5AOD, WT RED, and TIRES II—are more expensive than EIA's projected gasoline price in 2000, and thus they fail this test of cost effectiveness. As can be seen in Figure 9.7, this mix of fuel economy technologies and costs yields cost-effective fuel savings in the year 2000 of about

1.5 quads (quadrillion Btu). This is a 15 percent reduction in the fuel that would be consumed by automobiles, relative to a scenario in which new car fuel economy is held to its 1987 level of 28.3 mpg (21.7 mpg actual) in 2000.

Using the cost-effective technologies in Group 2 (Table 9.6) would result in a new car fuel economy level of 32.2 mpg (41.9 mpg EPA-rated). The cost and energy savings for each technology in Table 9.6 are plotted in Figure 9.8. As can be seen, all but the last three technologies are cost-effective. Fuel savings of 1.9 quads (20 percent) are achievable using cost-effective technologies. Again, this level of savings depends on how much fuel would be used if new car fuel economy were held at 1987 levels. All results are summarized in Table 9.8.

## POLICY ISSUES

We first consider the full range of policies to improve energy efficiency and reduce greenhouse gas emissions while providing daily access for people and maintaining or improving such values as safety

TABLE 9.8. SUMMARY OF RESULTS

| | Projected auto fleet mpg | QUADS/% REDUCTION | |
| | | Relative to frozen eff. | Relative to market |
| Scenario | | | |
|---|---|---|---|
| Frozen efficiency | | | |
| EPA | 28.3 | | |
| Actual | 21.6 | NA | NA |
| Market scenario | | | |
| EPA | 28.4 | | |
| Actual | 21.8 | 0/0% | 0/0% |
| Group 1 cost effective | | | |
| EPA | 38.3 | | |
| Actual | 29.3 | 2.0/26% | 2.0/26% |
| Group 2 cost effective | | | |
| EPA | 41.9 | | |
| Actual | 32.2 | 2.5/33% | 2.5/32% |

*Note:* Analysis assumes 1675 billion miles of automobile travel in the year 2000.

and environmental quality. Daily access means being able to reach places to work, shop, or engage in other activities. This does not, strictly speaking, mean mobility with its implication of expanded vehicle-miles or passenger-miles. In a given situation, improving access may involve enabling people to travel farther; but it may instead involve reconfiguring land-use patterns so that less travel is needed.

Certain policy areas bear directly on improved access and environmental quality:

- Land use
- Public transport
- Substitutes for transportation
- Traffic and parking management, road controls, and road design
- Driver behavior, including vehicle maintenance and driving style
- Improvements in light vehicles

Land-use and public transport policies are a major focus of those interested in improved access (Pushkarev and Zupan 1977; Burchell and Listakin, 1982; Holtzclaw, 1990). Their potential impact is suggested by the fact that per capita gasoline use in the Toronto metropolitan area is roughly half that in Houston, Phoenix, Denver, or Detroit (Newman and Kenworthy, 1988). Yet Toronto is not that different. It is a relatively affluent North American metropolis with a high quality of life. The key characteristics of Toronto that seem responsible for its low gasoline use are regional control of land use and a widely used public transport system.

A major insight into provision of access is that while almost all passenger-miles traveled (87 percent) are due to autos and light trucks, a relatively small increase in the use of public transport appears to be associated with driving many fewer miles. Recent research indicates that one added mass transit passenger-mile is associated with a reduction in personal vehicle-miles by a factor of 5 to 10 (Holtzclaw, 1990). This large leverage is argued to be a consequence of the synergism of mass transit and higher density of housing, work sites, shopping, and the like. With that kind of leverage, mass transit and appropriate land-use policies should eventually offer far more potential for reducing VMT than suggested by the small share of passenger-miles provided by transit. Substitutes for transportation, such as telecommunications, which enable some people to work and shop at home, and satellite places of work, which rely heavily on telecommunications, also have substantial potential.

Traffic management—such as high-occupancy vehicle lanes and car pooling assistance—has had some success in reducing travel demand and fuel consumption (Burke, 1990). A key to the success of many of these programs is charging full cost for parking privileges (Replogle, 1990). Road charges in congested areas have long been considered in Europe and Asia and have proved successful in Singapore, where they have been combined with provision of extensive modern public transport (Ang, 1991).

Highway controls, such as sophisticated signal management to encourage smooth traffic flow in congested areas, have been successfully developed, especially in Australia where resulting fuel savings of up to 20 percent have been estimated (Watson, 1990). Moreover, the rate of vehicles entering expressways can be controlled to limit congestion (Institute of Transportation Engineers, 1989). Enforcement of speed limits also contributes to efficiency.

While all these areas, including driving behavior, contribute to reducing the environmental impacts and energy consumption of light vehicles, the focus of this section is on improvements in light vehicles. Improving light vehicle technology will not be a sufficient means of resolving the environmental and energy problems created by light vehicles, but it may be the most important.

Several aspects of improving vehicle fuel economy are of interest to policymakers:

- Modest modifications to conventional vehicles
- Alternative fuels
- Radical vehicle technology, such as the fuel-cell vehicle or the small commuter car
- Interactions with vehicle safety
- Interactions with emissions of regulated pollutants

Our emphasis here will be on the first item in this list and its associated issues: near-term technological changes.

## FUEL PRICING

The market price of gasoline does not reflect its full social cost. Some studies estimate that the national security cost of importing oil amounts to at least 30 cents per gallon (Broadman and Hogan, 1986). The costs of air pollution and the risks of climate change make the social cost higher. Logically, these costs should be internalized through a tax on transportation fuels. But while higher fuel prices can play an important role in spurring fuel economy improvements, as they have in the past, increased fuel prices alone are not a strong enough motivator to improve fuel economy to levels that are cost-effective from a social perspective. This conclusion is based on two reasons.

First, fuel constitutes a relatively small part of the cost of driving a new car. At the price of gasoline that prevailed during the late 1980s, about $1 per gallon, annual fuel costs were only about 10 percent of the cost of driving the average new car. (See Figure 9.9.) If fuel prices were twice as high and the amount of driving remained the same, fuel costs would still only be about 20 percent of the cost of driving.

Second, buying a more fuel-efficient car has only a small effect on annual driving costs. Purchasing a 35-mpg car instead of a 30-mpg car, for example, will reduce annual fuel costs only $50 per year. If the two cars are identical in every respect except fuel economy, the more fuel-efficient car will be more expensive because of additional manufacturing cost. The extra up-front cost, converted into an annual cost, is about $25 per year, making the net savings to the buyer of the 35-mpg car only $25 per year. With fuel costing twice as much, the net annual saving would still be only $75. Most new car buyers do not do such calculations, but they may be aware that, for most cars, fuel economy performance does not greatly affect the economics of buying and owning a new car.

It is not surprising that new car buyers find fuel economy to be a secondary consideration. Many other attributes have higher priority: brand, safety, interior volume, trunk size, handling, price, reliability, and the like (McCarthy and Tay, 1989). Indeed, manufacturers have decided that fuel economy is of so little interest to buyers that they only offer it as part of a package in bottom-of-the-market vehicles (such as the Geo Metro), making it impossible for buyers simply to

FIGURE 9.9. COSTS OF OWNING AND OPERATING A CAR.

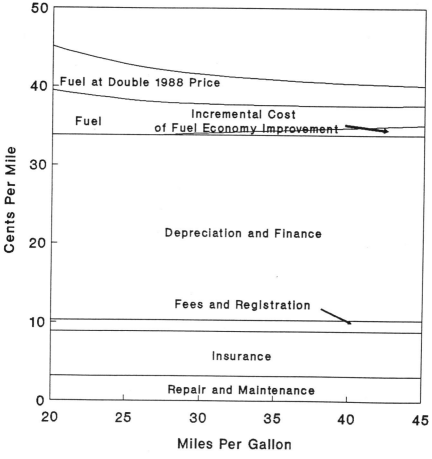

SOURCE: ACEEE, adapted from von Hippel and Levi (1983).

choose added fuel economy at extra cost while preserving the other vehicle attributes in which they are interested.

There is a different way of expressing these observations: new car buyers appear to place a low value on future fuel savings. That is, their implicit discount rate is high, perhaps 30 to 50 percent (as with other household energy conservation investments), rather than the 5 to 10 percent real interest on most new car loans. This implicit undervaluing of future fuel costs will probably continue to characterize new vehicle purchases, except perhaps in times of fuel crises.

Other evidence that higher fuel prices will not push passenger car fuel economy, rated in mpg, into the high 30s or 40s is found in industrialized countries with gasoline prices that are two to four times higher than U.S. prices. Vehicle ownership and use are different in these countries than in the United States, of course, so quantitative comparisons may be misleading. Nevertheless, it is impressive that much higher fuel prices are associated with new vehicle mpg values at most in the mid-30s.

While fuel pricing alone is not an adequate policy, it would be a useful part of a larger policy package. How large should the fuel price increases be? While small federal and state gas tax increases of approximately 5 cents a gallon have been adopted lately, large increases, especially those not earmarked for highway improvements, are strongly opposed. Some of this opposition is driven by questions of fairness. Imposing large, new fuel costs on people who earn their living driving cars and trucks, or those whose living arrangements require long-distance driving, would oblige them to shoulder a disproportionate share of the tax burden.

There is, however, at least one way to increase the apparent price of gasoline substantially without imposing new taxes or increasing the cost of driving: by restructuring the way people pay for automobile insurance. Instead of paying for all the insurance in independent contracts with insurance companies, we could pay for a large fraction at the gasoline pump. The price of gasoline at the pump could include a charge of perhaps 50 cents to $1 a gallon for basic, driving-related, automobile insurance that would be organized by state governments and auctioned in blocks to private insurance companies. Supplementary insurance could be independently arranged as at present (El-Gasseir, 1990).

In conclusion, we favor gradual introduction of fuel-tax increases totaling about 50 cents a gallon in 1991 dollars. Although this is not large by the standards of other OECD countries, it is substantial in the U.S. context. In addition we would like to see the "pay as you drive" insurance scheme developed and an equitable form adopted.

PERFORMANCE STANDARDS

The Motor Vehicle Information and Cost Savings Act of 1975 set corporate average fuel economy (CAFE) standards that required the fuel economy of new cars to increase from about 14 mpg in the early

1970s to 27.5 mpg by 1985. (See Figure 9.10.) The act gives manufacturers flexibility by applying the standard to the sales-weighted average for each corporation instead of individual vehicles. Further flexibility is provided by allowing manufacturers to earn credits for exceeding the standard in any year; these credits may be used to offset penalties in years when a manufacturer may fall short of the standard. Moreover, the secretary of transportation is given the discretion to set a lower standard, as was done for 1986 through 1989 on appeal from manufacturers (especially General Motors and Ford), and may also set standards for light trucks.

FIGURE 9.10. TRENDS IN FUEL PRICE AND DOMESTIC NEW CAR MPG.

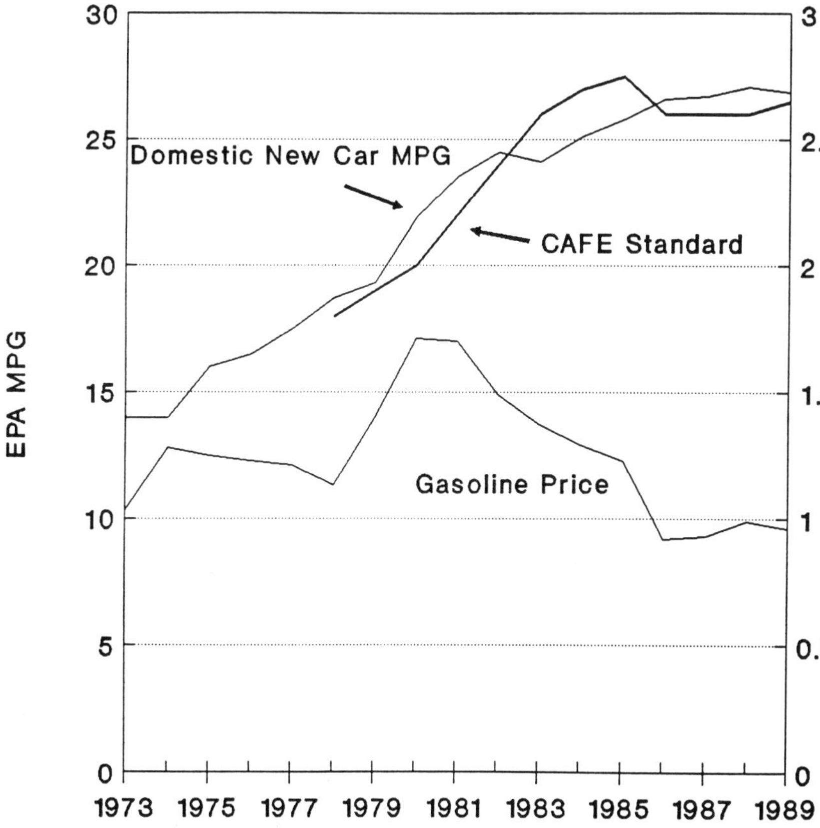

SOURCE: ACEEE, adapted from EPA and DOE.

In hearings on the 1975 act, the manufacturers stated that the technology to achieve 27.5 mpg was not available on the proposed time scale. The only way to achieve the standard, they said, would be by making the average car much smaller. They claimed that the act would "outlaw full-size sedans and station wagons" (Chrysler), "require all subcompact vehicles" (Ford), and "restrict availability of five and six-passenger cars regardless of consumer needs" (GM) (Energy Conservation Coalition, 1989). Indeed, there was some reduction in the ratio of maximum power to weight—though almost none in interior volume—in the early 1980s. By the mid and late 1980s, however, manufacturers were achieving the mandated standards with vehicles of interior volume and maximum power equal to and even higher than those of the early 1970s. The CAFE standards were thus an important example of successful "technology forcing" by regulation.

Some have claimed that the CAFE regulations were unnecessary—that the increased price of gasoline in the late 1970s and early 1980s was responsible for the fuel economy improvements (Mayo and Mathis, 1988; Crandall et al., 1986). This argument is unconvincing on two related grounds:

- The estimated fuel price elasticities for vehicle purchase are moderate (Bohi and Zimmerman, 1984), whereas the increase in fuel economy in that period was more rapid than that for fuel price (Figure 9.10).
- Statistical analysis of separate manufacturers' CAFE achievements show that "the CAFE standards were a significant constraint for many manufacturers and were perhaps twice as important an influence as gasoline prices" during that period (Greene, 1990:37).

GM and Ford have argued that the CAFE formulation placed them at a disadvantage because their mix of vehicles includes large cars while that of the Asian manufacturers does not. Recent fuel economy legislation introduced in the U.S. Congress addresses this problem by changing the basis of the standards so that each manufacturer is required to improve its fuel economy by the same percentage above its base-year fuel economy.

Depending on one's perspective, the fuel economy regulatory program can be viewed as ineffective or as a remarkable success. If one takes a static perspective—comparing the absolute level of fuel consumption today with the levels that existed when the regulatory programs were begun—one would not declare success. Even though a new car today has approximately twice the fuel economy (and 10 percent of the emissions) of cars built 10 to 15 years ago, the overall use of gasoline has actually grown somewhat (and air quality has improved only slightly). But from a dynamic perspective—where one asks what fuel consumption would have been without fuel economy improvement—the program has been very successful. As pointed out earlier, had average light-vehicle fuel economy not risen since 1973, the United States would be consuming an additional 4 million barrels of oil per day. Nonetheless, it is important to find out why the substantial improvement in fuel economy has not produced significant reductions in fuel use.

Vehicle-miles traveled on highways increased 59 percent from 1973 to 1989 (Figure 9.1). In the same period, the average fuel economy of all cars on the road also improved, but not quite enough to compensate for the higher VMT. Average fuel economy grew about 45 percent, much less than the 100 percent improvement in the new-car test value. This discrepancy is primarily due to three factors: the long time required for retirement of old, inefficient vehicles, the increasing share of light trucks with their poorer fuel economy, and the increasing gap between EPA-rated fuel economy and actual on-road fuel economy discussed in the introduction.

The disappointment as seen from a static perspective is not that fuel economy regulation has been unsuccessful. It is that new-vehicle fuel economy is only one aspect of the problem. As noted earlier, increased vehicle-miles of travel and changes in driving patterns also have important effects on fuel use. The conclusion for policymaking is the need for a package of policies that addresses all these problems, so that the gains in one aspect of the problem are not canceled by losses in another.

THE NEXT GENERATION OF STANDARDS

Regulatory performance standards are an important policy option for bringing motor vehicle fuel use under control. They have worked well in the past, and market conditions and technological opportunities are such that they are likely to work well again. Since substantially higher fuel economies are practical and cost-effective and since society has a major interest in reducing petroleum demand, it is not surprising that stronger regulatory standards for fuel economy—that is, mandating higher mpg—are actively being considered in Congress.

Automobile manufacturers strongly oppose such legislation. As before, they claim that it is not practical to improve fuel economy substantially except by moving to much smaller cars. Manufacturers are stonewalling on this point. There are in fact other, more compelling, reasons for their opposition:

1. Major additional tooling investments would be needed to make the changes, especially if, as proposed, a moderately rapid timetable is required.
2. The required rate of improvement in fuel economy would prevent manufacturers from fully exploiting sales opportunities for low-fuel-economy models already in production.
3. High fuel economy standards would somewhat restrict designers' options in developing new vehicles and markets—for example, there would be a premium on streamlining and on certain kinds of transmission shift management.

It is important to address such concerns by creating a schedule of strengthened standards allowing adequate time for manufacturers to adjust and by enacting a policy package (with components discussed elsewhere in this chapter) such that the burden of compliance would not fall entirely on the manufacturer. Policies should be enacted that motivate buyers to select high-fuel-economy vehicles. The underlying concept in these suggestions is that it will be difficult for the manufacturers to raise vehicle fuel economy substantially. Therefore an increase in the standards is not itself a sufficient policy for boosting average fuel economy to 40 mpg or higher.

## Fees and Rebates

Some incentives to encourage new-car buyers toward higher fuel economy are in place, and stronger incentives have been proposed.

THE GAS GUZZLER TAX

The gas guzzler tax, enacted as part of the Energy Tax Act of 1978, has been overlooked as an effective policy tool for improving fuel economy. There is strong evidence that this tax played an important role in improving fuel economy between 1983 and 1986.

Figure 9.11 shows the fuel economy of cars whose average fuel economy was below 21 mpg in 1980, the year before the gas guzzler tax took effect. Also plotted is the gas guzzler tax threshold. As can be seen, the fuel economy of low-mpg cars rose after the guzzler tax

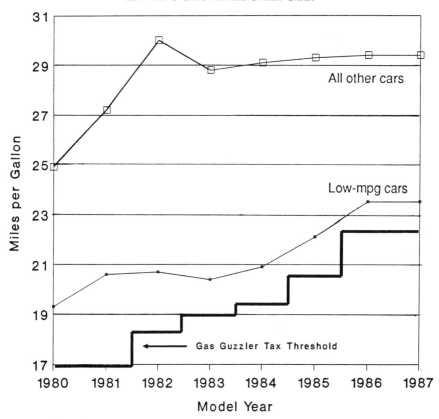

FIGURE 9.11. CHANGE IN THE FUEL ECONOMY OF
LOW-MPG CARS VS. ALL OTHER CARS.

SOURCE: The authors.

threshold was raised high enough to pose a tax threat. Manufacturers clearly decided it was better to improve fuel economy than to pay even a small gas guzzler tax. This improvement in fuel economy occurred during a period of sustained drops in the price of gasoline.

Gas guzzler taxes have a number of desirable features. Since the tax applies only to new cars, low-income people are largely unaffected. And since the tax is a large penalty imposed at the point of automobile purchase, instead of small sums stretched out over many years (as with a gasoline tax), it is likely to motivate car buyers to seek high-mpg cars. We recommend that a gas guzzler tax be established for light trucks, as well. And, as for passenger cars, the tax threshold should be increased by the same percentage as the light truck fuel economy standard.

DRIVE PLUS

An extension of the concept of the gas guzzler tax is a system of fees and rebates that would be levied on new cars according to whether they are above or below average fuel economy. Such an approach has been introduced in legislation in California, where it has been called "Drive Plus" (Gordon and Levenson, 1990). As proposed in California, fees and rebates would also be set according to whether a car's emissions were above or below an average level. Drive Plus would thus encourage the production of cars that are certified for emissions at levels below the legal limits. There is good evidence, from cars made by Volkswagen, Suzuki, and others, that such low emissions can be achieved at modest cost, at least by high-fuel-economy vehicles. The program is designed so that total rebates roughly equal total fees. The concept is just as appropriate at the federal level as it is at the state. The Drive Plus program was passed overwhelmingly by the California state legislature in 1990, but was vetoed by the governor.

Fees and rebates at the point of purchase of a new vehicle are an important tool to improve fuel economy and emissions (Geller, 1989). Given our society's sensitivity to first cost, it is easier and more effective to adjust for market imperfections and influence new-car fuel economy and emission levels at the point of capital equipment purchase than it is to adjust for imperfections in the course of operations, as, for example, with a gasoline tax.

TECHNOLOGY POLICIES

The policies just discussed indirectly encourage the creation of new technology to meet changed economic conditions or regulatory constraints. Experience shows, however, that a more direct focus on new technology can be highly effective. Before considering such policies, let us briefly suggest how new technology can help us meet our goals. By new technology we mean advanced vehicles and their energy supply systems that could radically reduce energy requirements and emissions. There are three potential types of vehicles:

1. Vehicles with much higher fuel economy but still based on gasoline or diesel fuel and still serving four or more passengers with, roughly, today's driving capabilities
2. Special-purpose vehicles requiring much less energy at the drive wheels, such as a small commuter car
3. Alternative-fuel vehicles including those that could flexibly operate on both gasoline and an alternative fuel

In Group 1 the engine could be an advanced, direct-injection diesel, now entering production in Europe, which is about one-third more efficient than corresponding conventional gasoline-powered engines. High-fuel-economy prototype vehicles incorporating advanced diesels have been built or partially developed by Volvo, Volkswagen, Renault, Peugeot, and Toyota. Fuel economies are estimated to be 70 mpg and higher (Bleviss, 1988).

In Group 2 are vehicles such as the proposed Lean Machine and the demonstration electric vehicle called Impact, both developed by General Motors. The Lean Machine is a two-seater with one passenger behind the driver. Both the Lean Machine and the Impact are small, have little air and tire drag, and require very little power to be delivered to the wheels in typical driving. (The fact that the Impact is an electric vehicle is incidental to this discussion.) Although both of these prototype vehicles happen to have strong acceleration performance, it is not clear whether this is an important attribute for marketing such a vehicle. Safety is a critical issue for such vehicles. It may be necessary to consider separate lanes on high-speed roadways.

In Group 3 there is an enormous range of possibilities. We mention only two of the most exciting: hybrid-electric and fuel-cell vehicles. The hybrid electric is powered both by batteries and by an internal combustion engine. A common configuration of the car uses the batteries (and electric motor) on short day trips and the internal combustion engine for longer trips. The batteries would be expected to be recharged overnight, when electricity demand is low and there is substantial unused electricity generating capacity. The hybrid overcomes the severe disability of electric vehicles: their short daily range and long battery recharge period.

The fuel cell, essentially a large battery, has the advantage of relying on a stored fluid fuel like methanol or hydrogen. The fuel cell converts the chemical energy of the fuel to electricity without combustion. Virtually no emissions, with the possible exception of carbon dioxide, are associated with fuel cell operation. Much higher efficiencies of conversion are possible than with the present kind of engine.

TECHNOLOGY PUSH-AND-PULL POLICIES

The U.S. government has been highly effective in encouraging new technology in certain sectors such as agriculture, commercial aircraft, and semiconductors. The tools used are, broadly, technology push and technology pull.

Technology push concerns the creation of technology: research, invention, development, and demonstration. This is not a linear sequence of activities, in which one follows the next, but a complex interaction in which new technologies are created. Technology-push policies involve government support for research, development, and demonstration (RD&D) and government encouragement of private-sector RD&D through tax incentives, patent law, and the like.

Technology pull concerns the demand for new technology—that is, demand after it has reached initial commercial status. It cannot be overemphasized that just as the existence of a likely market for a new or improved process or product strongly motivates development and production of new technologies, the apparent absence of a market strongly inhibits them. Government policies can provide technology pull through government purchases and by encouraging the private sector's propensity to purchase new technology (Ross and Socolow, 1990).

A major example of a technology-push policy is government-supported R&D on generic technologies that could form the basis for many new product developments. Modest government involvement is proving very beneficial in electrochemistry (new and improved batteries), combustion (understanding of knock and soot formation), and ceramic insulation (for the combustion chamber). We urge continued support in these areas as well as dramatic expansion of the government's efforts in, for example, engine friction, fuel cells, and systems control for hybrid-electric vehicles.

It may seem that it is the private sector's responsibility to conduct research on generic technologies such as those just mentioned. It is well known and well documented, however, that the private sector underinvests in research (Young, 1986). The roots of this underinvestment lie in a firm's inability to prevent its competitors from capturing many of the benefits of its research. Other contributing factors are the short time horizons and cyclical earnings patterns experienced by many firms. The private sector cannot support research leading to innovation in many socially useful areas of technology, at least not nearly at a level consonant with today's needs.

An attractive example of a technology-pull policy would be to give extra fuel economy credits to manufacturers who produce automobiles or light trucks that attain exceptionally high levels of fuel economy. Such a provision would reward manufacturers for boldly introducing new technology—an incentive for manufacturers to take a significant leap forward with fuel economy technologies as opposed to taking more conservative, incremental steps. The incentive could be made especially strong for improving the fuel economy of mid-size and large cars.[6] A schedule of fuel economy thresholds for the major EPA automobile size classes could be established as part of a strengthened fuel-economy-standards law. Present fuel economy regulations already include similar CAFE credits for alternative-fuel vehicles.

FUEL ECONOMY AND SAFETY

Opponents of efforts to improve automobile fuel economy have recently argued that the standards increase highway fatalities. Fuel-efficient cars are commonly equated with small, light cars. The record shows, however, that fuel economy can be substantially increased without reducing vehicle weight. The average new-car fuel economy began to improve sharply after 1974. Initially much of this improvement was due to reducing average vehicle weight. It was the easiest and cheapest way for manufacturers to improve fuel economy. The potential for making further improvements without reducing vehicle weight remains promising.

Figure 9.12 presents the weight and safety performance of 1984 to 1988 cars crash-tested by the National Highway Traffic Safety Administration. These cars were crashed into a fixed barrier at 35 mph. The measure of safety performance is the driver head injury criterion (HIC). The higher the number, the greater the potential for injury. As shown here, there is no relationship between automobile weight and head injury criteria. In fact, some heavy vehicles perform poorly (upper-right portion of the figure) and some light vehicles perform very well (lower left).

Crashing a car into a fixed barrier does not necessarily measure how weight affects a car's performance in a crash. Nonetheless, as a 1982 study pointed out, the differences in crash performance within weight classes are greater than the differences among weight classes (OTA, 1982). Evidence that fuel economy and automobile safety can be improved simultaneously is also found in the statistical record established in the United States. From 1975 to 1989, the average fuel economy for all cars on the road rose from 13.5 mpg to 20.3 mpg. During the same period, traffic fatalities fell from 3.45 per 100 million VMT to 2.25. Safer cars and highways, increased use of seatbelts, and campaigns against drunk driving are widely recognized as major reasons for the improvement.

Despite the evidence that improving fuel economy and automobile safety are compatible goals, some critics cite studies of actual automobile crash data demonstrating a relationship between car size and fatalities. A common problem, however, with studies based on actual · accident data is that it is difficult to separate the effects of driver behavior from vehicle characteristics. The bad fatality record of a few high-performance sports cars, for example, may lead one to conclude that these cars are inherently unsafe. But dangerous driving practices of the people who most commonly own and drive these cars may be responsible for their bad safety record. Similarly, the worse than

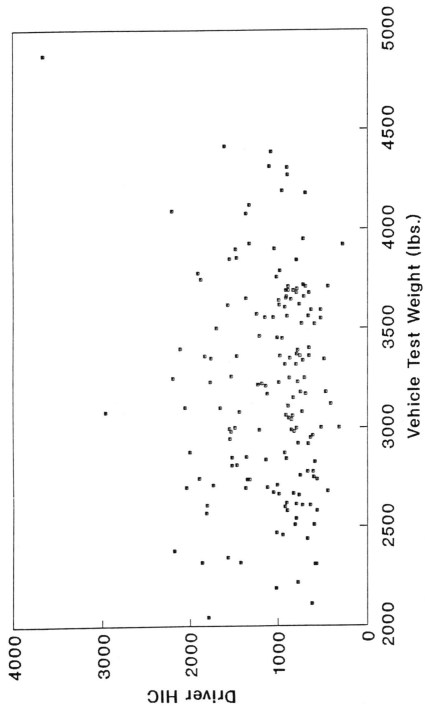

FIGURE 9.12. AUTO WEIGHT VS. DRIVER HEAD INJURY CRITERION (HIC).

SOURCE: ACEEE, based on NHTSA data.

average safety record of a few small, inexpensive, and fuel-efficient cars may be due to the atypical driving behavior of people who tend to buy these cars—for example, drivers of small cars tend to be young.

In summary, research has shown that with careful design cars can be both fuel efficient and safe. Nonetheless, the results of certain recent studies justify a close look at the effect of car size and weight on crash performance. If new research indicates that weight is indeed critical to safety, then future fuel economy improvements should be based on approaches other than weight reduction. If new research indicates that car size, such as wheelbase, is critical to safety, then future fuel economy improvements should focus on measures that do not reduce car size. With either approach, current and future technologies offer a broad range of ways to improve auto fuel economy substantially while simultaneously improving auto safety.

CONCLUSIONS

Federal policies have made a major contribution to the fuel economy improvements achieved since 1975. But because they require no further improvements, and because real gasoline prices have fallen, the long upward trend in fuel economy has stalled. This cessation in fuel economy improvement puts our nation's economy and security at risk. Among the options for reducing our oil imports and carbon dioxide emissions within the next 10 to 20 years, none will have greater effect than substantially improving light vehicle fuel economy. (See Figures 9.13 and 9.14.)[7] The market could be of great assistance in pushing fuel economy levels higher, but because of large externalities and other barriers, the market alone will not be sufficient.

Policies to reduce light vehicle fuel consumption will not only benefit the United States directly. Given the enormous influence U.S. policies and technologies have on other countries, these policies could leverage substantial international reductions in transportation fuel use.

In addition to fuel economy improvements, it is imperative that we slow growth in vehicle-miles of travel and offer attractive transportation alternatives to low-occupancy light vehicles. Otherwise, even with improvements in fuel economy, we could find ourselves looking back on the 15 years that follow 1990 with the same sense of running-in-place that we get when looking back on the 15 years since 1975.

FIGURE 9.13. PROJECTED U.S. LIGHT VEHICLE CARBON EMISSIONS WITH AND WITHOUT 40/30 MPG CAFE STANDARD BY 2001.

*The "no CAFE change" scenario assumes that new vehicle fuel economy does not rise above current levels.*

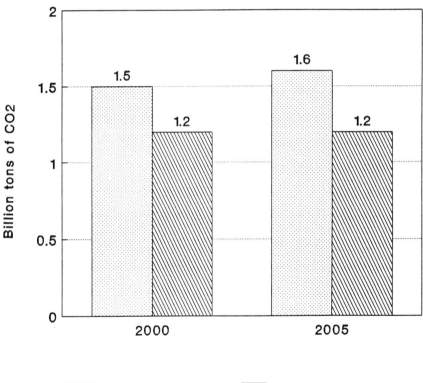

SOURCE: ACEEE.

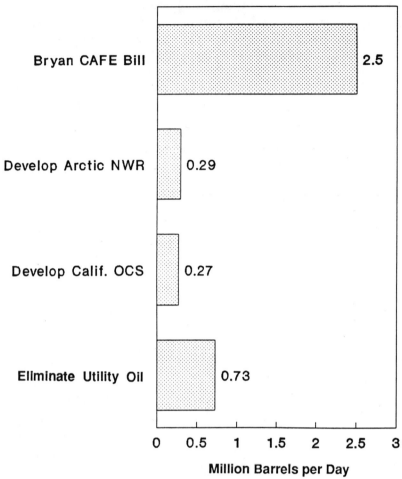

FIGURE 9.14. POTENTIAL DAILY PRODUCTION RATES FOR
SELECTED NEW U.S. OIL RESOURCES.

SOURCE: ACEEE.

## NOTES

1. The usual fraction of oil used by transportation, cited from the *Monthly Energy Review*, Energy Information Administration, U.S. Department of Energy, is 63 percent. Our calculation, however, excludes secondary petroleum-based fuels burned at oil refineries and in other industries from total U.S. oil consumption, because it would not be used if the primary uses of petroleum did not exist. We believe this approach better reflects the extent to which the transportation sector is responsible for oil consumption in the United States.

2. This calculation includes both the carbon emissions resulting directly from combustion in vehicles (20.2 kg $C/10^9$ J) and indirectly from production and transport of the fuel (2.7 kg $C/10^9$ J) (MacDonald, 1990).

3. The continuously variable transmission in Technology Group 1 would also change driving feel, but it was left in Group 1 to keep the list identical to that of Difiglio et al.

4. The mileage distribution was taken from the Department of Transportation's 1983–1984 Personal Transportation Study. Since cars are driven many more miles in their first years of use than later, capital recovery for technology improvements is accelerated, resulting in a lower annual capital charge. Using the DOT mileage distribution results in an annual capital charge equal to 96 percent of what it would be if capital were recovered in equal increments over 10 years.

5. A new car buyer typically has a much higher discount rate and might only be willing to consider the fuel savings achieved over 4 years, the typical period of time a new car buyer holds onto the car before reselling it.

6. As pointed out earlier, high fuel economy has been associated with bottom-of-the-market vehicles. One of the major policy challenges is to encourage manufacturers to create "green" cars in the middle of the market.

7. The 40/30 mpg CAFE standard referred to in Figure 9.13 is for cars and light trucks, respectively, and corresponds to EPA test composite fuel economy.

## REFERENCES

Amann, Charles A. 1989. *The Automotive Engine—A Future Perspective.* SAE 891666. Warrendale, Pa.: SAE.

An, Feng, and Marc Ross. 1991. *Automobile Energy Efficiency at Part-Load: The Opportunity for Improvement.* Washington: American Council for an Energy-Efficient Economy.

Ang, B. W. 1991. Traffic management systems and energy savings: The case of Singapore. In *Transportation and Energy in Developing Countries: Experiences and Options*, ed. D. Bleviss and M. Birk. New York: Quorum Books.

Asmus, T. W. 1991. *Perspectives on Applications of Variable Valve Timing.* SAE 910445. Warrendale, Pa.: SAE.

Bleviss, Deborah Lynn. 1988. *The New Oil Crisis and Fuel Economy Technologies.* New York: Quorum Books.

Bohi, Douglas, and Mary Beth Zimmerman. 1984. An update on econometric studies of energy demand behavior. In *Annual Review of Energy*, vol. 9. Palo Alto: Annual Reviews.

Bosch (Robert Bosch GmbH). 1986. *Automotive Handbook.* 2nd ed. Warrendale, Pa.: SAE.

Broadman, H. G., and W. W. Hogan. 1986. Oil tariff policy in an uncertain market. Energy and Environmental Policy Center, Harvard University.

Burchell, Robert W., and David Listakin (eds.). 1982. *Energy and Land Use.* Piscataway, N.J.: Center for Urban Policy Research, Rutgers University.

Bureau of the Census. 1990. Truck inventory and use survey. 1987 Census of Transportation, Washington.

Burke, Monica. 1990. *High Occupancy Vehicle Facilities: General Characteristics and Potential Fuel Savings.* Washington: American Council for an Energy-Efficient Economy.

Crandall, R. W., et al. 1986. *Regulating the Automobile.* Washington: Brookings Institution.

Davis, S. C., et al. 1989. *Transportation Energy Data Book.* 10th ed. Oak Ridge, Tenn.: Oak Ridge National Laboratory.

Demmelbauer-Ebner, Wolfgang, Alois Dachs, and Hans Peter Lenz. 1991. *Variable Valve Actuation Systems for the Optimization of Engine Torque.* SAE 910447. Warrendale, Pa.: SAE.

Department of Transportation (DOT). 1986. *1983–1984 Nationwide Personal Transportation Study.* Washington: Federal Highway Administration.

Difiglio, Carmen, K. G. Duleep, and David L. Greene. 1990. Cost effectiveness of future fuel economy improvements. *Energy Journal* 11(1):65–86.

El-Gasseir, Mohamed. 1990. The potential benefits of pay-as-you-drive automobile insurance. Submission to the California Energy Commission, docket 89-CR-90.

Energy and Environmental Analysis, Inc. (EEA). 1985. Documentation of the characteristics of technological improvements utilized in the TCSM. Arlington, Va.: EEA.

————. 1986. Analysis of the capabilities of domestic auto-manufacturers to improve corporate average fuel economy. Report to the Office of Policy, Planning, and Analysis, Department of Energy.

————. 1988. Developments in the fuel economy of light-duty highway vehicles. Report prepared for the Office of Technology Assessment.

————. 1990. An assessment of potential passenger car fuel economy objectives for 2010. Report to the Air and Energy Policy Division, Environmental Protection Agency.

Energy Conservation Coalition. 1989. *The Auto Industry on Fuel Efficiency: Yesterday and Today.* Fact Sheet. Washington: Energy Conservation Coalition.

Energy Information Administration. 1989. *Annual Energy Outlook 1989.* Washington: Department of Energy.

Federal Highway Administration. 1986. *1983–1984 Nationwide Personal Transportation Study.* Vol. 1: *Personal Travel in the U.S.* Washington: FHA.

Frame, Phil. 1990. GM upshifting to electronics. *Automotive News,* 17 Sept., p. 3.

Ganoung, David. 1990. *Fuel Economy and Drive-by Wire Control.* SAE 900160. Warrendale, Pa.: SAE.

Geller, Howard. 1989. *Financial Incentives for Reducing Auto Emissions: A Proposal for Clean Air Act Amendments.* Washington: American Council for an Energy-Efficient Economy.

Gordon, Deborah. 1991. *Steering a New Course: Transportation Energy and Environment.* Cambridge, Mass.: Union of Concerned Scientists.

Gordon, Deborah, and Leo Levenson. 1990. Drive +: Promoting cleaner and more fuel-efficient motor vehicles through a self-financing system of state sales tax incentives. *Journal of Policy Analysis and Management* 9(3):409–415.

Greene, David L. 1990. CAFE or price?: An analysis of the effects of federal fuel economy regulations and gasoline price on new car MPG, 1978–89. *Energy Journal* 11(3):37–57.

————. 1991. *Vehicle Use and Fuel Economy: How Big Is the Rebound Effect?* Oak Ridge, Tenn.: Oak Ridge National Laboratory.

Hamilton, William. 1989. Electric and hybrid vehicles: Technical background report for the DOE flexible and alternative fuels study. DOE/ID-10252. Report to EG&G, Idaho.

Heavenrich, Robert M., J. Dillard Murrell, and Karl H. Hellman. 1991. *Light*

*Duty Automotive Technology and Fuel Economy Trends Through 1991.* Ann Arbor: EPA Office of Mobile Sources.

Hellman, Karl, and J. Dillard Murrell. 1984. *Development of Adjustment Factors for the EPA City and Highway MPG Values.* SAE 8400496. Warrendale, Pa.: SAE.

Holtzclaw, John. 1990. Explaining urban density and transit impacts on auto use. Testimony before the California Energy Commission, docket 89-CR-90, 23 April.

Horton, E. J., and W. D. Compton. 1984. Technological trends in automobiles. *Science* 225:587–593.

Huber, Werner, Bernd Lieberoth-Leden, Wolfgang Maisch, and Andreas Reppich. 1991. *New Approaches to Electronic Throttle Control.* SAE 910085. Warrendale, Pa.: SAE.

Inoue, Kazuo, Osainu Kubota, Noriguki Kishi, and Shunji Yano. 1989. *A High Power, Wide Torque Range, Efficient Engine with a Newly Developed Variable Geometry Turbocharger.* SAE 890457. Warrendale, Pa.: SAE.

Institute of Transportation Engineers. 1989. *A Toolbox for Alleviating Traffic Congestion.* Washington: ITE.

Keebler, Jack, and Mary Ann Maskery. 1991. Mitsubishi, Honda claim gains with "lean burn" engines. *Automotive News,* 5 August.

Lashof, Daniel A., and Dennis Tirpak. 1989. *Policy Options for Stabilizing Global Climate.* Washington: Office of Policy, Planning, and Evaluation, Environmental Protection Agency.

Ledbetter, Marc, and Marc Ross. 1990a. *Supply Curves of Conserved Energy for Automobiles.* Washington: American Council for an Energy-Efficient Economy.

————. 1990b. A supply curve of conserved energy for automobiles. *Proceedings of the 25th Intersociety Energy Conversion Engineering Conference.* New York: American Institute of Chemical Engineers.

————. 1991. Light vehicles: Policies for reducing their energy use and environmental impacts. In *Energy and Environment.* Energy Policy Studies Series, vol. 6, ed. John Byrne and Dan Rich. New Brunswick, N.J.: Transaction Books.

Lenz, H. P., K. Wichart, and A. Grudent. 1988. *Variable Valve Timing—A Possibility to Control Engine Load Without Throttling.* SAE 880388. Warrendale, Pa.: SAE.

Ludema, Kenneth C. 1984. Mechanical friction and lubrication in automobiles. In John C. Hilliard and George S. Springer, eds., *Fuel Economy in Road Vehicles Powered by Spark Ignition Engines.* New York: Plenum Press.

MacDonald, Gordon. 1990. The future of methane as an energy source. *Annual Review of Energy,* vol. 15. Palo Alto: Annual Reviews.

MacKenzie, James, and Michael Walsh. 1990. *Driving Forces: Motor Vehicle*

*Trends and Their Implications for Global Warming, Energy Strategies, and Transportation Planning.* Washington: World Resources Institute.

Mayo, John, and John Mathis. 1988. The effectiveness of mandatory fuel efficiency standards in reducing the demand for gasoline. *Applied Economics* 20:211–219.

McCarthy, Patrick S., and Richard Tay. 1989. Consumer evaluation of new car attributes. *Transportation Research* 23A(5):367–375.

McCosh, Dan. 1990. Variable valve timing. *Popular Science*, March, p. 82.

McInerney, Larry. 1990. *Performance with Economy from High-Torque Backup for SI Engines.* SAE 900172. Warrendale, Pa.: SAE.

Motor Vehicle Manufacturers Association. 1990. *MVMA Facts and Figures '90.* Detroit: MVMA.

Muranaka, Shigeo, Yasuo Takagi, and Tokuhei Ishida. 1988. *Factors Limiting the Improvement in Thermal Efficiency of S.I. Engines at Higher Compression Ratios.* SAE 870548. Warrendale, Pa.: SAE.

Murrell, J. D. 1975. *Factors Affecting Fuel Economy.* SAE 750958. Warrendale, Pa.: SAE.

Murrell, J. D., and R. M. Heavenrich. 1990. Downward trend in passenger car fuel economy—A view of recent data. Technical Report EPA/AA/CTAB/90-01. Ann Arbor: EPA Motor Vehicle Emissions Laboratory.

Newman, Peter G., and Jeffrey R. Kenworthy. 1988. *Cities and Automobile Dependence: A Sourcebook.* Aldershot, England: Gower Technical.

Office of Technology Assessment (OTA). 1982. *Increased Automobile Fuel Efficiency and Synthetic Fuels.* Washington: OTA.

*Popular Science.* 1990. Jekyll & Hyde engines. Automotive Newsfront, Sept., p. 4.

Post, Richard. 1991. Lawrence Livermore Laboratory, personal communication.

Pushkarev, Boris S., and Jeffrey M. Zupan. 1977. *Public Transportation and Land Use Policy.* Bloomington: Indiana University Press.

Replogle, Michael. 1990. *U.S. Transportation Policy: Let's Make It Sustainable.* Washington: Institute for Transportation and Development Policy.

Ross, Marc. 1989. Energy and transportation in the United States. *Annual Review of Energy*, vol. 14. Palo Alto: Annual Reviews.

Ross, Marc, and Robert Socolow. 1990. Technology policy and the environment. Paper presented at "Toward 2000: Environment, Technology, and the New Century," Annapolis, Md.

Ross, Marc, Marc Ledbetter, and Feng An. 1991. *Options for Reducing Oil Use by Light Vehicles: An Analysis of Technologies and Policies.* Washington: ACEEE.

Schatz, Oskar. 1991. *Cold Start Improvements with a Heat Store.* SAE 910305. Warrendale, Pa.: SAE.

Seiffert, Ulrich, and Peter Walzer. 1991. *Automobile Technology of the Future.* Warrendale, Pa.: SAE.

Stedman, Donald. 1990. University of Denver, personal communication.

Stone, Richard. 1989. *Motor Vehicle Fuel Economy.* London: Macmillan.

Sztenderowicz, Mark L., and John B. Heywood. 1990. *Cycle-to-Cycle IMEP Fluctuations in a Stoichiometrically-Fueled S.I. Engine at Low Speed and Load.* SAE 902143. Warrendale, Pa.: SAE.

Tobias, Andrew. 1982. *The Invisible Bankers.* New York: Pocket Books.

von Hippel, Frank, and Barbara Levi. 1983. Automobile fuel efficiency: The opportunity and the weakness of existing market incentives. *Resources and Conservation* 10:103–124.

Ward, Daniel. 1990. Smaller engine, more power. *Automotive News,* 26 Feb.

Watson, H. C. 1990. University of Melbourne, personal communication.

Weber, Trudy. 1988. *Analysis of Unconventional Powertrain Systems.* SAE 885023. Warrendale, Pa.: SAE.

Westbrook, Fred, and Phil Patterson. 1989. Changing driving patterns and their effect on fuel economy. Paper presented at the 1989 SAE Government/Industry Meeting, 22 May, Washington.

Yagi, Shizuo, Yoichi Ichibasi, and Hiroshi Sono. 1990. *Experimental Analysis of Total Engine Friction in Four Stroke S.I. Engines.* SAE 900223. Warrendale, Pa.: SAE.

Young, J. A. 1986. Global competition—the new reality: Results of the President's Commission on Industrial Competitiveness. In *The Positive Sum Strategy,* ed. R. Landau and N. Rosenberg. Washington: National Academy Press.

CHAPTER 10

# ENERGY USE IN MANUFACTURING

*Dan Steinmeyer*

INDUSTRY CONSUMES 35 to 45 percent of the energy used in developed countries and a larger share in most of the lesser developed countries. Hence efficiency of use is a key issue in defining future global energy use and its impact on the environment. The good news is that there has been a drop in the developed world's energy use per unit of manufactured product of 1 to 3 percent a year for the last several decades. Rising industrial efficiency will soften the energy impacts over the next few decades, but it will not solve the energy supply for the twenty-first century. The rate of progress is inconsistent with aspirations of the developing world for goods, the energy required to make them, and the implications for the environment. The need for technological breakthroughs is clear—as is the contribution of fundamental science in enabling these breakthroughs.

This chapter is adapted from an article jointly authored with Marc Ross (Ross and Steinmeyer, 1990). It reflects many of his concepts, such as the importance of dematerialization and the distinction between technological change and cost optimization, and includes many of his quantified estimates for changes in the 1971 to 1985 period.

*319*

Rising industrial efficiency evolves naturally from the pressures of a competitive, growing, industrial society, our evolving consumption patterns, and fundamental thermodynamics. Rising industrial *energy* efficiency is guaranteed by the competitive process and will continue without government action. It includes "dematerialization" of the things we buy. It comes from technological progress much broader than energy—computers permit better designs, for example, and stronger plastics replace steel.

Scientists and engineers tend to make a big thing about energy efficiency. Industry does not—it cares about lowering costs and selling products. The desire for lower costs drives technological change. And as a by-product, it drives toward reduced energy use. And as a further by-product, it drives toward reduced environmental discharges—ash, sulfur dioxide, carbon dioxide, and others. When energy prices shoot up, as in the late 1970s, the efficiency improvement is much faster. This stems from "cost optimization"—choosing to trade capital for energy. This is a choice from *existing* technical options. It is distinct from new technology. Industry also recognizes that a better environmental image is in its own self-interest, and energy efficiency is beginning to gain support as an environmental issue.

Industrial energy use is dominated by the sectors that lie closest to the front end of the raw material processing cycle—those that deal with commodity-type materials. (See Table 10.1.) A separate break-

TABLE 10.1. BREAKDOWN OF U.S. INDUSTRIAL ENERGY USE IN 1988
(IN QUADRILLION BTU)

| Industry | ELECTRICITY | | Oil | Gas and LPG | Coal and coke | Other[b] | Total |
| | Direct use | Loss[a] | | | | | |
| --- | --- | --- | --- | --- | --- | --- | --- |
| Paper | 0.19 | 0.38 | 0.19 | 0.43 | 0.32 | 1.24 | 2.75 |
| Chemicals | 0.42 | 0.84 | 0.14 | 2.95 | 0.32 | 0.55 | 5.22 |
| Petroleum | 0.11 | 0.22 | 0.13 | 0.78 | 0.00 | 5.39 | 6.63 |
| Primary metals | 0.51 | 1.02 | 0.06 | 0.75 | 1.51 | 0.05 | 3.90 |
| Other | 1.17 | 2.34 | 0.29 | 1.95 | 0.64 | 0.45 | 6.84 |
| | | | | | | | 25.3 |

[a] Assumes that delivered electricity has one-third the Btu content of the fuel used to produce it—restated, 1 Btu of electrical supply to an industrial user represents 3 Btu of fuel consumed.
[b] This category is unevenly handled. In the case of petroleum, it is primarily feedstock used in "nonenergy products" such as asphalt. In the case of paper, it is the fuel value of biomass.

down (EIA, 1991) shows that 30 percent of the industrial fuel use is actually chemical feedstock. Here are the major contributors (in quadrillion Btu) in addition to the petroleum industry in 1988:

| | |
|---|---|
| Blast furnace coal | 1.06 |
| Plastics | 0.29 |
| Nitrogenous fertilizers | 0.26 |
| Other organic chemicals | 0.81 |

Often the mining and agricultural use of roughly 5 quadrillion Btu is combined with the manufacturing sector. The decision whether or not to include electrical generation losses, feedstock, and mining/agricultural explains the twofold variation in industrial energy use reported in various sources.

## DEMATERIALIZATION

One of the most important elements in reduced industrial energy use has nothing to do with the classic concept of energy efficiency. While we "consume" far more than our parents did, the things we consume are different and generally take less energy to make.

### DIFFERENT THINGS FOR BETTER LIVING

The composition of production inevitably changes in a way that drives toward less energy use. The first steps of manufacturing—the initial conversions of material—are much more energy-intensive than the rest of manufacturing. For each $1 spent on wages and capital:

- An aluminum plant spends $1 on energy.
- An inorganic chemical manufacturer spends $0.25 on energy.
- A frozen food manufacturer spends $0.05 on energy.
- A computer manufacturer spends only $0.01 on energy.

The shift in production has been toward more complex, information-laden products. People are buying more electronic equipment and fewer refrigerators. The completion of infrastructure in developed countries also means a lower requirement for energy-intensive materials like cement and steel.

## BETTER-MADE THINGS

We also use less material in the production of the same basic set of products. Part of this is due to materials substitution—for example, glass fiber replacing copper wire for telephone transmission. Part is due to improved design—for example, thinner sheets of higher-strength alloy steel substituting for thicker sheets of conventional steel in automobile bodies. A dramatic example of long-term design improvement is the fall in the weight/power ratio for railroad locomotives. Between 1810 and 1980 this ratio fell by a factor of 100. Moreover, the consumption of all major materials is declining relative to consumption as a whole (Williams et al., 1987). Since 1970 the relative contribution of basic materials to this country's GNP has dropped by nearly 30 percent.

The combined shift in the mix of consumption and less physical mass per unit of product caused a decline of 1.5 percent a year in manufacturing energy use in the period 1971–1985. A related impact is that our exports have become less energy-intensive than our imports. Data suggest that the energy embodied in imports rose about 2 quadrillion Btu more than that in exports between 1972 and 1984 (DOE, 1989).

## THINGS THAT ARE NOT DISCARDED: RECYCLING

Broadening the dematerialization picture to include recycling improves the efficiency prospect for the future. The thermodynamic requirement for turning a bottle into a bottle is zero. This underlies the commonsense case for the refillable recycling we formerly practiced.

When we look at recycling *nonrefillable* beverage containers, the practical energy requirements go up—but a substantial margin remains for recycling. It is not surprising that aluminum, which has the largest differential, is by far the leader in container recycling with approximately 60 percent of aluminum cans recycled. (The plastic is PET and the energy is net of combustion value.) Consider these recycling figures (reckoned in Btu/pound):

| Source | Glass | Steel | Plastic | Aluminum |
|---|---|---|---|---|
| Made from raw materials | 6500 | 35,000 | 35,000 | 100,000 |
| Made from recycled materials | 4500 | 15,000 | 15,000 | 25,000 |

Recycling in other areas also correlates with economics. Automobile hulk recycling approaches 95 percent, for example, because the mass is large enough to repay the effort. In contrast, recycling of plastics is only in the range of 1 to 2 percent, in large part because the individual pieces are small and hard to segregate.

An obvious problem is finding acceptable end uses for the recycled material. There is a limited quantity of cellulose insulation that can be made from old newspapers as well as a limited demand for substitute lumber made from recycled plastic. The major success story is the aluminum beverage can, a technologically demanding, high-value product that is made primarily from recycled material. The challenge is to create other recycling systems like this, systems that use recycled materials in high-value applications, such as recycling old plastic bottles into new plastic bottles. Recycling was recently highlighted as one of three initiatives for energy R&D in industry (NAS, 1990) and as the major short-term opportunity for improving industrial energy efficiency.

## ENERGY INTENSITY—AND TECHNOLOGICAL PROGRESS

Dematerialization is one part of rising energy efficiency. The other part is using less energy to make a given mass of a commodity. Fuel per ton of steel or electricity per kilogram of polyethylene is often referred to as energy intensity. Fuel intensity declined 3 percent a year in the 1971–1985 period while electricity intensity stayed roughly constant. Real energy intensity is an average of the two, calculated in a way that eliminates the effect of shifts in the mix of production. Its decline is estimated at roughly 2 percent a year for this period. The improvement in intensity was driven by two distinct forces: technological progress, which is the subject of this section, and cost optimization, which is the short-term response to price swings and is discussed later in a separate section.

### THERMODYNAMIC LIMITS: SETTING THE TARGETS

One reason engineers focus on energy is that energy efficiency improvement is uniquely easy to measure and calculate. The rules are given by thermodynamics. There are two sets.

Almost all the official rules count *energy* (the first law of thermodynamics). As an accounting exercise this is accurate, but not helpful in setting expectations or understanding progress. In contrast the rules that deal with *work* (the second law of thermodynamics) give an accurate measure of potential for improvement. When we speak of "using energy" what we really mean is "using the embedded work potential." What we value in energy is its ability to do work. A typical production process uses the work potential embedded in chemical fuels to transform a raw material into the desired product, such as iron ore into steel. Part of this work potential is retained in the product—steel has a higher potential than the ore it is derived from—but most of the work is lost to low-level heat.

For every process we can compute a theoretical work requirement. The ratio of this requirement to the actual work potential consumed is the true thermodynamic efficiency. This calculation tells us that even for the best processes, the thermodynamic efficiency is remarkably low. The efficiency of producing oxygen by separating it from air is 20 to 30 percent, for example, and the efficiency of producing ethylene, the prime petrochemical building block, is 15 to 25 percent. These low efficiencies provide the margin from which gains are carved.

The calculation yields an even sharper comment on other energy use by our society:

- The net work of moving something from one place to another at the same elevation is zero. Hence the thermodynamic efficiency of our entire transport industry is zero.
- The net work of consuming energy in a partially insulated, leaky enclosure—and letting the energy mix with the ambient

through openings or through the somewhat insulated walls—is zero. Hence the thermodynamic efficiency of our entire building sector (in space conditioning) is zero.

The second law of thermodynamics is often hidden behind abstract terms, but a simple, functional definition of the second law is: It takes work to change things. A useful corollary states: We can calculate the theoretical minimum work needed for a given change. This applies not only to a total process like making methanol from natural gas, but to the detail of the individual processing steps. This means we can find the loss points and focus on them directly (Steinmeyer, 1984; Kenney, 1984):

- It is a little helpful to know that a distillation step uses five times the theoretical energy. But it is far more helpful to know that 60 percent of the loss is due to driving the heat exchanges—that is, in the temperature difference between the hot and cold streams. This immediately changes the search. Now the question becomes: How can we change the temperature levels to cut this loss? (Answer: By redesigning the refrigeration system to operate closer in temperature to the process and by using bigger heat exchangers.)
- Similarly, it is a little helpful to know that the largest loss in a methanol plant is at the fired heater. But it is much more helpful to know that the losses can be broken down into the specific areas shown in Figure 10.1. Note that the combustion step accounts for one-third of the lost work. Again this changes the search. Now the question that emerges is: How can we cut the loss when we run the highly irreversible step of combustion? (Answer: By preheating the air used to burn the fuel.)

One message is that efficiencies will improve. Another is that we can pinpoint where losses occur and track our progress as we drive them down.

TRACKING IMPROVEMENTS: CAN WE EXTRAPOLATE THE PAST?

Estimates of efficiency intensity improvement for the coming decades vary so widely (Table 10.2) that at first glance one is left with skepticism about our ability to predict the future. Nevertheless, the

FIGURE 10.1. CONVENTIONAL HEAT BALANCE VS. SECOND LAW BALANCE.

*The second law of thermodynamics gives the designer a new set of eyes. The traditional heat balance suggests that the designer should focus on recovering heat from the hot gases leaving the furnace. But it would completely miss other losses that can be much greater. Combustion is customarily the largest loss in a process—it is highly irreversible. Typically 30 percent of the work potential is lost in the flame. Options for cutting this loss would be adding a heat exchanger to preheat the air or by feeding gas turbine exhaust as combustion air.*

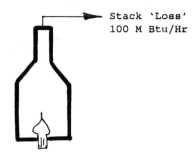

Stack 'Loss'
100 M Btu/Hr

Combustion    550 M Btu/hr

**(a)** Conventional "heat" balance

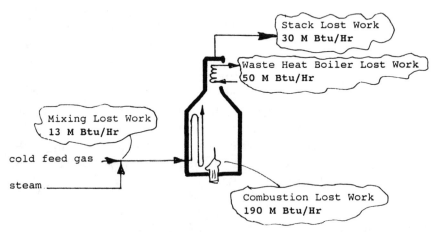

Stack Lost Work
30 M Btu/Hr

Waste Heat Boiler Lost Work
50 M Btu/Hr

Mixing Lost Work
13 M Btu/Hr

cold feed gas

steam

Combustion Lost Work
190 M Btu/Hr

**(b)** Second law balance

actual data align fairly well with thermodynamic efficiency. Those with more margin for improvement, like chemicals and paper, have shown more improvement (Carlsmith et al., 1990). The far right-hand column of Table 10.2 shows the actual annual improvement in energy intensity between 1972 and 1985. The forecasts are below the actual improvement because much of the 1972–1985 gain is due to the unusual energy price surges during this period.

## LEARNING CURVES: GROWING OUR WAY TO LOWER ENERGY INTENSITY

Most technological progress in the industries that dominate energy use, such as the ones cited earlier, comes as a series of small steps and small decisions:

- Deciding what operating pressure to use in air separation
- Deciding what kind of catalyst basket to use in an ammonia fertilizer plant
- Deciding how to control a distillation tower
- Installing sensors and on-line control systems for rolling metal sheet (yielding a reduction in scrap)
- Using variable-speed drives to control motors for pumps
- Finding a better catalyst producing fewer impurities or achieving the desired yield at less extreme pressures
- Introducing a more efficient type of lighting

This is broadly referred to as "learning." In most manufacturing processes, for each doubling of cumulative production, total process-

TABLE 10.2. FORECASTS FOR INTENSITY IMPROVEMENT (% PER YEAR)

| Industry | NAS (1990) | NAS (1979) | DOE (1989) | OTA (1983) | Actual improvement |
|---|---|---|---|---|---|
| Steel | 1.0 | 0.9 | 2.4 | 2.3 | 2.1 |
| Chemicals | 1.5 | 1.7 | 0.7 | 0.6 | 3.4 |
| Paper | 2.2 | 1.3 | 2.1 | 1.3 | 4.7 |
| Petroleum | 0.5 | NA | 0.9 | 0.6 | 2.5 |

SOURCE: Carlsmith et al. (1990).

ing costs, including energy, drop by about 20 percent. Often energy savings are merely a by-product of changes made to improve quality or safety, or to increase productivity, or to reduce emissions. This relation has been found to hold for processes as diverse as aircraft manufacture and the production of polyethylene. The learning curve does not depend on increasing energy costs. As shown by Figure 10.2, substantial gains in energy efficiency were made during the 1950s and 1960s with energy prices flat and in some cases falling.

Figure 10.2 follows a key commodity, ethylene, and tracks the energy efficiency of new plants offered by The Lummus Company, an engineering contractor (Sumner, 1990). The gains can be traced to a mix of sources:

- More efficient turbines and compressors from suppliers
- Adjustment of the purification sequence by engineering contractors
- Improvements in computerized control by operating companies
- More heat recovery everywhere—by common consent

The individual events are not exciting. Neither is the 3 percent annual improvement in energy efficiency. But the net result was a 60 percent drop in energy use of new facilities over a 35-year period, which is very significant. The chemical industry is the largest industrial user of energy, and ethylene production is the largest user of energy in this sector. Despite the large efficiency gains, a major opportunity for improvement remains: energy use is still about four times the thermodynamic minimum.

The driving force behind "learning curve" progress is the desire for a competitive advantage: lower cost—which yields economic efficiency. Energy efficiency is one component of economic efficiency, but it is rarely a dominant one. For the overall manufacturing sector, capital and labor costs are roughly twenty times as great. One result is that a spurt of federal funding for research focused purely on energy is unlikely to cause much of a change in practice. A few person-years of effort focused on one cost element simply does not have enough leverage to make a major change in a process that is based on thousands of person-years of work. This is the main reason why industry has been lukewarm about the DOE's programs and skeptical about its claims for contributions.

FIGURE 10.2. ETHYLENE FROM ETHANE: A THERMODYNAMIC SUCCESS STORY.

*Ethylene plant competition has traditionally been fought by the engineering contractors in the area of low energy per pound of ethylene. As shown here, this competition has produced a dramatic reduction in use.*

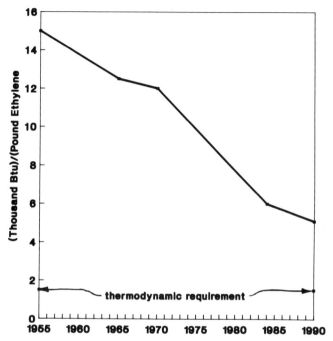

SOURCE: Lummus Company.

A key element in learning is a *growing* industry, one that replaces old facilities and offers continual opportunities to test improved technology. If one inputs the 20 percent learning rule to a spreadsheet and looks at expected yearly cost reduction, growth rate turns out to be as important as age (Table 10.3). A factor in the ethylene success story is its high growth rate (5 to 10 percent a year) through most of the 1955–1990 period.

### SLIGHTLY BIGGER STEPS ON THE CURVE

But the learning rule is only an approximation, and old industries occasionally find themselves on new learning curves as a result of technological breakthroughs. Two breakthroughs from recent

TABLE 10.3. COST REDUCTION (% PER YEAR)

| *Growth rate* | YEAR FROM START | | | |
|---|---|---|---|---|
| | *10* | *20* | *30* | *50* |
| 3% | 4.0 | 2.3 | 1.7 | 1.3 |
| 6% | 4.5 | 3.1 | 2.4 | 2.0 |
| 9% | 5.1 | 4.1 | 3.2 | 2.9 |
| 12% | 5.7 | 5.2 | 4.1 | 3.8 |

decades in older industries are the float-glass process to make flat glass on a pool of molten tin and the basic oxygen converter to make steel by blowing oxygen into the solution of carbon and iron that comes from a blast furnace. Both breakthroughs are probably due more to individual inventiveness than scientific discovery, though scientific discoveries often play a key role in *enabling* the breakthroughs.

An example of how scientific discovery and innovation interact with learning is polyethylene. Polyethylene began its commercial life in the early 1940s with a very high pressure (1200 atmospheres) process. The high-pressure process saw continual improvement such that the energy required to produce a pound of polyethylene was cut in half in about 25 years. Meanwhile, two European chemists made some fundamental discoveries that led to a radically new production process that utilized a solvent and operated at low pressure. This in turn led to Union Carbide's development in the 1970s of the low-pressure gas-phase process. It uses only 15 percent of the energy of the original high-pressure process. The new process is simpler, safer, and requires much less capital. It even yields a stronger polymer.

Sometimes scientific discovery and commercial application are so tightly linked that they appear to proceed almost side by side. An example is the interaction between polymer chemistry and development of synthetic fibers such as nylon and kevlar. And sometimes the progress is really the by-product of major scientific discovery in unrelated areas. Quantum physics and the invention of the transistor led to microprocessors and modern computers. The impact of this hardware shows up very broadly in the way we live. The industrial fallout too is extremely broad. Low-cost microelectronics permits more efficient industrial control systems as diverse as robots in laboratory analysis and excess-air measurement in furnaces. Perhaps more important, it has revolutionized industry's ability to explore for better processes. In the ethylene process cited earlier, engineering contractors assigned much more credit to enhanced calculation capability than to all bench-scale R&D.

## BIGGER BREAKTHROUGHS: ENTIRELY NEW CURVES

While the evolutionary improvements are important, major breakthroughs are more important in the long run. The learning curve process would have run its course long ago were it not for fresh starts on new curves due to fresh inputs of new science and radical innovations. Breakthroughs are the reason why the future rarely turns out the way we foresee it. It is usually much more exciting, and often happier.

Some economists believe that inventions and innovations come when they do in response to a generally perceived constraint (Haustein and Neuwirth, 1982). An extrapolation would suggest that if the scientific/industrial community sees an imperative for more energy, they will find it—and will probably find it in unexpected places. Convince them that global warming is real, and they will find a way to control the global heat balance. This may sound utopian, but consider our history. Energy and material constraints have been critical in the past. An example is widespread doubt about the availability of sufficient wood for the masts of sailing ships—leading to the development of steel ships and alternative propulsion. Other historic examples are the concern about the ability to move coal (leading to railroads) and the concern about sufficient waterwheel power to drive machinery (leading to the steam engine). Breakthroughs have shaped our industrial development and will continue to cause our future to be different from our expectations. We rarely bump into the limits we fear—whether the limit is the availability of wood for our sailing ships in the 1800s or today's apparent set of limits: oil and gas in the ground, locations for our domestic waste, and releases of greenhouse gases to the atmosphere.

Scientific discovery does not guarantee commercial innovation, but any discovery offers a set of possibilities that did not exist before. What breakthrough is likely to contribute to a *sustainable* global economy? What are some of the blue sky possibilities? One can guess at some breakthroughs from scientific progress, recent events, and trends. If some of these lack an immediate, obvious tie to industrial energy use, so did the developments in microelectronics.

As a start, some recent events are:

- Over two dozen species of crop plants have been transformed by molecular engineering to achieve an altered characteristic.
- Measured superconductivity has dramatically moved toward room temperature.
- Diamond films have been produced at moderate temperature and pressure.
- Photovoltaic conversion efficiencies of 35 percent have been achieved.
- Unexplained fusionlike phenomena have been encountered at low temperatures in electrolytic cells.

Some current technology trends are:

- Computers have made theoretical prediction of chemical reaction pathways feasible and are being used for such things as molecular design of systems that mimic photosynthesis.
- Life cycles and recyclability are beginning to affect product design.

The discovery that appears most clearly latent with possibilities today is the understanding of molecular biology and the ability to insert desired genetic traits into plants. This breakthrough should have practical consequences for industry in the first half of the twenty-first century through the development of new and modified agricultural plants. For example:

- Genetically engineered plant systems allowing crops to fix their own nitrogen from the air (as legumes do now via a symbiotic process with bacteria) could eliminate the need for nitrogen fertilizers. Not only does the manufacture of these fertilizers consume 2 percent of all industrial energy, but their use is believed to be responsible for the major share of human-derived emissions of nitrous oxide, a gas implicated in the greenhouse effect.

- Plants that can better tolerate environmental stress could double the fraction of land available for crops. (Pasture and forest could be transformed into cropland by increased plant tolerance of drought and temperature.)
- High-yield plants readily convertible to liquid fuels could support our transportation energy needs.
- Bioprocesses could recover fuels from municipal and agricultural wastes.
- Agricultural plants and bioprocessing could yield chemical products at costs lower than at present and with no petroleum feedstock.

## COST OPTIMIZATION

A final reason to expect falling energy intensity is that energy prices are expected to rise faster than those of other cost components. This means that cost optimization will drive the substitution of capital equipment for energy. Industry chooses its processes and products largely on the basis of cost. Energy is just one of the cost components, and capital equipment is a component it can be traded against. If one spends more on equipment, less energy is needed. The curves that plot the overall cost of a production process versus energy use are "gentle" in the region of the optimum. (See Figure 10.3.) Thus a great deal of trading can occur with relatively small impact on total cost. As the price of energy goes up, industry is driven by logic to substitute capital. This is a major cause of the efficiency gains we saw in 1975–1985. When energy costs rise in the future, this pattern will repeat. This is optimization within existing technology.

### TRADING CAPITAL FOR ENERGY

The trade between capital and energy is surprisingly clearly defined for certain kinds of operations. A fourfold rise in the relative price of

FIGURE 10.3. ENERGY USE VS. TOTAL COST—PIPING.

*The slope of the total cost curve is very gentle near the optimum—the low point on the curve. The shape of this curve depends on the ratio of energy costs to capital costs. For piping the ratio is extremely low (about 1 to 7). The result, as shown here, is the low impact of energy optimization on total cost.*

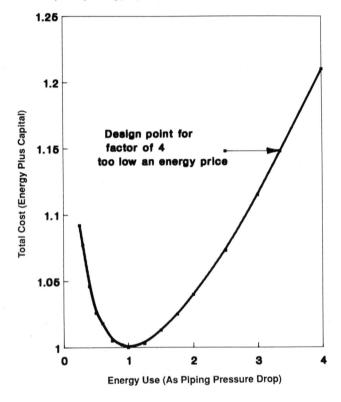

Energy Use (As Piping Pressure Drop)

energy, for example, will cause the following reduction in energy use—assuming the owners adjust their design to the economic optimum (Steinmeyer, 1982):

| | |
|---|---|
| Unrecovered energy in heat exchange | 50–75% |
| Distillation reflux above the minimum | 50–60% |
| Piping pressure drop | 70% |

There is wide variation in the ease of capital and energy trading (Carlsmith et al., 1990). Much of the energy use and much of the capital use is not subject to the marginal interchange shown here—

for example, much of the capital is fixed by the decision to build a plant of a certain size or efficiency. Examples are control rooms and laboratories. Conversely, endothermic reactions require a minimum level of fuel firing.

### TIME LAGS

At first glance Figure 10.3 may seem puzzling. Note that between half the optimum pressure drop and twice the optimum, the total cost stays within a few percent of that at the optimum. The reason for this is that for an optimum piping system, the annualized energy cost for piping pressure drop is only about one-seventh the annualized cost of the pipe. Although piping is an extreme case of capital cost dominating energy cost, it illustrates one of the reasons why there is a long time lag in industry's adjustment to energy price increases. The energy saving would easily justify the incremental cost of going to a larger pipe size in a new facility—but it would not justify the cost of replacing the existing piping system in an existing plant with a slightly larger one. Industry is constrained by existing facilities and can only change them economically as part of an orderly capital replacement.

### COST OPTIMIZATION AND CAPITAL LIMITS

We have recently seen a flattening of energy efficiency. The Chemical Manufacturers Association index,

$$\frac{Energy}{Value\ added}$$

showed a drop of 42 percent from 1974 to 1985 but only a 1 percent fall from 1985 to 1989. Part of this is due to falling energy prices, but part is due also to a rise in the price of capital for cost reduction projects. The price of capital is subject to variance between types of project (Carlsmith et al., 1990), as some firms "discriminate" against cost reduction efforts (like energy conservation) in favor of market-share projects. The rationale is that an energy project can be executed this year or next with little impact on the project's net present value. In contrast, a market-share project sees a window of commercial opportunity. In other words, the projects are sorted into "strategic" and "nonstrategic." The nonstrategic projects are used to "level" capital expenditure—that is, to utilize any leftover capital in the yearly ration.

### THE UNITED STATES VERSUS OTHER COUNTRIES

The gentleness of the curves also supports major differences in energy intensity between comparable plants throughout the world. An ethylene plant on the Houston ship channel and one in Korea can differ appreciably even though the same technology is accessible to both. The cause, again, is a difference in the ratio of price of energy/price of capital. Both numerator and denominator are subject to great variance. Natural gas is a premium fuel when viewed by environmentalists—but it is difficult to ship and store and hence sells into a very inelastic market. Regions that have gas (and the pipelines to distribute it) tend to have low energy prices and low efficiency.

State subsidy of energy cost has the same impact. Data on Eastern Bloc facilities are less readily available, but they appear to suggest a use as much as 40 percent higher than the United States. For example, a fired heater designed for the same process, by the same engineering contractor, for the same startup time, runs a stack temperature 100°C hotter because of less heat exchange equipment installed. This reflects a much lower cost of energy input to the design optimization—not a difference in technology level.

That capital is less costly in some world areas than others is difficult to explain. Perhaps the good news is that a Japanese firm operating a U.S. facility may evaluate the capital versus energy trade on a basis more favorable to conservation. Directly comparable energy intensity data are normally kept secret by competitive firms, but a company that performs comparative operating analyses of similar petroleum and petrochemical units throughout the world has estimated the relative energy usage as:

| | |
|---|---|
| United States | 100 |
| Japan | 91 |
| Germany | 83 |
| Scandinavia | 75 |

SOURCE: Capers (1991).

Other sources suggest that a 20 percent reduction may be available in the United States with current technology and good economic return.


ENVIRONMENTAL IMPACTS: A NEW DIMENSION

The optimization framework adjusts when society recognizes a new set of costs and values—for example, the benefit of cleaner air and water. The environmental framework is a bit harder for industry to see clearly, but industry has learned that failure to respond carries a host of indirect costs such as delays in approval for new products—as well as in more directly measurable costs like lawsuits and legal fees. The voluntary movement today by many companies to reduce wastes is a consequence.


POLICY OPTIONS: WHAT HAVE WE LEARNED?


Policy, too, should be subject to a learning curve. The extensive experience with energy, economic, and environmental policies in the past two decades points to actions that should be taken. It also suggests that we should be patient. Further, the relatively low success rate of major initiatives suggests that we should be skeptical of the benefits from radical shifts in R&D and regulatory policies.

Perhaps the first thing to learn is that much of the future change in energy efficiency will happen with or without government policy:

• Dematerialization
• Learning-curve improvements in efficiency

Perhaps the second thing to learn is that some of the best programs are the older ones:

• Support for education
• Support for broad-based research

Perhaps the third thing to learn is that patience and persistence are virtues:

- Breakthroughs take decades to move through the cycle of scientific discovery, to invention, to innovation, to commercialization.
- The scientific base on which we build has literally millions of person-years.

## What Can We Do to Encourage Breakthroughs?

One of the important policy elements in our drive for a sustainable economy and environmental quality is a broadened scientific base and allowance of enough time for technology and inventions to develop. There will be surprises, and many will be positive. A long time horizon and patience are both necessary. The average interval between first invention in a new area and peak inventive activity is 10 years; between the first invention and the subsequent industrial surge the interval is 60 years (Haustein and Neuwirth, 1982). The basic research funding that underlies an innovation is rarely visible when the innovation enters commercial practice.

## Learning Curves: Are They Really Inevitable?

It is easy to recognize the sources of learning-curve improvement:

- A vigorous technology base is needed—both in terms of people and equipment.
- Good profit levels are needed to fund research but even more important to exploit good ideas in demonstrations—which inevitably carry risk.
- The ability to reap the benefits of the successful innovation is needed—either in a growing internal market or in technology sale abroad.
- Strong competition between firms within each industry, and between industries, is needed.

- Finally, a long time horizon is needed—for operators of the plants, for equipment and engineering suppliers, as well as for the government regulatory bodies.

Policies that explicitly recognize these needs are not difficult to envision nor necessarily new:

- Education: The reinforcement of science education in high schools is fundamental to a supply of scientists and engineers— and the understanding of their contribution by others.
- Depreciation/tax codes: Plant operators, equipment suppliers, engineering firms, and government regulatory bodies must all have long time horizons. The financial policies that have pushed American firms to think in terms of fiscal quarters instead of decades must be changed. Favorable depreciation treatment of productivity investments would be a start toward redressing the balance.
- International focus: Active participation in the international areas of growth is required. To return to the ethylene example, the growth in production will be overseas. The most recent innovation in the falling energy usage is gas turbine preheat of the furnace combustion air. It has been implemented by Lummus in two Korean plants but not in the United States. This says something about the relative costs of energy and capital in the two locations—but even more about risk aversion and growth prospects. An operating company that does not plan on a series of additional plants in the next 10 years has little incentive to gamble on innovations.

RECYCLING

Recycling is a less dramatic topic, perhaps, but it is a current opportunity for dematerialization that is particularly open to enabling government action—such as requirements in product design, taxes to align economics better with social benefits, and implementation of a system of local collection/separation stations.

Recycling is also potentially ripe for research contribution—for example, how can one efficiently segregate the fifteen major varieties of plastics? Part of the solution may be to develop end-use products that can accept a lower level of purity. Part of the solution may be to develop marking systems enabling automatic separation.

## LINKAGES AND LONG PERSPECTIVES: SOME SCENARIOS

The forecasts shown in Figure 10.4 range from flat energy use to a 75 percent increase over the years between 1990 and 2020. The difference—about 2 percent compounded over the 30 years—is driven by forecasts for three numbers:

- Population growth rate
- Manufacturing product/population
- Energy use/manufactured product

The third number is a summation of impacts:

- Shift in the mix of manufactured product
- Improvement in processing efficiency (intensity reduction) through technological change
- Improvement in processing efficiency as a result of capital substitution (cost optimization driven by rising energy prices)
- Shift in the *net* material and energy content of a given item (including material substitution, improved design, and increased recycling of material)

The first two numbers are outside the scope of this chapter. Perhaps all that can be said here is that growth is down in the United States. What will happen to the third number—energy use/manufactured product—is the interesting story. We can postulate several futures.

### Scenario I: Things Continue

The simplest scenario is that energy continues to be available at a slightly falling cost, both in cost of production and environmental externalities. We continue to build, to buy, and to move material

FIGURE 10.4. INDUSTRIAL ENERGY USE IN THE UNITED STATES.

*Projections of U.S. energy consumption for the industrial sector show great uncertainty.*

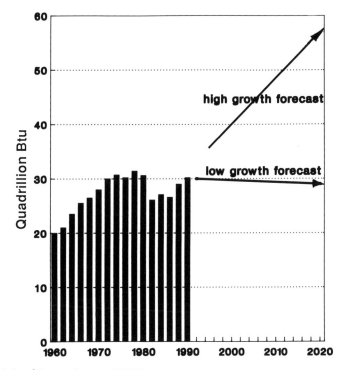

SOURCE: National Energy Strategy (1991).

from mines to new landfills with increasing efficiency. If one chooses a time perspective of the first part of this century, this is our history. Nonbiomass energy use in the United States grew at a compound rate of 3.2 percent a year from 1890 to 1938 (Kennedy, 1987).

## SCENARIO II: THINGS CONTINUE TO CHANGE, SLOWLY

If one chooses a time perspective of only the last two decades, things look very different. Energy prices are rising in surges. Consumption of essentially all commodities is falling on a per capita basis. Industrial energy usage is flat.

We have already seen major shifts in energy capability as well as in our behavior relative to our energy supply and use. Some indicators:

- Reduction of oil and gas production in the United States
- Energy import costs pushing our balance of payments deficit to an unsustainable level
- Environmental shutdown of products such as freon
- Drive for understanding the impact of greenhouse gases
- Wars justified by preserving access to oil
- Industrial capital allocation to environmental projects becoming a major share of the total capital expended

Industrial energy consumption is about the same as in 1973. To forecast that it will resume its historical rise is to argue with the data trend as well as the detail behind the trend discussed in this chapter.

This is neither a happy scenario nor an unhappy one. We simply define progress differently.

## SCENARIO III: A FUTURE FULL OF BIGGER SURPRISES

As individuals (particularly engineers), we tend to think of energy use as something uniquely important. Manufacturers see it as just one item to be balanced in the overall drive for minimum cost. Other factors are skilled labor, capital, raw materials, and public acceptance. We worry that we will hit a barrier in energy. This could well be true, but there are other barriers we might hit sooner:

- We could exhaust our capital in environmental improvement projects. Capital is limited and represents a major share of material and energy usage.
- We could exhaust our labor fighting plagues like AIDS and crime—or we could expend our surplus productivity in education or care for an aging population.
- We could exhaust our ability to manufacture goods with high material content due to lack of raw materials.

A happier set of surprises would involve technological break-throughs in fusion, biomass, or other unsuspected areas. Considering the long course of our history, this is the result our great-grandchildren are likely to see if our children encounter the energy barriers we currently fear.

## REFERENCES

Capers, David (Solomon Associates). 1991. Personal communication.
Carlsmith, R. G., W. U. Chandler, J. E. McMahon, and D. J. Santini. 1990. *Energy Efficiency: How Far Can We Go?* Oak Ridge, Tenn.: Oak Ridge National Laboratory.
Haustein, H., and E. Neuwirth. 1982. Long waves in world industrial production, energy consumption, innovations, inventions, and patents, and their identification by spectral analysis. *Technological Forecasting and Social Change* 22:53–89.
Kennedy, P. 1987. *The Rise and Fall of the Great Powers.* New York: Random House.
Kenney, W. F. 1984. *Energy Conservation in the Process Industries.* Orlando: Academic Press.
NAS. 1990. *Confronting Climate Change: Strategies of Energy Research and Development.* Washington: National Academy Press.
Ross, M. H., and D. E. Steinmeyer. 1990. Energy for industry. *Scientific American* 263 (Sept.):88–98.
Steinmeyer, D. E. 1982. Take your pick: Capital or energy. *Chemtech* (March):188–192.
_____. 1984. Process energy conservation. In Kirk Othmer, *Encyclopedia of Chemical Technology,* supp. vol. New York: Wiley.
Sumner, Charles (Lummus Company). 1990. Personal communication.
U.S. Department of Energy (DOE). 1989. *Energy's Role in International Trade: Structural Changes and Competitiveness.* Washington: DOE.
_____. Energy Information Administration (EIA). 1991. *Manufacturing Energy Consumption Survey: 1988.* Washington: DOE.
Williams, R. H., E. D. Larson, and M. H. Ross. 1987. Materials, affluence and industrial energy use. *Annual Review of Energy* 12:99–144.

# PART III

---

# ENERGY
# IN TRANSITION

CHAPTER 11

# ECONOMICS, ETHICS, AND THE ENVIRONMENT

*Richard B. Norgaard and Richard B. Howarth*

AMONG THE MANY COMPLEX, long-term, global environmental issues—biodiversity conservation, ecosystem management, resource scarcity, population growth, and the control of toxic substances—the issue of global climate change is relatively tractable to conventional economic query. Unlike the other issues, controlling climate change has a clear starting point: managing the release and absorption of greenhouse gases. Economists are beginning to quantify the costs of mitigating climate change through greenhouse gas reduction and absorption. Dollar estimates are appearing in research reports and academic journals. To be sure, the benefits of mitigating climate change are more difficult to quantify, but in light of the importance of the controversy, economists have begun to stake a position in the debate and have spoken out publicly in the press (Nordhaus, 1990; Becker, 1991).

While there has been some variation in their positions, economists have generally argued that:

- Only a few mitigation measures such as the reduction of chloro-fluorocarbons (CFCs) are justified now.
- For most mitigation measures, the benefits over the long run, discounted to the present, do not justify the costs that will occur in the next few decades.
- Perhaps a little global warming would be good.
- Little or no action should be taken until the uncertainties are reduced and a rational decision can be made.
- Since future peoples will be materially wealthier than people of today regardless of climate change, it would be inequitable for current peoples to bear the burden of mitigating climate change.

The positions taken by economists to date, however, have been reached through the conventional frame of analysis incorporating traditional economic assumptions about technological progress and natural resources.

The responses identify a flaw in how economists conceptually frame the future and inform the political process. People are asking whether our children and our children's children are not entitled to a climate that supports the species, soils, and agricultural technologies and communities we enjoy today. The majority of people in the United States no longer think their children will be better off than they are themselves (Mishan, 1986a, 1986b). Economists, on the other hand, are asking whether mitigation is a good investment. Framing the question as an investment decision implicitly assumes that current peoples have the right to exploit the atmosphere and that doing so will not leave future generations worse off.

Economists have questioned how this generation might most efficiently exploit its environment; the political discourse has questioned whose environment it ought to be. The critical question is not simply "Will climate mitigation provide a suitable return as an investment for us?" but "What kind of a world do we want to leave to our children and what is the best way to provide it?" We present an approach that is better suited to the questions being addressed and to the environmental controversy behind the sustainability debate.

In this chapter we show why we should make moral decisions

about the environmental rights of future generations and then seek economic efficiency. Equally important, we show that it is logically fallacious to construct economic arguments for or against such moral decisions using efficiency analyses which implicitly assume that the current generation has the right to fully exploit the climate and using data generated by an economy running on the same assumption. Correctly formulating how economic reasoning should address questions concerning the future means reframing the literature on resource allocation, the valuation of nonmarket benefits, and the discount rate controversy.

## THE RELATIONSHIP BETWEEN EQUITY AND EFFICIENCY

The relationship between the distribution of rights to resources and the efficient allocation of resources should be explicated at the outset since it is the core of the argument. Imagine two countries with identical populations and identical resources allocated by markets. In the first country, rights to resources are distributed between people approximately equally, people have similar incomes, and they consume similar products. In the second country, rights are concentrated among a few people who can afford luxury goods while those who have few rights to resources, living largely on their labor alone, consume only the most basic of goods. Here is the critical point of the argument. In each country, markets efficiently allocate resources to the production of goods, but the types of goods produced (and the people who consume each good) depend on how rights to resources are distributed. For each distribution of rights, there is a different efficient allocation of resources.

Economists typically invoke the term "efficiency" as if there were only one best way of using resources. They reduce the number of efficient allocations to one, ignoring other efficient allocations associated with other distributions of rights, by implicitly taking the existing distribution of rights to resources as given. Existing practice assumes that efficiency, not equity, questions are being asked. The benefit-cost analyses of alternative strategies for abating climate change that are beginning to appear in professional reports and the popular press implicitly follow the convention of assuming that the critical question is how the current generation should efficiently exploit the environment. This approach may have been reasonable during the nineteenth century when the United States had many resources untapped. In the crowded, tightly knit, global economy at the end of the twentieth century, however, it is clearly inappropriate.

Economic reasoning can incorporate alternative distributions of rights. Yet it is important to keep in mind that economic reasoning cannot determine which distribution of rights should prevail. The people—directly, through their elected representatives, and through administrative and judicial processes—must address the moral question as to whether future generations should have rights to a climate approximately like the climate of today. When that decision is made, economic reasoning can be used to efficiently allocate resources between uses.

## SUSTAINABILITY: THE CHALLENGE TO ECONOMIC THOUGHT

While economists have yet to frame questions of sustainability correctly, the widespread concern that development must be sustainable is challenging the conventions of economic thought. Economists are asking how the inclusion of the value of environmental goods and services not traded in the market affects mitigation strategies (Hall, 1991), whether the practice of discounting future benefits and costs is consistent with concern for the future (Markandya and Pearce, 1988), how the uncertainties of climate change affect a rational response (Barbier and Pearce, 1990), whether economies operating at higher rates of resource use will invest in more research and technological development, facilitating access to new resources that might offset environmental problems like climate change (Pezzey, 1989; d'Arge et al., 1991), or whether the use of resources and environmental services must be constrained to ensure sustainability (Daly and Cobb, 1989). These are well-reasoned responses to the challenge. They result, however, from a line of reasoning that is at odds with the moral question driving the political discourse on climate change and sustainable development. The moral question is concerned with whether future generations have rights to environmental services.

Neoclassical economic theory is fairly malleable and has been successfully applied to a wide range of problems. Thus economists initially assumed that by further consideration of environmental interrelations and the long run, an economics addressing the problems of sustainability would readily unfold. Efforts to date at applying economic reasoning to the questions raised by sustainability reasoning, however, have been less than satisfying. In other applications, the concept of efficiency has helped explain how an objective can best be achieved. But efficiency, both as understood by lay people and even now by economists, frequently seems to conflict with sustainability. In a working document of the World Bank, for example, we find this quote: "A fundamental question is whether (or under what conditions) 'sustainability' is on the whole economically justified" (World Bank, 1990:17).

Many economists have argued that the apparent conflict between economic efficiency and sustainability results from efficiency being too narrowly defined. The inclusion of environmental goods and services that are not traded in markets frequently brings efficient environmental management closer to sustainable environmental management (Batie, 1989; Pearce and Turner, 1990). For this reason, economists have promoted the valuation of nonmarket goods and services as a way of improving environmental management (Krutilla, 1972) and economic valuation is now advocated by environmentalists as well (Myers, 1983; McNeely, 1988). And yet there is still a fundamental contradiction. Most of the valuation techniques now utilized measure the preferences of *current* generations, and any benefits accruing to future generations are discounted in net present value calculations to reflect what they are *currently* worth. Sustainability, on the other hand, is concerned with the future.

This tension between the current orientation of economic reasoning and the future orientation of sustainability reasoning, in turn, has led economists to rethink the question of discounting future benefits and costs (Markandya and Pearce, 1988). Under current economic reasoning, it is not efficient to mitigate climate change if other investments have higher returns. It may be better, by current reasoning, to invest in space travel, for example, than to try to stabilize the climate. In an effort to reconcile the concept of efficiency with sustainability, economists also have argued that we do not understand all of the possible consequences of a warmer climate, that such changes would have irreversible consequences, and that we do not know what sort of environments people in the future might prefer. Given this uncertainty and irreversibility, mitigation may be the safer option. And yet, what if we did know? Would we opt for efficiency as the concept is narrowly understood today? Or for sustainability, whatever this term is coming to mean? The economic issues presented by global climate change must be framed correctly before empirical research can be pursued properly.

## REFRAMING THE FUTURE

The apparent conflict between sustainability reasoning and efficiency reasoning is resolved by thinking of sustainability as a matter of intergenerational equity. Different intergenerational distributions of rights result in new efficient allocations of resources and environmental services, different patterns of consumption and investment, and different factor and commodity prices including different interest rates. The appearance of a conflict is an artifact of a long history of not incorporating equity in economic thinking. The overlap between economic and equity reasoning is illustrated in Figure 11.1. Economists to date have basically tried to work only in the area that is diagonally striped, ignoring the horizontally striped area that includes equity considerations. Clearly the overlap between economic reasoning and sustainability reasoning is greater when equity considerations are included.

While economists have concentrated their efforts on the efficient use of resources, environmentalists have consistently argued that societies must consider how much resources in aggregate they are leaving for future generations. The political debate is over the distribution of rights to resources and environmental services between generations, not over how efficiently this generation exploits its current rights. By acknowledging that the efficient intertemporal allocation of resources depends on the intergenerational distribution of rights to resources, this apparent conflict disappears.

The relation between intertemporal allocative efficiency and the intergenerational distribution of resource and environmental rights is illustrated in Figure 11.2 (see Bator, 1957). The utility frontier ($U$)

FIGURE 11.1. OVERLAP OF ENVIRONMENTAL, ECONOMIC, AND EQUITY ISSUES.

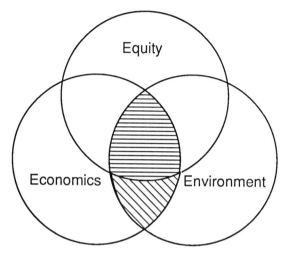

indicates the highest utility possible for people in future generations for any given utility of people in the current generation and vice versa. Each point on this frontier results from an efficient allocation of resources associated with different distributions of resource rights of caring between generations. Points within the frontier represent inefficient allocations of resources. Clearly there are many possible efficient allocations. A society's location on the frontier is determined by the initial distribution of rights to productive assets, including natural assets. While Figure 11.2 is limited to only the present and the future, the relationships between distributional equity and allocative efficiency are more fully elaborated in the theoretical literature through mathematical models of overlapping generations (Howarth, 1990, 1991a, 1991b; Howarth and Norgaard, 1990; Norgaard, 1991; Norgaard and Howarth, 1991a, 1991b). These models, of course, are also simplifications, but they explicitly document that while efficiency is important in that it puts society on the utility frontier, the sustainability of development is entirely a matter of whether the economy is operating above the 45-degree line—that is, a matter of the distribution of productive assets or of caring across generations.

While Figure 11.2 is very simple, it illustrates an important point.

FIGURE 11.2. THE UTILITY FRONTIER.

*The meaning of sustainability can be clarified by considering an intergenerational utility frontier (U).*

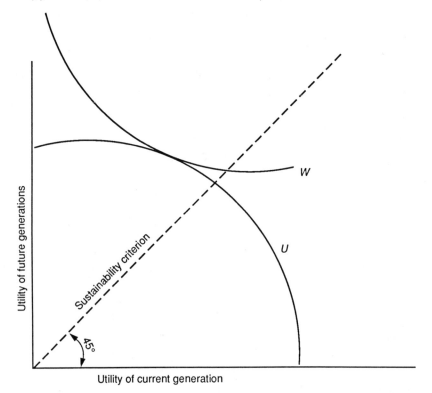

Utility of current generation

Much of the economic literature to date on sustainability stresses the importance of internalizing externalities. Development is conceived as a process of spurring economies to go faster; sustainability is conceived as a process of perfecting how economies work. Perfecting how economies work, however, will move the economy toward the efficiency frontier but may not make it any more sustainable. Thus we have the unfortunate situation where economies are still being stimulated, even the already developed economies, while sustainability waits for a perfection of market performance that has never been achieved and is unlikely to lead to sustainability if it is.

The best point on the utility frontier in Figure 11.2 would be at the tangency with an intergenerational welfare function (*W*) as illustrated. Such a welfare function, of course, has never been revealed to economists. When it comes to equity decisions, economists must work with politics. The tenor of the political discourse certainly indicates that sustainability is at least a minimum intergenerational criterion on which there is broad consensus. While economists cannot determine how resource and environmental rights should be distributed across generations, they can more effectively engage in policy dialogue and assist society in making decisions if they understand sustainability as a matter of assuring that assets are available to future generations.

## RESOURCE ALLOCATION, DISCOUNTING, AND ENVIRONMENTAL VALUATION

Acknowledging that sustainability is a matter of both equity and efficiency does not solve the moral question of how the rights of future generations and responsibilities of current generations should be defined. Nor does it reduce the uncertainties of the interplay between climate, biodiversity, resource scarcity, and the many other complexities of sustainable development. It does, however, provide a framework for thinking about the future that gives new insights into the efficient use of resources over time, clarifies certain issues in nonmarket valuation, and solves the contradictions of discounting the future.

### THE ALLOCATION OF RESOURCES

Economists have developed an extensive literature on the "optimal" depletion of stock resources over time based on the model elaborated by Hotelling (1931). In this now classic paper, Hotelling explored how resource producers would behave in a world of perfect competition and perfect knowledge of future market conditions. Profit-maximizing producers would equate the return realized through holding units of the resource for future extraction to the return available from extracting the resource and investing the net revenue earned from the sale in the capital market. Accordingly, under a wide variety of conditions, the resource royalty—the difference between price and extraction cost—would increase over time at the rate of interest (Dasgupta and Heal, 1979).

Since the energy crisis of 1973–1974 alerted the profession to the importance of stock resources, economists have written roughly a thousand articles elaborating how this basic result differs under various combinations of different assumptions. Well-respected economists have pondered the equity implications of this line of reasoning but have never built formal models showing that there are different efficient solutions for different intergenerational distributions of assets and that the equity implications of alternative efficient allocations can be significant.

In part, economists implicitly assumed that progress would take care of the interests of future generations by assuring them access to resources through new technology; in part, the profession has become so accustomed to not addressing equity issues that a whole generation of economists has been incompletely trained. In any case, the economics of intertemporal resource use was reduced to questions of efficient allocation as if this generation had all the rights to resources. Economists have argued, for example, that energy resource markets work *socially optimally* so long as externalities are internalized. Similarly, they have argued that the extinction of certain species may be optimal without considering that the policy decision under consideration is whether future generations might have rights to biological diversity. And by staying within existing conventions of the discipline, economists are treating climate change as a matter of how this generation should most efficiently exploit the atmosphere. By sustainability reasoning, these are equity issues. We suspect that future generations, if they could speak for themselves, would express the same concern.

Our overlapping-generations models of economies based on both renewable and nonrenewable resources show how alternative assignments of resource rights and different ways of caring about the future affect the allocation of resources. The research by others to date has been entirely directed at moving an economy from a position represented by point A in Figure 11.3 to a position such as point B, which is only one among many on the efficiency frontier. If the economy is unsustainable, as it is at point A, making it more efficient without changing the rights of future generations or otherwise exercising care for the future simply moves the economy toward some point B, which is also likely to be unsustainable.

## THE DISCOUNT RATE CONTROVERSY

Economists have long recognized the apparent perversity of discounting the benefits received and costs borne in the future (Ramsey, 1928; Markandya and Pearce, 1988). How can one discount the effects borne by future generations while being concerned with their welfare or with the sustainability of development? Are not lower discount rates more consistent with sustainability than high ones since a zero discount rate treats generations equally? Low discount rates, for example, favor letting trees grow longer and planting trees that take longer to grow. Perhaps a lower social rate of discount would lead to sustainable development.

The net effect of lower discount rates, however, has led to a tangled thicket of reasoning within the efficiency half of economic theory. If interest rates were low, more investments would be made—including investments in the capital necessary for mineral extraction or in projects that transform environmental systems on which the well-being of future peoples may depend. For these reasons, environmentalists opposed to water development projects and other major investments have argued against subsidized discount rates.

By reframing questions of the future in terms of the intergenerational distribution of rights to natural and other assets, the case for using lower discount rates to protect future generations becomes moot. If societies want to protect future generations, they should assure their rights or otherwise care for them more. Our models show that if rights to the services of resources and environmental systems, or any other asset, are transferred to future generations, both the supply of savings and the demand for investment funds in the present change, determining a new rate of interest. While the discount rate has been treated as an instrumental variable, it is inappropriate to think of the rate of interest as anything other than simply another price—as something that equilibrates savings and investment on the margin. What is important is the extent to which current and future generations have rights to resources, environmental services, and human and physical capital. The distribution of these rights over generations determines the kinds and levels of investments made within generations. The emphasis must be on the kinds and levels of investments and hence what is passed between generations, not the level of the equilibrating mechanism between the dollar flows of savings and investments.

Transfers of rights to future generations and other decisions to implement greater concern for the future are equity, not efficiency, decisions. This distinction can also be illustrated with the help of Figure 11.3. Imagine an economy operating at point $A$. The economy can be moved toward the efficiency frontier at point $B$ through the corrections of market failures—for example, by internalizing environmental externalities. Equity decisions, however, can be illustrated as movements *between* different objectives on the efficiency frontier, such as a movement from point $B$ to point $B^*$. Such decisions are made in accordance with equity criteria. Efficiency criteria cannot be used because at both positions $B$ and $B^*$ resources are being allocated perfectly efficiently. Since efficiency criteria cannot be used to choose between efficient points, the benefits to future generations from shifting from one level of concern for the future to another are not discounted.

When rights are redistributed to achieve the objective of point $B^*$, new investment opportunities and a new discount rate arise to guide the economy to a position such as point $A^*$, equivalent to point $A$ under the initial distribution of rights. The long-standing concern over the rate of discount and the welfare of future generations only becomes manifest within an efficiency framework that has been separated from the distribution of rights between generations. The concern has been identified and elaborated in the framework of investment analysis, a very partial approach, rather than in a general equilibrium framework. Bringing equity back into neoclassical economics and using general equilibrium analyses resolves the apparent contradiction between our concern for the future and our concern for efficiency.

ENVIRONMENTAL VALUATION

For the same reasons that the rate of interest or discount is a function of how environmental and other rights are distributed across generations, the values of environmental services are also a function of how

FIGURE 11.3. REFRAMING THE ISSUE OF SUSTAINABILITY.

*Assuring sustainability is a matter of transferring assets to future generations, moving from point* B *to* B\*, *or even from point* A *to* A\*, *not simply improving efficiency through moving from point* A *to point* B.

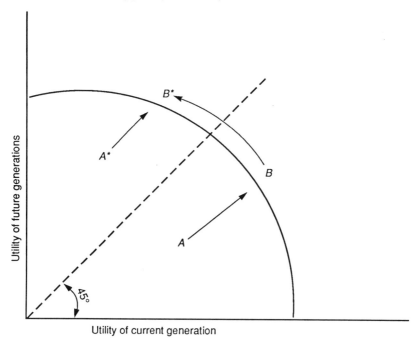

rights are distributed. If the present generation has complete rights to the environment, then clearly the marginal value of environmental services will be lower than it would be if the present generation had more restricted access to environmental services and had to exercise environmental responsibilities to future generations.

While this argument is theoretically straightforward, economists have fostered a rather roundabout alternative. They have argued, and in part rightly, that environments are misused and degraded because their full value is not reflected in markets. For example, too much fossil fuel is used because the market cost of using fossil fuels does not include the health and aesthetic costs of pollution. The valuation of nonmarket goods and services makes considerable sense with respect to more efficient use of hydrocarbons and the air we breathe subject to the current distribution of rights. Knowing nonmarket values can help us move the economy from point *A* to point *B*.

But economists have also advocated the valuation of nonmarket goods to reach sustainability objectives (Pearce and Turner, 1990). To be sure, if short-run environmental and health costs of burning fossil fuels were included in their price, less carbon dioxide and other greenhouse gases would be released to the atmosphere. With fuel prices reflecting their true costs, Darwin Hall (1990 and 1991) effectively argues that energy conservation and the use of alternative technologies would be significantly enhanced, substantially reducing the costs of mitigating climate change per se. There will frequently be cases where internalizing externalities from a point such as *A* in Figure 11.3 will move the economy closer to sustainability. But internalizing the environmental costs of developing oil in the Arctic National Wildlife Refuge could very well hasten its depletion and contribute further to climate change.

In any case, redistributing rights to future generations will move the economy to a point such as $A^*$ in Figure 11.3. This point, like point *A* before it, is not on the efficiency frontier because of external costs that can be internalized through further refinements in how rights are assigned, through taxes or subsidies, or government allocation. But the market failures that occur at point $A^*$ are different failures than those that occur at point *A* because different activities dominate. If climate is stabilized by reducing the combustion of fossil hydrocarbons, for example, the environmental and health externalities associated with air pollution from combustion would be substantially reduced (Hall, 1991). The elimination of these externalities might be thought of as a benefit of moving to point $A^*$, but new externalities are likely to exist at point $A^*$ and the values people place on goods and services external to the market will change due to the redistribution of wealth between generations in moving from points *A* to $A^*$ and due to the new mix of market goods and services that may complement or substitute for the nonmarket goods and services.

With respect to protecting the welfare of future generations, environmental valuation as it is now undertaken, to make the point strikingly clear, resembles an effort to determine how men really value women in a society that condones rape. Where moral issues loom large, first ethical decisions must be made and widely accepted; valuation can only follow. We contend that climate change as well as many other issues of sustainable development must be tackled first as issues of intergenerational equity and then as questions of economic efficiency.


## RESOLVING THE DEBATE

A multitude of individual proposals to prohibit certain practices, enhance energy efficiency, develop new technologies, and plant trees are being advanced to mitigate climate change. The policy debate to date has consisted of opposing analyses of the benefits and costs of these individual proposals. All of these analyses, however, have started with a pattern of reasoning with implicit assumptions that make it inherently biased against protecting the future. The analyses are based on data generated by an economy of a society trying to go from point $A$ to point $B$ rather than a society trying to go from point $A$ to point $A^*$. The relative efficiency of different projects for mitigating climate change and the best time to initiate them depends on the future we are trying to achieve. The efficient timing and mix of energy conservation, solar technologies, and tree planting will depend on the rights we assign to future generations, not vice versa. If we decide that future generations are entitled to a climate approximately like the one we enjoy, it will be efficient to conserve more fossil fuels relative to the current situation where their rights are not protected. Similarly, it would be efficient to make a more rapid shift to alternative technologies such as geothermal, solar, and wind and to undertake greater efforts to absorb carbon in living biomass.

The appropriate framework eliminates the myopia long associated with economic reasoning. It does so by distinguishing between decisions to redistribute rights and decisions concerning the efficient use of resources given a distribution of rights. The decision to protect the rights of future generations will surely benefit them, but the benefits they experience in the future are not discounted and weighed against the costs this generation will experience by granting such rights. It makes no sense to discount the objective we are trying to reach. This is exactly analogous to a decision with respect to civil rights. Economic calculations and weighing are inadmissible during political debates over civil rights because equity or moral criteria are considered not only to be more appropriate but are thought to be "higher" values.

And thus we conclude as promised. It is better economics for society to make moral decisions about the climate rights of future generations and let the economy follow than to construct arguments for or against such moral decisions using reasoning and data rooted in an implicit assumption that current generations hold all rights.

## REFERENCES

Barbier, Edward B., and David W. Pearce. 1990. Thinking economically about climate change. *Energy Policy* (Jan./Feb.):11–18.

Batie, Sandra S. 1989. Sustainable development: Challenges to the profession of agricultural economics. *American Journal of Agricultural Economics* 71:1083–1101.

Bator, Francis. 1957. The simple analytics of welfare maximization. *American Economic Review* 47:22–59.

Becker, Gary S. 1991. The hot air inflating the greenhouse effect. *Business Week*, 17 June, p. 16.

Daly, Herman E., and John B. Cobb, Jr. 1989. *For the Common Good: Redirecting the Economy Toward Community, the Environment, and a Sustainable Future.* Boston: Beacon Press.

d'Arge, Ralph C., Richard B. Norgaard, Mancur Olson, Jr., and Richard Sommerville. 1991. Economic growth, sustainability, and the environment. *Contemporary Policy Issues* 9:1–23.

Dasgupta, Partha S., and Geoffrey M. Heal. 1979. *Economic Theory and Exhaustible Resources.* Cambridge: Cambridge University Press.

Hall, Darwin C. 1990. Preliminary estimates of cumulative private and external costs of energy. *Contemporary Policy Issues* 8:283–307.

———. 1991. Social cost of $CO_2$ abatement from energy efficiency and solar power in the United States. Paper presented at the Western Economics Association meeting in Seattle, 2 July.

Hotelling, Harold. 1931. The economics of exhaustible resources. *Journal of Political Economy* 39:137–175.

Howarth, Richard B. 1990. Economic theory, natural resources, and intergenerational equity. Ph.D. dissertation, Energy and Resources Program, University of California at Berkeley.

————. 1991a. Intergenerational equity under technological uncertainty and an exhaustible resource constraint. *Journal of Environmental Economics and Management* 21(3):225–243.

————. 1991b. Intertemporal equilibria and exhaustible resources: An overlapping generations approach. *Ecological Economics* 4(3):237–252.

Howarth, Richard B., and Richard B. Norgaard. 1990. Intergenerational resource rights, efficiency, and social optimality. *Land Economics* 66(1): 1–11.

Krutilla, John V. (ed). 1972. *Natural Environments: Studies in Theoretical and Applied Analysis.* Baltimore: Johns Hopkins University Press.

Markandya, Anil, and David Pearce. 1988. Environmental considerations and the choice of the discount rate in developing countries. Environment Department Working Paper no. 3. Washington: World Bank.

McNeely, Jeffrey A. 1988. *Economics and Biological Diversity: Developing and Using Economic Incentives to Conserve Biological Resources.* Gland, Switzerland: International Union for Conservation of Nature and Natural Resources.

Mishan, Ezra J. 1986a. The future is worse than it was. In *Economic Myths and the Mythology of Economics.* Atlantic Highlands, N.J.: Humanities Press International.

————. 1986b. The mystique of economic expertise. In *Economic Myths and the Mythology of Economics.* Atlantic Highlands, N.J.: Humanities Press International.

Myers, Norman. 1983. *A Wealth of Wild Species.* Boulder: Westview.

Nordhaus, William D. 1990. Greenhouse economics: Count before you leap. *Economist,* 7 July, pp. 21–24.

————. 1991. The cost of slowing climate change: A survey. *Energy Journal* 12:37–66.

Norgaard, Richard B. 1991. Sustainability as intergenerational equity: The challenge to economic thought and practice. World Bank Discussion Paper. Washington: World Bank.

Norgaard, Richard B., and Richard B. Howarth. 1991a. Sustainability and discounting the future. In *The Ecological Economics: The Science and Management of Sustainability.* New York: Columbia University Press.

————. 1991b. Sustainability and intergenerational environmental rights: Implications for benefit-cost analysis. In *Economic Issues in Global Climate Change: Agriculture, Forestry, and Natural Resources,* ed. John Reilly and Margot Anderson. Boulder: Westview.

Pearce, David W., and R. Kerry Turner. 1990. *Economics of Natural Resources and the Environment.* Baltimore: Johns Hopkins University Press.

Pezzey, John. 1989. Economic analysis of sustainable growth and sustainable development. Environment Department Working Paper no. 15. Washington: World Bank.

Ramsey, F. P. 1928. A mathematical theory of saving. *Economics Journal* 38:543–559.

World Bank. 1990. Initiating memorandum, Malaysia forestry sector review, 23 Oct.

CHAPTER 12

# TOWARD A SUSTAINABLE WORLD

## William D. Ruckelshaus

THE DIFFICULTY of converting scientific findings into political action is a function of the uncertainty of the science and the pain generated by the action. Given the current uncertainties surrounding just one aspect of the global environmental crisis—the predicted rise in greenhouse gases—and the enormous technological and social effort that will be required to control that rise, it is fair to say that responding successfully to the multifaceted crisis will be a difficult political enterprise. It means trying to get a substantial proportion of the world's people to change their behavior in order to (possibly) avert threats that will otherwise (probably) affect a world most of them will not be alive to see.

The models that predict climatic change, for example, are subject to varying interpretations as to the timing, distribution, and severity of the changes in store. Also, whereas models may convince scientists, who understand their assumptions and limitations, as a rule projections make poor politics. It is hard for people—hard even for the groups of people who constitute governments—to change in response to dangers that may not arise for a long time or that just might not happen at all.

How, then, can we make change happen? Other chapters in this book document the reality of the global ecological crisis and point to some specific ameliorative measures. This chapter is about how to shape the policies, launch the programs, and harness the resources that will lead to the adoption of such measures—and that will actually convince ordinary people throughout the world to start doing things differently.

Insurance is the way people ordinarily deal with potentially serious contingencies, and it is appropriate here as well. People consider it prudent to pay insurance premiums so that if catastrophe strikes, they or their survivors will be better off than if there had been no insurance. The analogy is clear. Current resources forgone or spent to prevent the buildup of greenhouse gases are a kind of premium. Moreover, as long as we are going to pay premiums, we might as well pay them in ways that will yield dividends in the form of greater efficiency, improved human health, or more widely distributed prosperity. If we turn out to be wrong on greenhouse warming or ozone depletion, we still retain the dividend benefits. In any case, no one complains to the insurance company when disaster does not strike.

That is the argument for some immediate, modest actions. We can hope that if shortages or problems arise, there will turn out to be a technological fix or set of fixes, or that technology and the normal workings of the market will combine to solve the problem by product substitution. Already, for example, new refrigerants that do not have the atmospheric effects of the chlorofluorocarbons are being introduced; perhaps a cheap and nonpolluting source of energy will be discovered.

It is comforting to imagine that we might arrive at a more secure tomorrow with little strain, to suppose with Dickens's Mr. Micawber that something will turn up. Imagining is harmless, but counting on such a rescue is not. We need to face up to the fact that something enormous may be happening to our world. Our species may be pushing up against some immovable limits on the combustion of fossil fuels and damage to ecosystems. We must at least consider the possibility that, besides those modest adjustments for the sake of prudence, we may have to prepare for far more dramatic changes, changes that will begin to shape a sustainable world economy and society.

Sustainability is the nascent doctrine that economic growth and development must take place, and be maintained over time, within the limits set by ecology in the broadest sense—by the interrelations of human beings and their works, the biosphere, and the physical and chemical laws that govern it. The doctrine of sustainability holds too that the spread of a reasonable level of prosperity and security to the less developed nations is essential to protecting ecological balance and hence essential to the continued prosperity of the wealthy nations. It follows that environmental protection and economic development are complementary rather than antagonistic processes.

Can we move nations and people in the direction of sustainability? Such a move would be a modification of society comparable in scale to only two other changes: the agricultural revolution of the late Neolithic and the industrial revolution of the past two centuries. Those revolutions were gradual, spontaneous, and largely unconscious. This one will have to be a fully conscious operation, guided by the best foresight that science can provide—foresight pushed to its limit. If we actually do it, the undertaking will be absolutely unique in humanity's stay on the earth.

The shape of this undertaking cannot be clearly seen from where we now stand. The conventional image is that of a crossroads: a forced choice of one direction or another that determines the future for some appreciable period. But this does not at all capture the complexity of the current situation. A more appropriate image would be that of a canoeist shooting the rapids: survival depends on continually responding to information by correct steering. In this case the information is supplied by science and economic events; the steering is the work of policy, both governmental and private.

Taking control of the future therefore means tightening the connection between science and policy. We need to understand where the rocks are in time to steer around them. Yet we will not devote the appropriate level of resources to science or accept the policies mandated by science unless we do something else. We have to understand that we are all in the same canoe and that steering toward sustainability is necessary.

Sustainability was the original economy of our species. Preindustrial peoples lived sustainably because they had to; if they did not, if they expanded their populations beyond the available resource base, then sooner or later they starved or had to migrate. The sustainability of their way of life was maintained by a particular consciousness regarding nature: the people were spiritually connected to the animals and plants on which they subsisted; they were part of the landscape, or of nature, not set apart as masters.

The era of this "original sustainability" eventually came to an end. The development of cities and the maintenance of urban populations called for intensive agriculture yielding a surplus. As a population grows, it requires an expansion of production, either by conquest or colonization or improved technique. A different consciousness, also embodied in a structure of myth, sustains this mode of life. The earth and its creatures are considered the property of humankind, a gift from the supernatural. Man stands outside of nature, which is a passive playing field that he dominates, controls, and manipulates. Eventually, with industrialization, even the past is colonized: the forests of the Carboniferous are mined to support ever-expanding populations. Advanced technology gives impetus to the basic assumption that there is essentially no limit to humanity's power over nature.

This consciousness, this condition of "transitional unsustainability," is dominant today. It has two forms. In the underdeveloped, industrializing world, it is represented by the drive to develop at any environmental cost. It includes the wholesale destruction of forests, the replacement of sustainable agriculture by cash crops, the attendant exploitation of vulnerable lands by people such cash cropping forces off good land, and the creation of industrial centers that are also centers of environmental pollution.

In the industrialized world, unsustainable development has generated wealth and relative comfort for about one-fifth of humankind, and among the populations of the industrialized nations the consciousness supporting the unsustainable economy is nearly universal. With a few important exceptions, the environmental-protection movement in those nations, despite its major achievements in passing legislation and mandating pollution-control measures, has not had a substantial effect on the lives of most people. Environmentalism has been ameliorative and corrective—not a restructuring force. It is encompassed within the consciousness of unsustainability.

Although we cannot return to the sustainable economy of our distant ancestors, in principle there is no reason why we cannot create a sustainability consciousness suitable to the modern era. Such a consciousness would include the following beliefs:

1. *The human species is part of nature. Its existence depends on its ability to draw sustenance from a finite natural world; its continuance depends on its ability to abstain from destroying the natural systems that regenerate this world.* This seems to be the major lesson of the current environmental situation as well as being a direct corollary of the second law of thermodynamics.

2. *Economic activity must account for the environmental costs of production.* Environmental regulation has made a start here, albeit a small one. The market has not even begun to be mobilized to preserve the environment; as a consequence an increasing amount of the "wealth" we create is in a sense stolen from our descendants.

3. *The maintenance of a livable global environment depends on the sustainable development of the entire human family.* If 80 percent of the members of our species are poor, we cannot hope to live in a world at peace; if the poor nations attempt to improve their lot by the methods we rich have pioneered, the result will eventually be world ecological damage.

This consciousness will not be attained simply because the arguments for change are good or because the alternatives are unpleasant. Nor will exhortation suffice. The central lesson of realistic policymaking is that most individuals and organizations change when it is in their interest to change, either because they derive some benefit from changing or because they incur sanctions when they do not—and the shorter the time between change (or failure to change) and benefit (or sanction), the better. This is not mere cynicism. Although people will struggle and suffer for long periods to achieve a goal, it is not reasonable to expect people or organizations to work against their immediate interests for very long—particularly in a democratic system, where what they perceive to be their interests are so important in guiding the government.

To change interests, three things are required. First, a clear set of values consistent with the consciousness of sustainability must be articulated by leaders in both the public and the private sector. Next, motivations need to be established that will support the values. Finally, institutions must be developed that will effectively apply the motivations. The first is relatively easy, the second much harder, and the third perhaps hardest of all.


Values similar to those I described above have indeed been articulated by political leaders throughout the world. In the past year the president and the secretary of state of the United States, the leader of the Soviet Union, the prime minister of Great Britain, and the presidents of France and Brazil have all made major environmental statements. In July the leaders of the Group of Seven major industrialized nations called for "the early adoption, worldwide, of policies based on sustainable development." Most industrialized nations have a structure of national environmental law that to at least some extent reflects such values, and there is even a small set of international conventions that begin to do the same thing.

Mere acceptance of a changed value structure, although it is a prerequisite, does not generate the required change in consciousness, nor does it change the environment. Although diplomats and lawyers may argue passionately over the form of words, talk is not action. In the United States, which has a set of environmental statutes second to none in their stringency, and where for the past 15 years poll after poll has recorded the American people's desire for increased environmental protection, the majority of the population participates in the industrialized world's most wasteful and most polluting style of life. The values are there; the appropriate motivations and institutions are patently inadequate or nonexistent.

The difficulties of moving from stated values to actual motivations and institutions stem from basic characteristics of the major industrialized nations—the nations that must, because of their economic strength, preeminence as polluters, and dominant share of the world's resources, take the lead in any changing of the present order. These nations are market-system democracies. The difficulties, ironically, are inherent in the free-market economic system on the one hand and in democracy on the other.

The economic problem is the familiar one of externalities: the environmental cost of producing a good or service is not accounted for in the price paid for it. As the economist Kenneth E. Boulding has put it: "All of nature's systems are closed loops, while economic activities are linear and assume inexhaustible resources and 'sinks' in which to throw away our refuse." In willful ignorance, and in violation of the core principle of capitalism, we often refuse to treat environmental resources as capital. We spend them as income and are as befuddled as any profligate heir when our checks start to bounce.

Such "commons" as the atmosphere, the seas, fisheries, and goods in public ownership are particularly vulnerable to being overspent in this way, treated as either inexhaustible resources or bottomless sinks. The reason is that the incremental benefit to each user accrues exclusively to that user, and in the short term it is a gain. The environmental degradation is spread out among all users and is apparent only in the long term, when the resource shows signs of severe stress or collapse. Some years ago the biologist Garrett Hardin called this the tragedy of the commons.

The way to avoid the tragedy of the commons—to make people pay the full cost of a resource use—is to close the loops in economic systems. The general failure to do this in the industrialized world is related to the second problem, the problem of action in a democracy. Modifying the market to reflect environmental costs is necessarily a function of government. Those adversely affected by such modifications, although they may be a tiny minority of the population, often have disproportionate influence on public policy. In general, the much injured minority proves to be a more formidable lobbyist than the slightly benefited majority.

The Clean Air Act of 1970 in the United States, arguably the most expensive and far-reaching environmental legislation in the world, is a case in point. Parts of the act were designed not so much to cleanse the air as to protect the jobs of coal miners in high-sulfur coal regions. Utilities and other high-volume consumers were not allowed to substitute low-sulfur coal to meet regulatory requirements but instead had to install scrubbing devices.

Although the act expired 7 years ago, Congress found it extraordinarily difficult to develop a revision, largely because of another set of contrary interests involving acid rain. The generalized national interest in reducing the environmental damage attributable to this long-range pollution had to overcome the resistance of both high-sulfur-coal mining interests and the Midwestern utilities that would incur major expenses if they were forced to control sulfur emissions. The problem of conflicting interests is exacerbated by the distance between major sources of acid rain and the regions that suffer the most damage. It is accentuated when the pollution crosses state and national boundaries: elected representatives are less likely to countenance short-term adverse effects on their constituents when the immediate beneficiaries are nonconstituents.

The question, then, is whether the industrial democracies will be able to overcome political constraints on bending the market system toward long-term sustainability. History provides some cause for optimism: a number of contingencies have led nations to accept short-term burdens in order to meet a long-term goal.


War is the obvious example. Things considered politically or economically impossible can be accomplished in a remarkably short time, given the belief that national survival is at stake. World War II mobilized the U.S. population, changed work patterns, manipulated and controlled the price and supply of goods, and reorganized the nation's industrial plant.

Another example is the Marshall Plan for reconstructing Europe after World War II. In 1947 the United States spent nearly 3 percent of its gross domestic product on this huge set of projects. Although the impetus for the plan came from fear that Soviet influence would expand into Western Europe, the plan did establish a precedent for massive investment in increasing the prosperity of foreign nations.

There are other examples. Feudalism was abandoned in Japan, as was slavery in the United States, in the nineteenth century; this century has seen the retreat of imperialism and the creation of the European Economic Community. In each case important interests gave way to new national goals.

If it is possible to change, how do we begin to motivate change? Clearly, government policy must lead the way, since market prices of commodities typically do not reflect the environmental costs of extracting and replacing them, nor do the prices of energy from fossil fuels reflect the risks of climatic change. Pricing policy is the most direct means of ensuring that the full environmental cost of goods and services is accounted for. When government owns a resource, or supplies it directly, the price charged can be made to reflect the true cost of the product. The market will adjust to this as it does to true scarcity: by product substitution and conservation.

Environmental regulation should be refocused to mobilize rather than suppress the ingenuity and creativity of industry. For example, additional gains in pollution control should be sought not simply by increasing the stringency or technical specificity of command-and-control regulation but also by implementing incentive-based systems. Such systems magnify public-sector decisions by tens of thousands of individual and corporate decisions. To be sure, incentive systems are not a panacea. For some environmental problems, such as the use of unacceptably dangerous chemicals, definitive regulatory measures will always be required. Effective policies will include a mixture of incentive-based and regulatory approaches.


Yet market-based approaches will be a necessary part of any attempt to reduce the greenhouse effect. Here the most attractive options involve the encouragement of energy efficiency. Improving efficiency meets the double-benefit standard of insurance: it is good in itself, and it combats global warming by reducing carbon dioxide emissions. If the world were to improve energy efficiency by 2 percent a year, the global average temperature could be kept within 1°C of present levels. Many industrialized nations have maintained a rate of improvement close to that over the past 15 years.

Promoting energy efficiency is also relatively painless. The United States reduced the energy intensity of its domestic product by 23 percent between 1973 and 1985 without much notice. Substantial improvement in efficiency is available even with existing technology.

Something as simple as bringing all U.S. buildings up to the best world standards could save enormous amounts of energy. Right now more energy passes through the windows of buildings in the United States than flows through the Alaska pipeline.

Efficiency gains may nevertheless have to be promoted by special market incentives, because energy prices tend to lag behind increases in income. A "climate protection" tax of $1 per million Btu on coal and 60 cents per million Btu on oil is an example of such an incentive. It would raise gasoline prices by 11 cents a gallon and the cost of electricity an average of 10 percent, and it would yield $53 billion annually.

Direct regulation by the setting of standards is cumbersome, but it may be necessary when implicit market signals are not effective. Examples are the mileage standards set in the United States for automobiles and the efficiency standards for appliances that were adopted in 1986. The appliance standards will save $28 billion in energy costs by the year 2000 and keep 342 million tons of carbon out of the atmosphere.

Over the long term it is likely that some form of emissions-trading program will be necessary—and on a much larger scale than has been the case heretofore. (Indeed, the president's new Clean Air Act proposal includes a strengthened system of tradeable permits.) In such a program all major emitters of pollutants would be issued permits specifying an allowable emission level. Firms that decide to reduce emissions below the specified level—for example, by investing in efficiency—could sell their excess "pollution rights" to other firms. Those that find it prohibitively costly to retrofit old plants or build new ones could buy such rights or could close down their least efficient plants and sell the unneeded rights.

Another kind of emissions trading might reduce the impact of carbon dioxide emissions. Companies responsible for new greenhouse-gas emissions could be required to offset them by improving overall efficiency or closing down plants, or by planting or preserving forests that would help absorb the emissions. Once the system is established, progress toward further reduction of emissions would be achieved by progressively cranking down the total allowable levels of various pollutants, on both a national and a permit-by-permit basis.

The kinds of programs I have just described will need to be supported by research providing a scientific basis for new environmental-protection strategies. Research into safe, nonpolluting energy sources and more energy-efficient technologies would seem to be particularly good bets. An example: in the mid-1970s the U.S. Department of Energy developed a number of improved-efficiency technologies at a cost of $16 million; among them were a design for compact fluorescent lamps that could replace incandescent bulbs and window coatings that save energy during both heating and cooling seasons. At current rates of implementation, the new technologies should generate $63 billion in energy savings by the year 2010.

The motivation of change toward sustainability will have to go far beyond the reduction of pollution and waste in the developed countries, and it cannot be left entirely to the environmental agencies in those countries. The agencies whose goals are economic development, exploitation of resources, and international trade—and indeed foreign policy in general—must also adopt sustainable development as a central goal. This is a formidable challenge, for it touches the heart of numerous special interests. Considerable political skill will be required to achieve for environmental protection the policy preeminence that only economic issues and national security (in the military sense) have commanded.

But it is in relations with the developing world that the industrialized nations will face their greatest challenges. Aid is both an answer and a perpetual problem. Total official development assistance from the developed to the developing world stands at around $35 billion a year. This is not much money. The annual foreign-aid expenditure of the United States alone would be $127 billion if it spent the same proportion of its gross national product on foreign aid as it did during the peak years of the Marshall Plan.

There is no point, of course, in even thinking about the adequacy of aid to the undeveloped nations until the debt issue is resolved. The World Bank has reported that in 1988 the seventeen most indebted countries paid the industrialized nations and multilateral agencies $31.1 billion more than they received in aid. This obviously cannot go on. Debt-for-nature swapping has taken place between such major lenders as Citicorp and a number of countries in South America: the bank forgives loans in exchange for the placing of land in conservation areas or parks. This is admirable, but it will not in itself solve the problem. Basic international trading relations will have to be redesigned in order to eliminate, among other things, the ill effects on the undeveloped world of agricultural subsidies and tariff barriers in the industrialized world.

A prosperous rural society based on sustainable agriculture must be the prelude to future development in much of the developing world, and governments there will have to focus on what motivates people to live in an environmentally responsible manner. Farmers will not grow crops when governments subsidize urban populations by keeping prices to farmers low. People will not stop having too many children if the labor of children is the only economic asset they have. Farmers will not improve the land if they do not own it; it is clear that land-tenure reform will have to be instituted.

Negative sanctions against abusing the environment are also missing throughout much of the undeveloped world; to help remedy this situation, substantial amounts of foreign aid could be focused directly on improving the status of the environmental ministries in developing nations. These ministries are typically impoverished and ineffective, particularly in comparison with their countries' economic-development and military ministries. To cite one small example: the game wardens of Tanzania receive an annual salary equivalent to the price paid to poachers for two elephant tusks—one reason the nation has lost two-thirds of its elephant population to the ivory trade in the past decade.

To articulate the values and devise the motivations favoring a sustainable world economy, existing institutions will need to change and new ones will have to be established. These will be difficult tasks, because institutions are powerful to the extent that they support powerful interests—which usually implies support of the status quo.

The important international institutions in today's world are those concerned with money, with trade, and with national defense. Those who despair of environmental concerns ever reaching a comparable level of importance should remember that current institutions (for example, NATO, the World Bank, multinational corporations) have fairly short histories. They were formed out of pressing concerns about acquiring and expanding wealth and maintaining national sovereignty. If concern for the environment becomes comparably pressing, comparable institutions will be developed.

To further this goal, three things are wanted. The first is money. The annual budget of the United Nations Environment Program (UNEP) is $30 million, a derisory amount considering its responsibilities. If nations are serious about sustainability, they will provide this central environmental organization with serious money, preferably money derived from an independent source in order to reduce its political vulnerability. A tax on certain uses of common world resources has been suggested as a means to this end.

The second thing wanted is information. We require strong international institutions to collect, analyze, and report on environmental trends and risks. The Earthwatch program run by the UNEP is a beginning, but there is need for an authoritative source of scientific information and advice that is independent of national governments. There are many nongovernmental or quasi-governmental organizations capable of filling this role; they need to be pulled together into a cooperative network. We need a global institution capable of answering questions of global importance.

The third thing wanted is integration of effort. The world cannot afford a multiplication of conflicting efforts to solve common problems. On the aid front in particular, this can be tragically absurd: Africa alone is currently served by eighty-two international donors and more than 1700 private organizations. In 1980, in the tiny African nation Burkina Faso (population about 8 million) 340 independent aid projects were under way. We need to form and strengthen coordinating institutions that combine the separate strengths of nongovernmental organizations, international bodies, and industrial groups and to focus their efforts on specific problems.

Finally, in creating the consciousness of advanced sustainability, we shall have to redefine our concepts of political and economic feasibility. These concepts are, after all, simply human constructs; they were different in the past, and they will surely change in the future. But the earth is real, and we are obliged by the fact of our utter dependence on it to listen more closely than we have to its messages.

CHAPTER 13

# ENERGY AGENDA
# FOR THE 1990s

*John P. Holdren*

CIVILIZATION IS NOT RUNNING OUT of energy resources in an absolute sense, nor is it running out of technological options for transforming these resources into the particular forms that our patterns of energy use require. We are, however, running out of the cheap oil and natural gas that powered much of the growth of modern industrialized societies, out of environmental capacity to absorb the impacts of burning coal, and out of public tolerance for the risks of nuclear fission. We seem to be lacking as well the commitment to make coal cleaner and fission safer, the money and endurance needed to develop long-term alternatives, the astuteness to embrace energy efficiency on the scale demanded, and the consensus needed to fashion any coherent strategy at all.

These deficiencies suggest that civilization has entered a fundamental transition in the nature of the energy/society interaction without any collective recognition of the transition's character or its implications for human well-being. The transition is from convenient but ultimately scarce energy resources to less convenient but more abundant ones, from a direct and positive connection between energy and economic well-being to a complicated and multidimensional one, and from localized pockets of pollution and hazard to impacts that are regional and even global in scope.

The subject is also being transformed from one of limited political interest within nations to a focus of major political contention between them, from an issue dominated by decisions and concerns of the Western world to one in which the problems and prospects of all regions are inextricably linked, and from one of concern to only a small group of technologists and managers to one where the values and actions of every citizen matter.

Understanding this transition requires a look at the two-sided connection between energy and human well-being. Energy contributes positively to well-being by providing such consumer services as heating, lighting, and cooking as well as serving as a necessary input to economic production. But the costs of energy—including not only the money and other resources devoted to obtaining and exploiting it but also the environmental and sociopolitical impacts—detract from well-being.

For most of human history, the dominant concerns about energy have centered on the benefit side of the energy/well-being equation. Inadequacy of energy resources or (more often) of the technologies and organizations for harvesting, converting, and distributing those resources has meant insufficient energy benefits and hence inconvenience, deprivation, and constraints on growth. Energy problems in this category remain the principal preoccupation of the least developed countries, where energy for basic human needs is the main issue; they are also an important concern in the intermediate and newly industrializing countries, where the issue is energy for production and growth.

Aside from having too little energy, it is possible to suffer by paying too much for it. The price may be paid in excessive diversion of capital, labor, and income from nonenergy needs (thereby producing inflation and reducing living standards), or it may be paid in excessive environmental and sociopolitical impacts. For most of the past

100 years, however, the problems of excessive energy costs have seemed less threatening than the problems of insufficient supply. Between 1890 and 1970 the monetary costs of supplying energy and the prices paid by consumers stayed more or less constant or declined, and the environmental and sociopolitical costs were regarded more as local nuisances or temporary inconveniences than as pervasive and persistent liabilities.

All this changed in the 1970s. The oil-price shocks of 1973–1974 and 1979 doubled and then quadrupled the real price of oil on the world market. In 1973 oil constituted nearly half of the world's annual use of industrial energy forms (oil, natural gas, coal, nuclear energy, and hydropower as opposed to the traditional energy forms of fuelwood, crop wastes, and dung). Inevitably, the rise in oil prices pulled the prices of the other industrial energy forms upward. The results illustrate vividly the perils of excessive monetary costs of energy: worldwide recession, spiraling debt, a punishing blow to the development prospects of the oil-poor countries of the Southern Hemisphere, and the imposition in the industrialized nations of disproportionate economic burdens on the poor.

The early 1970s also marked a transition in coming to grips with the environmental and sociopolitical costs of energy. Problems of air

TABLE 13.1. ENERGY-USE TRENDS

TRENDS IN POPULATION AND ENERGY USE PER PERSON ACCOUNT FOR THE PAST CENTURY'S RAPID GROWTH OF WORLD ENERGY DEMAND. INDUSTRIAL ENERGY FORMS ARE MAINLY COAL, OIL, AND NATURAL GAS, WITH SMALLER CONTRIBUTIONS FROM HYDROPOWER AND NUCLEAR ENERGY. TRADITIONAL FUELS ARE WOOD, CROP WASTE, AND DUNG. A TERAWATT IS EQUAL TO A BILLION TONS OF COAL OR 5 BILLION BARRELS OF OIL PER YEAR. DATA WERE COMPILED BY THE AUTHOR.

|  | 1890 | 1910 | 1930 | 1950 | 1970 | 1990 |
|---|---|---|---|---|---|---|
| World population (billions) | 1.49 | 1.70 | 2.02 | 2.51 | 3.62 | 5.32 |
| Traditional energy use per person (kilowatts) | 0.35 | 0.30 | 0.28 | 0.27 | 0.27 | 0.28 |
| Industrial energy use per person (kilowatts) | 0.32 | 0.64 | 0.85 | 1.03 | 2.04 | 2.30 |
| Total world energy use (terawatts) | 1.00 | 1.60 | 2.28 | 3.26 | 8.36 | 13.73 |
| Cumulative industrial energy use since 1850 (terawatt-years) | 10 | 26 | 54 | 97 | 196 | 393 |

and water pollution, many of them associated with energy supply and use, were coming to be recognized as pervasive threats to human health, economic well-being, and environmental stability. Consciousness of the sociopolitical costs of energy grew when overdependence on oil from the Middle East created foreign-policy dilemmas and even a chance of war, and when India's detonation of a nuclear bomb in 1974 emphasized that spreading competence in nuclear energy can provide weapons as well as electricity.

The 1970s, then, represented a turning point. After decades of constancy or decline in monetary costs—and of relegation of environmental and sociopolitical costs to secondary status—energy was seen to be getting costlier in all respects. It began to be plausible that excessive energy costs could pose threats on a par with those of insufficient supply. It also became possible to think that expanding some forms of energy supply could create costs exceeding the benefits.

The crucial question at the beginning of the 1990s is whether the trend that began in the 1970s will prove to be temporary or permanent. Is the era of cheap energy really over, or will a combination of new resources, new technologies, and changing geopolitics bring it back? One key determinant of the answer is the staggering scale of energy demand brought forth by 100 years of unprecedented population growth, coupled with an equally remarkable growth in per capita demand for industrial energy forms. Supplying energy at rates in the range of 10 terawatts (one terawatt is one billion watts), first achieved in the late 1960s, is an enterprise of enormous scale. The way it was done in 1970 required the harvesting, processing, and combustion of some 3 billion metric tons of coal and lignite, some 17 billion barrels of oil, more than a trillion cubic meters of natural gas, and perhaps 2 billion cubic meters of fuelwood. It entailed the use of dirty coal as well as clean; undersea oil as well as terrestrial; deep gas as well as shallow; mediocre hydroelectric sites as well as good ones; and deforestation as well as sustainable fuelwood harvesting.

The greatest part of the past century's growth in industrial energy forms was supplied by oil and natural gas—the most accessible, versatile, transportable, and inexpensive chemical fuels on the planet. The century's cumulative consumption of some 200 terawatt-years of oil and gas represented perhaps 20 percent of the ultimately recoverable portion of the earth's endowment of these fuels. If the cumulative consumption of oil and gas continues to double every 15 to 20 years, as it has done for a century, the initial stock will be 80 percent depleted in another 30 or 40 years.

Except for the huge pool of oil underlying the Middle East, the cheapest oil and gas are already gone. The trends that once held costs at bay against cumulative depletion, that is, new discoveries and economies of scale in processing and transport, have played themselves out. Even if a few more giant oil fields are discovered, they will make little difference against consumption on today's scale. Oil and gas will have to come increasingly, for most countries, from smaller and more dispersed fields, from offshore and Arctic environments, from deeper in the earth, and from imports whose reliability and affordability cannot be guaranteed.

There are, as the preceding chapters have shown, a variety of other energy resources that are more abundant than oil and gas. Coal, solar energy, and fission and fusion fuels are the most important ones. But they all require elaborate and expensive transformation into electricity or liquid fuels in order to meet society's needs. None has very good prospects for delivering large quantities of fuel at costs comparable to those of oil and gas prior to 1973 or large quantities of electricity at costs comparable to those of the cheap coal-fired and hydropower plants of the 1960s. It appears, then, that expensive energy is a permanent condition, even without allowing for its environmental costs.

The capacity of the environment to absorb the effluents and other impacts of energy technologies is itself a finite resource. The finitude is manifested in two basic types of environmental costs. "External" costs are those imposed by environmental disruptions on society but not reflected in the monetary accounts of the buyers and sellers of the energy. "Internalized" costs are increases in monetary costs imposed by measures, such as pollution-control devices, aimed at reducing the external costs.

Both types of environmental costs have been rising for several reasons. First, the declining quality of fuel deposits and energy-conversion sites to which society must now turn means more material must be moved or processed, bigger facilities must be constructed, and longer distances must be traversed. Second, the growing magnitude of effluents from energy systems has led to saturation of the environment's capacity to absorb such effluents without disruption.

Third, the monetary costs of controlling pollution tend to increase with the percentage of pollutant removed. The combination of higher energy-use rates, lower resource quality, and an already stressed environment requires that an increasing percentage be removed just to hold damages constant. Consequently, internalized costs must rise. And fourth, growing public and political concern for the environment has lengthened the time required for siting, building, and licensing energy facilities and has increased the frequency of mid-project changes in design and specifications, forcing costs still further upward.

It is difficult to quantify the total contribution of all these factors to the monetary costs of energy supply, in part because factors not related to the environment are often entwined with environmental ones. For example, construction delays have been caused not just by regulatory constraints but also by problems of engineering, management, and quality control. Nevertheless, it seems likely that in the United States actual or attempted internalization of environmental impacts has increased the monetary costs of supplying petroleum products by at least 25 percent during the past 20 years and the costs of generating electricity from coal and nuclear power by 40 percent or more.

Despite these expenditures, the remaining uninternalized environmental costs have been substantial and in many cases are growing. Those of greatest concern are the risk of death or disease as a result of emissions or accidents at energy facilities and the impact of energy supplies on the global ecosystem and on international relations.

The impacts of energy technologies on public health and safety are difficult to pin down with much confidence. In the case of air pollution from fossil fuels, in which the dominant threat to public health is thought to be particulates formed from sulfur dioxide emissions, a consensus on the number of deaths caused by exposure has proved impossible. Widely differing estimates result from different assumptions about fuel composition, air-pollution control technology, power-plant siting in relation to population distribution, meteorological conditions affecting sulfate formation, and, above all, the relation between sulfate concentrations and disease.

Large uncertainties also apply to the health and safety impacts of nuclear fission. In this case, differing estimates result in part from differences among sites and reactor types, in part from uncertainties about emissions from fuel-cycle steps that are not yet fully operational (especially fuel reprocessing and management of uranium-mill tailings), and in part from different assumptions about the effects of exposure to low-dose radiation. The biggest uncertainties, however, relate to the probabilities and consequences of large accidents at reactors, at reprocessing plants, and in the transport of wastes.

Altogether the ranges of estimated hazards to public health from both coal-fired and nuclear-power plants are so wide as to extend from negligible to substantial in comparison with other risks to the population. There is little basis, in these ranges, for preferring one of these energy sources over the other. For both, the very size of the uncertainty is itself a significant liability.

Often neglected, but no less important, is the public health menace from traditional fuels widely used for cooking and water heating in the developing world. Perhaps 80 percent of global exposure to particulate air pollution occurs indoors in developing countries, where the smoke from primitive stoves is heavily laden with carcinogenic benzopyrene and other dangerous hydrocarbons. A disproportionate share of this burden is borne, moreover, by women (who do the cooking) and small children (who are indoors with their mothers).

The ecological threats posed by energy supply are even harder to quantify than the threats to human health and safety from effluents and accidents. Nevertheless, enough is known to suggest they portend even larger damage to human well-being. This damage potential arises from the combination of two circumstances.

First, civilization depends heavily on services provided by ecological and geophysical processes such as building and fertilizing soil, regulating water supply, controlling pests and pathogens, and maintaining a tolerable climate; yet it lacks the knowledge and the resources to replace nature's services with technology. Second, human activities are now clearly capable of disrupting globally the processes that provide these services. Energy supply, both industrial and traditional, is responsible for a striking share of the environmental impacts of human activity. The environmental transition of the past 100 years—driven above all by a twentyfold increase in fossil-fuel use and augmented by a tripling in the use of traditional energy forms—has amounted to no less than the emergence of civilization as a global ecological and geochemical force.

Of all environmental problems, the most threatening and in many respects the most intractable is global climate change. Climate governs most of the environmental processes on which the well-being of 5.3 billion people critically depends. And the greenhouse gases most responsible for the danger of rapid climate change come largely from human endeavors too massive, widespread, and central to the functioning of our societies to be easily altered: carbon dioxide ($CO_2$) from deforestation and the combustion of fossil fuels; methane from rice paddies, cattle guts, and the exploitation of oil and natural gas; and nitrous oxides from fuel combustion and fertilizer use.

The only other external energy cost that might match the devastating impact of global climate change is the risk of causing or aggravating large-scale military conflict. One such threat is the potential for conflict over access to petroleum resources. The danger is thought to have declined since the end of the 1970s, but circumstances are easily imagined in which it could reassert itself—particularly given the current resurgence of U.S. dependence on foreign oil. Another threat is the link between nuclear energy and the spread of nuclear weapons. The issue is hardly less complex and controversial than the link between carbon dioxide and climate; many analysts, including me, think it is threatening indeed.

What are the prospects for abating these impacts? Clearly, the choices are to fix the present energy sources or to replace them with others having lower external costs.

As for fixing fossil fuels, it appears that most of their environmental impact (including the hazards of coal mining and most of the emissions responsible for health problems and acid precipitation) could be substantially abated at monetary costs amounting to additions of 30 percent or less to the current U.S. costs of fossil fuels or electricity generated from them. Still, a massive investment in retrofitting or replacing existing facilities and equipment would be needed, representing a particular barrier in parts of the world where capital is scarce and existing facilities and equipment are far below current U.S. standards. The carbon dioxide problem is much harder:

replacing coal with natural gas, which releases less $CO_2$ per giga-joule, is at best a short-term solution, and capturing and sequestering the $CO_2$ from coal and oil would require revamping much of the world's fuel-burning technology, at huge cost.

Nuclear energy is incomparably less disruptive climatologically and ecologically than fossil fuels are, but its expanded use is unlikely to be accepted unless a new generation of reactors with demonstrably improved safety features is developed, unless radioactive wastes can be shown to be manageable in the real world and not just on paper, and unless the proliferation issue is decisively resolved. I believe the first two conditions could be met, at least for nonbreeder reactors, without increasing the already high costs of nuclear electricity by more than another 25 percent. I think the third can be accomplished only by internationalizing a substantial part of the nuclear-energy enterprise, an approach blocked much more by political difficulties than by monetary costs. Fusion can, in principle, reduce the safety, waste, and proliferation hazards of fission, but it is not yet clear how soon, by how much, and at what monetary cost.

Biomass energy, if replaced continuously by new growth, avoids the problem of net $CO_2$ production, but the costs of controlling the other environmental impacts of cultivation, harvesting, conversion, and combustion of biomass will be substantial. Just bringing the consequences of today's pattern of biomass energy use under control, given its contribution to deforestation and air-pollution problems, will require huge investments of time and money. The tripling or quintupling of biomass supplies foreseen by some would be an even more formidable task, fraught with environmental as well as economic difficulties.

The superabundant long-term option whose external environmental costs are most clearly controllable is direct harnessing of sunlight, but it is now the most expensive of the long-term options and may remain so. The decision to pay the monetary costs of solar energy, if it is made, will represent the ultimate internalization of the environmental costs of the options that solar energy would displace.

There is much reason to think, then, that the energy circumstances of civilization are changing in fundamental rather than superficial ways. The upward trend in energy costs is solidly entrenched, above all because of environmental factors. It is quite plausible, in fact, given existing energy-supply systems, end-use technologies, and end-use patterns, that most industrialized nations are near or beyond the point where further energy growth will create greater marginal costs than benefits. "Full speed ahead" is no longer a solution.

Instead we will need transitions in energy-supply systems and patterns of end use just to maintain current levels of well-being; without such transitions, cumulative consumption of high-grade resources and the diminished capacity of the environment to absorb energy's impacts will lead to rising total costs even at constant rates of use. Providing for economic growth without environmental costs that undermine the gains will require even faster transitions to low-impact energy-supply technologies and higher end-use efficiency.

Although the situation poses formidable challenges, it is likely that the most advanced industrialized nations are rich enough and technologically capable enough to master most of the problems. The richest countries could, if they chose, live with low or even negative energy growth by milking increases in economic well-being from efficiency increases, and they could pay considerably higher energy prices to finance the transition to environmentally less disruptive energy-supply technologies. But so far there is little sign of this actually happening. And whether it could be managed in the former Soviet Union and Eastern Europe, even in principle, without massive help from the West is more problematic.

Still more difficult is the situation in the less developed countries (LDCs). They would like to industrialize the way the rich did, on cheap energy, but they see the prospects of doing so undermined by high energy costs—whether imposed by the world oil market or by a transition to cleaner energy options. An acute shortage of capital accentuates their tendency to choose options that are cheapest in terms of monetary costs, and they see the local environmental impacts of cheap, dirty energy as a necessary trade for meeting basic human needs (with traditional energy forms) and generating economic growth (with industrial ones).

Although the LDC share of world energy use is modest today, the demographics and economic aspirations of these countries represent a huge potential for energy growth. If this growth materializes and comes mainly from fossil fuels, as most of these countries now anticipate, it will add tremendously to the atmospheric burdens of

$CO_2$ and other pollutants both locally and globally. And while they resent and resist the go-slow approach to energy growth that global environmental worries have fostered in many industrialized nations, the LDCs are, ironically, more vulnerable to global environmental change: they have smaller food reserves, more marginal diets, poorer health, and more limited resources of capital and infrastructure with which to adapt.

Global climate change could have profound consequences for the nations of the Southern Hemisphere: more dry-season droughts, more wet-season floods, more famine and disease, perhaps hundreds of millions of environmental refugees. Even if the North suffered less from the direct effects of climate change because of the greater capacities of industrialized societies to adapt, the world is too interconnected by trade, finance, resource interests, politics, porous borders, and possibilities for venting frustrations militarily.

How should society respond to the changing and increasingly alarming interaction between energy and human well-being? How can the energy transition on which civilization has embarked, largely unaware, be steered consciously toward a more supportive and sustainable relation among energy, the economy, and the environment?

The first requirement is to develop an improved and shared understanding of where we are, where we are headed, and where we would like to go. There needs to be an extended public and indeed international debate on the connections between energy and well-being, supported by a greatly expanded research effort to clarify the evolving pattern of energy benefits and costs. Of course, study and debate will take time. Large uncertainties attend many of the important issues, and some of these will take decades to resolve.

Perhaps, with more information, the situation will seem less threatening and difficult than I have suggested; on the other hand, it could be even more threatening and difficult. In any case, we face the dilemma of action versus delay in an uncertain world: if we wait, our knowledge will improve, but the effectiveness of our actions may shrink; damage may become irreversible, dangerous trends more entrenched, our technologies and institutions even harder to steer and reshape.

The solution to the dilemma is a two-pronged strategy consisting of "no-regrets" and "insurance-policy" elements. No-regrets actions are those that provide leverage against the dangers we fear but are beneficial even if the dangers do not fully materialize. In contrast, insurance-policy actions offer high potential leverage against uncertain dangers in exchange for only modest investment, although some of that investment may later turn out to have been unnecessary.

One essential no-regrets program is to internalize and reduce the environmental and sociopolitical costs of existing energy sources. High priority should be given to abating emissions of sulfur and nitrogen oxides from fossil fuels and emissions of hydrocarbons and particles from fossil fuels and traditional fuels alike. Technologies for controlling these emissions exist and will more than repay their costs by reducing damage to health, property, and ecosystems. Another part of the program should be a carbon tax, the revenues from which could be used to develop and finance technologies for reducing fossil-fuel dependence worldwide. More effort is also needed to increase the safety and decrease the weapon-proliferation potential of contemporary (nonbreeder) nuclear reactors, including the development of better reactor designs and placing the most vulnerable fuel-cycle steps under international control.

Increasing the efficiency of energy use (another no-regrets approach) is the most effective way of all to abate environmental impacts. Fossil fuels and uranium saved through efficiency generate no emissions and create no fission products or proliferation hazards. (Efficiency, too, can have environmental impacts, but they are usually smaller, or can be made smaller, than those of the energy sources displaced.) Increased efficiency is also the most economical option in monetary terms and the most rapidly expandable, and its ultimate potential is both enormous and sustainable. The main obstacle is educating the vast numbers of individual energy consumers, whose actions hold the key to many of the potential gains, and then providing them with the capital to take advantage of more efficient technologies.

Also crucial to a sensible energy strategy is the acceleration of research and development on long-term energy alternatives: sunlight, wind, ocean heat, and biomass; the geothermal energy that is ubiquitous in the earth's crust at great depth; fission breeder reactors; fusion; and advanced approaches to energy efficiency. The research should emphasize not only the attainment of economical ways to harness these resources but also the prospects for minimizing their environmental costs. Investing in such research qualifies as an insurance approach in that we do not yet know which of the options will be needed or how soon. Some of the money will be wasted, in the sense that some of the options will never be exploited. But the funding required to develop these alternatives to the point that we can choose intelligently between them is modest compared with the potential costs of having too few choices.

Building East–West and North–South cooperation on energy and environmental issues, a no-regrets strategy that will help no matter how the future unfolds, might begin with increased cooperation on energy research. Such collaboration could alleviate the worldwide funding squeeze for such research by eliminating needless duplication, sharing diverse specialized strengths, and dividing the costs of large projects. (Until now nuclear fusion has been the only area of energy research that has enjoyed major international cooperation.) It is especially important that cooperation in energy research include North–South collaborations on energy technologies designed for application in developing countries.

International cooperation on understanding and controlling the environmental impacts of energy supply is also extremely important, because many of the most threatening problems are precisely those that respect no boundaries. Air and water pollution from Eastern Europe and the former Soviet Union reach across Western Europe and into the Arctic, and the environmental impacts of energy supply in China and India, locally debilitating at today's levels of energy use, could become globally devastating at tomorrow's. But pleas from the rich countries to solve global environmental problems through global energy restraint will fall on deaf ears in the least developed and economically intermediate countries unless the first group can find ways to help the last two achieve increased economic well-being and environmental protection at the same time.

Concerning carbon dioxide, the best hope is that a no-regrets approach to energy efficiency—together with reforestation and afforestation efforts that also fall in the no-regrets category—will be sufficient to stabilize $CO_2$ emissions even as we wait for the necessarily slower transition to environmentally, economically, and politically acceptable noncarbon-based energy sources. But it would be imprudent to assume that no-regrets approaches will suffice. We need more insurance, beyond the research advocated above, to protect us against the possibility that rapid and severe climate change might necessitate an accelerated retreat from fossil fuels. We ought to have a contingency plan—carefully researched, cooperatively developed, and continuously updated—for reducing global carbon emissions at a rate of 20 percent per decade or more if that proves necessary and if the no-regrets strategies already in place are not adequate.

None of the preceding measures, nor all of them together, will be enough to save us from the folly of failing to stabilize world population. The growth of population aggravates every resource problem, every environmental problem, and most social and political problems. Short of catastrophe, world population probably cannot be stabilized at less than 9 billion people; without a major effort to limit its growth, the number of human beings on the planet could soar to 14 billion or more.

Supplying 5.3 billion people in 1990 with an average of 2.6 kilowatts per person—a total of 13.7 terawatts—is severely straining the planet's technological, managerial, and environmental resources, and crucial human needs are going unmet. Let us suppose, optimistically, that tremendous progress in energy efficiency makes it possible to provide an acceptable standard of living at an average of 3 kilowatts per person (half the figure for West Germany today). Then 9 billion people would use 27 terawatts and 14 billion would use 42; the lower energy-use figure is twice today's, the higher one more than triple today's. Can we expect to achieve even the lower one at tolerable costs? As hard as controlling population growth may be, it is likely to be easier than providing increasing numbers of people with energy (and food and water and much else).

The foregoing prescriptions for taking positive control over the energy transition constitute a demanding and ambitious agenda for national and international action. Little of it will happen unless there is widespread consensus about the nature of the problem, the size of the stakes, and the possibilities for action.

# ABOUT THE CONTRIBUTORS

Duncan Brown is a Washington, D.C., area science writer and editor who specializes in technology and the environment. Brown served for six years as a staff officer of the National Research Council. He writes for magazines and the publications of research institutions, including the congressional Office of Technology Assessment, the National Research Council, and The Ohio State University. He holds a B.A. degree from St. Johns College in Annapolis, Maryland.

Umberto Colombo is the chairman of ENEA (Italian National Agency for Atomic and Alternative Energy Sources). A physical chemist with a doctorate from the University of Pavia, he held a Fulbright postdoctoral fellowship in chemical engineering at MIT. Among the positions he has held in his distinguished career are chairman of the Italian Atomic Energy Commission and director general for research and corporate strategies of the Montedison Co. He has been president of the Aurelio Peccei Foundation, a member of the Councils of the Club of Rome and the United Nations University, president of the European Science Foundation, and a member and chairman of a number of United Nations and OECD advisory bodies. Colombo is the author of more than 150 scientific papers and works on energy, science, and technology policy and coauthor of several books on science and technology.

Robert L. Goble is a research professor of environment, technology, and society at Clark University. A theoretical physicist who has worked for more than a decade on problems of energy, the atmospheric environment, and risk assessment, he recently participated in the Environmental Protection Agency's critical assessment of acid deposition. Currently his interests include policy responses to the threat of global warming and uncertainties in assessing human health risk.

John Harte holds a joint professorship in the Energy and Resources Group and the Department of Soil Science at the University of California, Berkeley. He received a B.A. in physics from Harvard University and a Ph.D. in theoretical physics from the University of Wisconsin. Harte is the author of three books and more than 100 publications in theoretical physics, ecology, energy and water resources, acid precipitation, and human and ecosystem toxicology. The recipient of a Pew Scholars Prize in Conservation and the Environment, Harte is a fellow of the American Physical Society and has served in an advisory capacity to the California Air Resources Board and the National Academy of Sciences. He now serves on the boards of directors of the Rocky Mountain Biological Laboratory, the Point Reyes Bird Observatory, and the Denali Foundation.

Christoph Hohenemser is professor of physics and chair of the Program on Environment, Technology, and Society at Clark University. He holds B.A. and Ph.D. degrees in physics from Swarthmore College and Washington University, respectively. For the past thirty years, Hohenemser has worked in experimental and solid-state physics, as well as environmental and risk analysis. Since spring 1986 he has been actively involved in analysis of the Chernobyl nuclear accident, on which he conducted early fallout measurements in Europe. Hohenemser is a fellow of the American Physical Society and recipient of a United Nations bronze medal for distinguished contributions to the cause of the environment.

John P. Holdren is Class of 1935 Professor of Energy and Resources at the University of California, Berkeley. He received his B.S. and M.S. degrees in aeronautics and astronautics from MIT and his Ph.D. in aeronautics and astronautics and theoretical plasma physics from

Stanford University. Holdren has authored over 200 papers and coauthored or coedited eleven books on plasma physics, fusion energy technology, international security, and energy, resources, and environmental subjects. He is a member of the National Academy of Sciences and a fellow of the American Academy of Arts and Sciences, the American Physical Society, the American Association for the Advancement of Science, and the California Academy of Sciences. In 1981, he received a five-year MacArthur Foundation Prize Fellowship. Holdren is chairman of the executive committee of the Pugwash Conferences on Science and World Affairs and a former chair of the Federation of American Scientists.

Jack M. Hollander is professor emeritus of energy and resources at the University of California, Berkeley. He received his B.S. degree in chemistry from Ohio State University and his Ph.D. in nuclear chemistry from the University of California, Berkeley. His multifaceted career has included basic research in nuclear-structure physics, energy and environment research, and academic research administration. The author of over 100 research publications and seventeen edited books, Hollander was coauthor of one of the ten most-cited review papers in physics during the thirty-year period 1955–1985. He was the first director of the Energy and Environment Division at Lawrence Berkeley Laboratory (1973–1976), director of the National Academy of Sciences CONAES energy study (1976–1978), cofounder of the American Council for an Energy-Efficient Economy (1980), chairman of the Beijer Institute of Energy and Human Ecology in Stockholm (1976–1988), and editor of the international book series *Annual Review of Energy* (1975–1992). From 1983 to 1989 Hollander served as vice-president for Research and Graduate Studies at Ohio State University. Recipient of two Guggenheim fellowships (1958 and 1966), he is a fellow of the American Physical Society, the American Association for the Advancement of Science, and the World Academy of Arts and Sciences and a member of the Royal Swedish Academy of Sciences (Uppsala).

Richard B. Howarth is a staff scientist in the Energy and Environment Division of Lawrence Berkeley Laboratory. He holds a Ph.D. from the Energy and Resources Program at the University of California, Berkeley, with undergraduate and master's degrees from

Cornell University and the University of Wisconsin–Madison. His research interests include energy and environmental economics, especially the analysis of energy utilization trends and the implications of the principle of sustainable development for resource policy analysis.

Mary Hope Katsouros is staff director of the Ocean Studies Board of the National Academy of Sciences/National Research Council. She participated in the two major studies on oil spills carried out by the National Research Council and has advised the congressional Office of Technology Assessment on oil spills. Katsouros received a B.A. degree in education and an M.S. in chemistry from George Washington University and has attended the Georgetown Center of Law.

Lars Kristoferson is currently vice-director of the Stockholm Environment Institute, an international environmental management and technology research institute based in Sweden. His interests focus on the interrelationships among energy, technology, environment, and society. He has held positions in the Swedish Energy Research Commission and the Board for Energy Supply Research and has worked closely with the World Bank, the Swedish International Development Agency (SIDA), the UN Food and Agriculture Organization (FAO), and a number of governments on energy, technology, and environmental policy issues. The author of several books and many publications, Kristoferson is a member of the editorial board of the journals *Ambio* and *Tomorrow*.

Marc Ledbetter is deputy director of the American Council for an Energy-Efficient Economy and has been with the organization since 1986. He received a B.A. in economics from the University of Montana. Ledbetter has extensive experience in energy analysis and policy, including solar energy, coal mining, power-plant siting, and federal energy-efficiency programs. He has written extensively on energy conservation policy and testifies frequently before Congress on this subject. He is coauthor of the book *Energy Efficiency: A New Agenda* (1988). Recently he has been assisting new energy-efficiency centers in Poland and Czechoslovakia.

Richard B. Norgaard is professor of energy and resources at the University of California, Berkeley. He received his undergraduate degree in economics at Berkeley, a master's in agricultural economics at Oregon State University, and a Ph.D. in economics at the University of Chicago. Norgaard's research interests in the area of energy have emphasized the economics of technological change and petroleum scarcity, environmental problems associated with outer continental shelf petroleum leasing and development, and, currently, intergenerational equity and intertemporal resource allocation. He helped initiate the International Society of Ecological Economics and has served on committees of the National Academy of Sciences and the congressional Office of Technology Assessment.

Janos Pasztor is an energy/environment analyst who is currently a senior officer with the United Nations Conference on Environment and Development (UNCED), based in Geneva. His responsibilities include UNCED's information system and its work on atmosphere, energy, and transport. He holds B.S. and M.S. degrees from MIT. He has held positions with the Stockholm Environment Institute, the United Nations Environment Program (Nairobi), and the World Commission on Environment and Development (Geneva).

Arthur H. Rosenfeld is a professor of physics at the University of California, Berkeley, and director of the Center for Building Science at Lawrence Berkeley Laboratory (LBL). He received his Ph.D. from the University of Chicago, the last student of the great physicist Enrico Fermi. Working until 1974 with Luis Alvarez's Nobel-prize winning particle-physics group at LBL, he codiscovered several elementary particles, codeveloped the bubble chamber data-processing system, and established the International Particle Data Center. In 1974 he founded the LBL Energy-Efficient Buildings Program, which has developed numerous energy-efficient technologies including low-E windows and advances in compact fluorescent lamps. In 1979 he cofounded the American Council for an Energy-Efficient Economy with Jack Hollander. In 1986 he received the Leo Szilard Prize of the American Physical Society for physics in the public interest. Coauthor of over 300 scientific papers and four books, he recently coauthored the "Mitigation Panel Report" of the NAS-NRC Panel on Policy Implications of Greenhouse Warming.

Marc Ross is professor of physics at the University of Michigan and senior scientist in the Energy and Environmental Systems Division,

Argonne National Laboratory. He received his Ph.D. in physics from the University of Wisconsin and worked in theoretical physics until 1972, when his interests turned to energy and environmental problems. His research currently focuses on energy and environmental implications of transportation. In 1974 he codirected the American Physical Society's study of efficient energy use. He is the coauthor of a book on energy policy, *Our Energy: Regaining Control* (1981). Recently Ross has worked on issues of technology and energy demand for the congressional Office of Technology Assessment and the American Council for an Energy-Efficient Economy.

William D. Ruckelshaus is chairman and chief executive officer of Browning-Ferris Industries, one of the nation's largest waste-disposal companies. He holds a B.A. degree from Princeton University and received a law degree from Harvard University. Ruckelshaus began a distinguished public service career in his home state of Indiana, where he served as deputy attorney general and chief counsel of the Office of Attorney General and was a member and majority leader of the Indiana House of Representatives. In 1970 he became the first administrator of the U.S. Environmental Protection Agency. He served again in that capacity from 1983 to 1985 and has served also as acting director of the Federal Bureau of Investigation and deputy U.S. attorney general. At present, Ruckelshaus serves as a member of The Trilateral Commission, the president's Council on Environmental Quality, and the Recycling Advisory Council. He is a director of several major corporations, including Monsanto Company and Weyerhaeuser Company.

Stephen H. Schneider is head of the Interdisciplinary Climate Systems Section at the National Center for Atmospheric Research. He received his Ph.D. in mechanical engineering and plasma physics from Columbia University. His current research interests include human impacts on climate (including global warming), public policy issues in environment and science, and modeling of paleoclimates. A prolific writer, Schneider is author of *Global Warming: Are We Entering the Greenhouse Century?* (1989), *The Genesis Strategy: Climate and Global Survival* (with L. Mesirow), and *The Coevolution of Climate and Life* (with R. Londer) and has authored or coauthored more than 175 scientific papers. Schneider was selected by *Science Digest* (1984) as one of the "Hundred Outstanding Young Scientists in America" and has been awarded the Louis J. Battan Author's Award from the American Meteorological Society. He is a fellow of the American Association for the Advancement of Science.

Paul Slovic is president of Decision Research in Eugene, Oregon, and a professor of psychology at the University of Oregon. During the past fifteen years, Slovic and his associates have pioneered in the development of methods for describing risk perceptions and measuring their impacts on individuals, industry, and society. They have created a taxonomic system that enables one to understand and predict perceived risk, attitudes toward regulation, and the impacts resulting from accidents or failures. Slovic has been a consultant to numerous companies and government agencies. In 1983–1984 he served as president of the Society for Risk Analysis.

Dan Steinmeyer is a senior fellow with Monsanto Company. His assignments with Monsanto have included process design manager on the largest project in its history, director of engineering in Brazil, and manager of process energy technology. He was selected in 1987 as one of the decade's ten leading energy contributors by the Association of Energy Engineers. He served as the industrial technical advisor to the National Academy of Science's committee on alternative energy R&D strategies (1989–1990). His current work focuses on process modification to eliminate waste production.

Ellen Ward is a science writer for the Center for Building Science at Lawrence Berkeley Laboratory. She is also communications manager/technology transfer coordinator for the California Institute for Energy Efficiency.

# INDEX

# ALSO AVAILABLE FROM ISLAND PRESS

*Balancing on the Brink of Extinction: The Endangered Species Act and Lessons for the Future*
Edited by Kathryn A. Kohm

*Better Trout Habitat: A Guide to Stream Restoration and Management*
By Christopher J. Hunter

*Beyond 40 Percent: Record-Setting Recycling and Composting Programs*
The Institute for Local Self-Reliance

*Coastal Alert: Ecosystems, Energy, and Offshore Oil Drilling*
By Dwight Holing

*The Complete Guide to Environmental Careers*
The CEIP Fund

*Death in the Marsh*
By Tom Harris

*Farming in Nature's Image*
By Judith Soule and Jon Piper

*The Global Citizen*
By Donella Meadows

*Healthy Homes, Healthy Kids*
By Joyce Schoemaker and Charity Vitale

*Holistic Resource Management*
By Allan Savory

*Inside the Environmental Movement: Meeting the Leadership Challenge*
By Donald Snow

*Last Animals at the Zoo: How Mass Extinction Can Be Stopped*
By Colin Tudge

*Learning to Listen to the Land*
Edited by Bill Willers

*Lessons from Nature: Learning to Live Sustainably on the Earth*
By Daniel D. Chiras

*The Living Ocean: Understanding and Protecting Marine Biodiversity*
By Boyce Thorne-Miller and John G. Catena

*Making Things Happen*
By Joan Wolfe

*Media and the Environment*
Edited by Craig LaMay and Everette E. Dennis

*Nature Tourism: Managing for the Environment*
Edited by Tensie Whelan

*The New York Environment Book*
By Eric A. Goldstein and Mark A. Izeman

*Our Country, the Planet: Forging a Partnership for Survival*
By Shridath Ramphal

*Overtapped Oasis: Reform or Revolution for Western Water*
By Marc Reisner and Sarah Bates

*Plastics: America's Packaging Dilemma*
By Nancy Wolf and Ellen Feldman

*Race to Save the Tropics: Ecology and Economics for a Sustainable Future*
Edited by Robert Goodland

*Rain Forest in Your Kitchen: The Hidden Connection Between Extinction and
Your Supermarket*
By Martin Teitel

*The Rising Tide: Global Warming and World Sea Levels*
By Lynne T. Edgerton

*The Snake River: Window to the West*
By Tim Palmer

*Steady-State Economics: Second Edition with New Essays*
By Herman E. Daly

*Taking Out the Trash: A No-Nonsense Guide to Recycling*
By Jennifer Carless

*Trees, Why Do You Wait?*
By Richard Critchfield

*Turning the Tide: Saving the Chesapeake Bay*
By Tom Horton and William M. Eichbaum

*War on Waste: Can America Win Its Battle with Garbage?*
By Louis Blumberg and Robert Gottlieb

*Western Water Made Simple*
From *High Country News*

For a complete catalog of Island Press publications, please write:
Island Press, Box 7, Covelo, CA 95428, or call: 1-800-828-1302